T0182094

The Science of Musical Sound

William Ralph Bennett, Jr.

Andrew C. H. Morrison
Editor

The Science of Musical Sound

Volume 1: Stringed Instruments, Pipe Organs, and the Human Voice

Foreword by Christy K. Holland

Author
William Ralph Bennett, Jr.
Yale University
New Haven, CT, USA

Editor
Andrew C. H. Morrison
Joliet Junior College
Joliet, IL, USA

This is a revised edition of the original publication, The Science of Musical Sound: Volume 1 by William R. Bennett, Jr. (Rose Lane Press). Copyright 2008 by William R. Bennett, Jr., Haverford, Pennsylvania, USA. All Rights Reserved.

ISBN 978-3-030-06519-5   ISBN 978-3-319-92796-1   (eBook)
https://doi.org/10.1007/978-3-319-92796-1

Library of Congress Control Number: 2018944607

This Springer imprint is published by the registered company Springer Nature Switzerland AG
The registered company address is: Gewerbestrasse 11, 6330 Cham, Switzerland

The ASA Press

The ASA Press imprint represents a collaboration between the Acoustical Society of America and Springer dedicated to encouraging the publication of important new books in acoustics. Published titles are intended to reflect the full range of research in acoustics. ASA Press books can include all types of books published by Springer and may appear in any appropriate Springer book series.

The Acoustical Society of America

On 27 December 1928 a group of scientists and engineers met at Bell Telephone Laboratories in New York City to discuss organizing a society dedicated to the field of acoustics. Plans developed rapidly and the Acoustical Society of America (ASA) held its first meeting on 10–11 May 1929 with a charter membership of about 450. Today ASA has a worldwide membership of 7000.

The scope of this new society incorporated a broad range of technical areas that continues to be reflected in ASA's present-day endeavors. Today, ASA serves the interests of its members and the acoustics community in all branches of acoustics, both theoretical and applied. To achieve this goal, ASA has established technical committees charged with keeping abreast of the developments and needs of membership in specialized fields as well as identifying new ones as they develop.

The Technical Committees include acoustical oceanography, animal bioacoustics, architectural acoustics, biomedical acoustics, engineering acoustics, musical acoustics, noise, physical acoustics, psychological and physiological acoustics, signal processing in acoustics, speech communication, structural acoustics and vibration, and underwater acoustics. This diversity is one of the Society's unique and strongest assets since it so strongly fosters and encourages cross-disciplinary learning, collaboration, and interactions.

ASA publications and meetings incorporate the diversity of these Technical Committees. In particular, publications play a major role in the Society. *The Journal of the Acoustical Society of America* (JASA) includes contributed papers and patent reviews. *JASA Express Letters* (JASA-EL) and *Proceedings of Meetings on Acoustics* (POMA) are online, open-access publications, offering rapid publication. *Acoustics Today*, published quarterly, is a popular open-access magazine. Other key features of ASA's publishing program include books, reprints of classic acoustics texts, and videos. ASA's biannual meetings offer opportunities for attendees to share information, with strong support throughout the career continuum, from students to retirees. Meetings incorporate many opportunities for professional and social interactions and attendees find the personal contacts a rewarding experience. These experiences result in building a robust network of fellow scientists and engineers, many of whom became lifelong friends and colleagues.

From the Society's inception, members recognized the importance of developing acoustical standards with a focus on terminology, measurement procedures, and criteria for determining the effects of noise and vibration. The ASA Standards Program serves as the Secretariat for four American National Standards Institute Committees and provides administrative support for several international standards committees.

Throughout its history to present day, ASA's strength resides in attracting the interest and commitment of scholars devoted to promoting the knowledge and practical applications of acoustics. The unselfish activity of these individuals in the development of the Society is largely responsible for ASA's growth and present stature.

To my grandchildren Sarah, Katie, Max, Danny, and Michael, with the hope that they will continue to enjoy playing music and studying science.

WRB

Foreword

Harmonizing principles from physics, engineering, and music, William Ralph Bennett, Jr. hit the right notes for undergraduate students in a popular semester-long course on the physics of musical instruments, which he taught at Yale University in the 1970s and 1980s. The course appealed to students with diverse talents and backgrounds, including musicians and scientists. His lectures were filled with demonstrations that helped students visualize complex concepts in wave analysis and train their sense of hearing to appreciate harmonics and historical tuning of musical instruments, and his classes were always infused with humor. In the mid-1970s, Bennett was selected by students as one of the "Ten Best Teachers" at Yale University 3 years in a row. The material for this course was based on Bennett's encyclopedic research of the physics of musical instruments. I was drawn to one of Bennett's lectures as a Yale graduate student when I heard strange noises and loud music emanating from gigantic electrostatic speakers in a lecture hall in Becton Center. As I slipped into the back of the hall, I was struck by his enthusiasm and the engagement of his students; his approach resonated with me. I became his teaching assistant on the spot. Lecture notes and homework assignments from Bennett's course served as the foundation for this book.

While at Bell Labs in 1960, Bennett invented the first gas laser, the helium-neon laser, along with Ali Javan and Donald Harriott. The word "LASER" originated as an acronym for light amplification by stimulated emission of radiation. Bennett went on to invent nearly a dozen other lasers using electron impact excitation in each of the noble gases, dissociative excitation transfer in the first chemical laser,

---

Musica et Lux - William Ralph Bennett, Jr. (January 30, 1930–June 29, 2008).

the neon-oxygen laser, and collision excitation in several metal vapor lasers. He was awarded 12 patents for his work in this area. Laser technology spawned many applications, including compact disc players, of relevance to the enjoyment of high-fidelity recorded music. Bennett authored 8 books and over 130 research papers.

Outside of his applied physics research on optical pumping, Bennett was also an accomplished musician. He enjoyed listening to and playing chamber music as Head of Silliman College at Yale and played clarinet with several amateur symphony orchestras. I personally enjoyed playing four-hand classical works with him on his beautifully restored Steinway concert grand piano. My hands worked hard to keep up with Bennett's, who hit the keys hard with his large hands. Before Bennett died in 2008 from esophageal cancer, he asked me to help Frances Commins Bennett, his wife of 55 years, see that the material in this book make it into print. Frances and their daughter Jean Bennett, a geneticist at the University of Pennsylvania, carefully organized and curated the contents of this book. Through my academic connections, and later as President of the Acoustical Society of America (ASA), I promised Frances to help find an editor for this book who is as passionate about teaching as William Ralph Bennett, Jr. was. Andrew C.H. Morrison was the perfect choice, an active member of the musical acoustics technical committee at the ASA, a society that generates, disseminates, and promotes the knowledge and practical applications of acoustics. Morrison is an assistant professor at Joliet Junior College, Northern Illinois University. Morrison's primary area of research is the physics of musical instruments, and he teaches acoustics, astronomy, general physics, and modern physics.

Bennett's intellectual legacy is partly immortalized in this first installment of a two-volume set on the science underpinning musical sounds. Wave motion and the physics of the propagation of disturbances in strings, membranes, and pipes are introduced with elegant descriptions and minimal algebraic and trigonometric mathematics. The history of spectral analysis and Fourier series is presented with delightful tidbits about significant scientists in the footnotes. Discrete Fourier analysis is described and applied to musical sounds and is employed to elucidate how human hearing works. Bennett provides sampled waveforms of many musical instruments and quirky household items, such as the garden hose, which can be played like a trumpet, complete with computer programs for the student to explore. While in my laboratory at the University of Cincinnati, Jason L. Raymond, currently a postdoctoral research assistant in the Department of Engineering Science at the University of Oxford, translated Bennett's original BASIC programs to MATLAB code, found in Appendix C. The scientific concepts of modes of vibration and resonance frequency are translated into the language of music, so that the nonscientist can understand the origins of overtones and pitch. Detailed descriptions of the inherent spectra of plucked, struck, and bowed strings are treated in separate chapters. The source of sounds made by the human voice is given special attention, and even the sampled spectrum of one of Bennett's beloved German Shepherd dogs named Mozart appears in one figure. The diagrams, photos, and anecdotes in the chapter on pipe organs reveal Bennett's fascination and familiarity with the construction of this intricate instrument. Bennett spent many hours building a pipe organ into a nook of his summer home in Colrain, Massachusetts. Students reading this book will be treated to Bennett's witty personality and love of music while learning about the instruments that contribute to this art form.

Professor, Internal Medicine, Division of Cardiovascular   Christy K. Holland
Health and Disease and Biomedical Engineering
Scientific Director, Heart, Lung, and Vascular Institute
University of Cincinnati
Cincinnati, OH, USA

Preface

Just as aging musicians often find the leap to the podium irresistible, aging scientists often turn to the study of musical sound. Examples start with the beginning of physics itself, from Aristotle and Pythagoras to Brook Taylor (Newton's student, who is best known for his invention of the *Taylor Series* in calculus, studied the physics of plucked vibrating strings). The more famous names of the nineteenth century include Felix Savart, John William Strutt (*alias* Lord Rayleigh), and Herman Helmholtz. In the early twentieth century, the roster contains molecular spectroscopist Chandrasekhara Raman (who investigated the physics of bowed violin strings), French nuclear physicist H. Bouasse (who wrote on resonances in tubular cavities), and that well-known British expert on the dynamical theory of gases, Sir James Jeans. More recent examples have included English scientist Alexander Wood, *RCA* engineer Harry F. Olson (1952), physicist John Backus, nuclear physicist Arthur Benade, solid-state physicist Thomas Rossing, former director of electronics research at the Bell Telephone Laboratories John Pierce, and, most humbly, the laser physicist who wrote this volume. Nuclear physicist Charles Kavalovski didn't wait for *his* "golden years" after teaching physics for 10 years, he became Principal Hornist in the Boston Symphony Orchestra. Interestingly, the French mathematician, Baron Jean-Baptiste-Joseph von Fourier, whose analysis is so frequently used to study the sound of instruments, seemed to have had no interest in music at all. (His revolutionary 1822 paper on Fourier Series arose from a study of heat flow in the boring of cannons for Napoleon's army. His other interests were Egyptology and draining swamps.) It is much rarer to find musicians who have turned to physics in their later years. The lone example of NBC Symphony violinist Norman Pickering comes to mind. (Although he conducted research on the physics of bowed strings, he was more widely known for his work in electrical engineering, especially the invention of the Pickering Phonograph Cartridge in the 1940s.)

In addition to physicists, many electrical engineers became enamored with the subject—a natural development in the early days of electronic sound recording and reproduction. The period from the late 1920s through the 1940s became a heyday in acoustical research at the Bell Laboratories, and, being the son of one of their electrical engineers, I was privileged to meet many famous people in that world at an early age. Harry Nyquist (sometimes thought of as the "Einstein" of the communication field), Homer Dudley (inventor of the Vocoder), John Pierce (the father of satellite communication), and many others were frequent guests in our home. As a result, I acquired a strong interest in the early developments in multi-channel high-fidelity stereophonic sound reproduction, audio technology, and the digital encoding of sound—which, in those days, went under the heading PCM (for pulse code modulation.) My father showed me how to construct electronic circuits at home, ranging from audio amplifiers to FM tuners, and the very smell of rosin smoke (probably poisonous) curling up in thin wisps from a hot soldering iron became a source of nostalgia. Later, when I worked at the Bell Laboratories myself, I came with sad reverence across some of the historic relics of that age gathering dust in the basement of Building One. When we got the first gas laser to work back in December of 1960,[1] Harry Nyquist, who was still at the Bell Labs as a consultant, came around to see how the device satisfied his famous criterion for oscillation. (Nyquist was one of many people at the Labs who had built an electronic organ at home. In those days, you had to make everything yourself if you wanted anything good, and the living rooms of many Lab employees were often cluttered with breadboard circuits and relay racks—much to the disgust of their wives.) (Fig. 1)

The wide availability of live classical music over FM in those days played an important role in my own musical education, as well as in that of many others. On a typical weekend in the New York City area, one could hear live concerts from the

Fig. 1 The author with one of the first helium-neon lasers

---

[1]Javan et al. (1961).

NBC Symphony under Toscanini, the Philadelphia Orchestra under Stokowski, the N.Y. Philharmonic under Metropolis, the Radio City Music Hall Orchestra under Rapeé, the CBS Symphony, miscellaneous chamber music recitals over WNYC and the WQXR String Quartet—not to mention the Longine Symphonette, and even the "Bell Telephone Hour." The decrease in live (and recorded) classical music over the radio since that time has been appalling.[2]

For me (in contrast to Baron von Fourier), life without music would be unimaginably bleak, and I am not surprised that many scientists turned to study the subject in their later years. Of course, some, including me, were amateur musicians themselves. I studied the clarinet with Simeon Bellison (Principal Clarinetist in the New York Philharmonic from 1925 to 1948) and clarinetist David Weber (then Principal in the New York Metropolitan Opera Orchestra) during my college years and was a soloist with several orchestras in the Princeton, NJ area.

At the Institute for Advanced Study, Einstein's mathematical assistant, physics professor Valentine Bargmann, was an excellent pianist. I once played the Brahms clarinet trio, along with chemist (and cellist) Peter Koerber, while Bargmann sight-read the difficult piano part. Also at the Institute, Swiss theoretical physicist Dominic Rivier was a good cellist with whom I spent many evenings alternately playing chamber music and talking about David Bohm's "new" book on quantum theory. ("You must work all the problems!") Then, of course, the Director of the Institute himself, physicist J. Robert Oppenheimer, was a great devotee of chamber music.

The music scene at the Columbia Physics department was also active among the student population. When I first arrived, Professor Charles Townes (of "maser-laser" fame and then chairman of the department) sternly advised that I would have to cut down on musical activity if I wanted to obtain a PhD in physics. I, of course, took his advice, although I learned later that Charlie was taking voice lessons then at the neighboring Juilliard School of Music.[3] One of the faculty members at Columbia was violinist Erwin Hahn, an expert in magnetic resonance spectroscopy who could also play tunes by beating the top of his head with spoons. Another, Jack Steinberger, played the flute; at our New Year's Eve party in 1956, he made the welcome suggestion to bring in a string quartet "to raise the tone of the event." But many students tended to conceal their musical abilities, probably fearing that they wouldn't be taken seriously as physicists.

A casual glance through the *Amateur Chamber Music Players Directory*, confirms my belief that there are lots of scientists who do love music. But I don't especially share the view expressed by some (e.g., Rothstein 1995) that it is primarily the mathematical structure of music that attracts them to the subject.

---

[2]Lebrecht (2001, p. 96) infers that the end of classical music broadcasts occurred when Toscanini's memory failed due to a transient ischemic attack during a televised concert. "He floundered in mid-Wagner and America listened aghast." The NBC engineers quickly shifted to a recording of the Maestro conducting Brahms. This was shortly followed by Toscanini's resignation and the disbandment of the NBC Symphony. The other networks quickly followed suit.

[3]Schawlow (1996, p. 4).

As a number of composers have demonstrated, it is possible to write music with very complex mathematical structure that sounds simply dreadful. For example, Hindemith himself decided to abandon a composition based on the wavelengths in the hydrogen atom, and, as once demonstrated by hornist Willie Ruff and geologist John Rodgers at Yale, "the music of the spheres" is not all that it is cracked up to be either.[4] I think the allure of both music and art is more on a tonal, partly subconscious level. Obviously, form plays a very important part. But the main thing is whether it sounds good. Similarly, in great writing, it is the *sound* of words that stimulates emotional interest. If Lincoln's *Gettysburg Address* had started off, "Eighty-seven years ago, our predecessors set up a new type of government," few people would have remembered it. Similar thoughts apply to changes from the King James version of the *Bible* and from Shakespeare's English. (In one pirated version of *Hamlet*, the famous soliloquy from Act III began, "To be or not to be; ay, there's the rub...") Some remarkable analysis must go on behind the scenes in the human brain for most people while listening to music. For example, the late Beethoven string quartets have a special, near-mystical appeal. One almost feels that they are the result of such intense emotional concentration that they could *not* have been written by someone who had not become totally deaf. Of course to the scientist, trying to understand the physical basis of acoustic phenomenon is as irresistible as the analysis of musical structure must be to a composer, and most physicists also have a well-developed fondness for problems involving wave motion.

In reviewing previous books on the physics of music, one sees a familiar problem in presenting the material. One wants to hold the attention of two rather different groups of readers: those who play musical instruments and have only an amateur interest in physics or mathematics and those who are quite at home with math and physics, but also play musical instruments as a hobby. The problem of how to deal with both groups simultaneously is one that I faced over a decade of teaching this subject at Yale University. One could see frustration and boredom develop among the musicians during the presentation of elegant mathematical proofs. Similarly, the scientists in the audience would become restive without some analytic treatment of the subject. Starting with Helmholtz, different authors have handled this problem in different ways. Helmholtz had a traumatic experience in 1855 while serving as Professor of Anatomy and Physiology at Bonn: he was severely criticized (nearly fired) for having had the temerity to write a cosine function on the blackboard during a lecture on physiological optics.[5] It is probably for that reason that he divided his famous opus *On the Sensations of Tone* into two major parts: the first and longest section was a qualitative discussion of the material in which equations were spelled out in words; the second was a series

---

[4]Physicist Henry Margeneau (private conversation) had computed the hydrogen wavelengths at Hindemith's request while the composer was at Yale. Margeneau also told me that, although he was a violinist, he had been forced to pass an examination in playing the pipe organ before he could receive his undergraduate degree in Germany! The opinion of the "music of the spheres" (based on the orbital periods of the planets in the solar system) is the author's own.

[5]Henry Margeneau in the Preface to the Dover (1954) edition of the Helmholtz (1885) work.

of very concise mathematical appendices. The basic trouble with that approach for the physicist is the necessity to keep jumping back and forth between the two parts of the book. Lord Rayleigh's 1877 book, *The Theory of Sound*, is strictly for physicists and for those having a good deal of patience at that. He is already deeply involved in the stability of solutions to nonlinear differential equations by page 80. Jeans in his 1937 volume, *Science and Music*, solved the problem by leaving out the equations altogether. But that results in such a watered-down treatment that the book would scarcely be useful in a "physics for poets" course. Various other approaches have been tried. John Backus (1969) in his opus entitled *The Acoustical Foundations of Music* used lots of illustrations, and almost no math. C.A. Taylor (1965) in *The Physics of Musical Sounds* mixes the equations in with the text in a fairly traditional way. Arthur Benade (1976) in *Fundamentals of Musical Acoustics* has lots of verbal description and relatively few equations. He fools the reader slightly by omitting equation numbers and referring instead to numbered statements in sections of the text that really are equations spelled out in words. His book goes into much practical detail (e.g., on the geometry of mouthpieces) that could be of value to a professional musician, but tends to be too detailed for the nonspecialist. His study of clarinet mouthpieces is a remarkable tour de force.

Many other authors have written about music and science, including Thomas Rossing (1990), Rossing and Fletcher (1991), Bouasse and Fouché (1929), and Cremer (1981). Many of these are excellent reference books for advanced students, but in some cases the mathematics can obscure physical meaning for a general college audience. An interesting experiment conducted by the late John R. Pierce (1983) only used equations derived by dimensional analysis. In that approach, one figures out what units the answer should have and then plays around with likely quantities until the right dimensions pop out. For example, if the answer is to be a velocity, you probably want to divide length by time, but you might also decide to multiply acceleration by time, so the process is ambiguous. In addition, numerical factors are both dimensionless and important. With so many books already written on the subject, one might ask, "Why still another?" It's a fair question. I found in teaching a course that used books by most of the authors listed above that no one of them seemed adequate for my own purposes. In addition to the problem of treating the mathematics of the subject, there were many interesting questions that simply weren't mentioned at all: those things ranged from modern methods and examples of spectral analysis to digital recording and computational models for the mechanism of sound production in various instruments. I also prefer a more historical approach to the subject than given in most texts. In addition, I have presented some of my own research results here. My approach in teaching the course was to make the undiluted mathematical material available in handouts, but not as a compulsory part of the reading. I did assume the students at least had an introduction to trigonometry and calculus. But I added footnotes in the reading to clarify those things when they were important. I also provided lots of illustrations in the course, sometimes based on "video clips" or "computer movies" that showed things like the motion of a plucked harpsichord string or the vibrational modes of a circular drumhead.

These were determined either by direct measurement or from numerical calculations based on reasonable models of the things being studied. Those demonstrations were supplemented by weekly laboratory experiments that permitted easy observation of real examples. I have tried to carry over some of those concepts here by putting most of the routine math in page footnotes and at the end of chapters and by referring the reader to more involved calculations in mathematical appendices. I have also provided numerous illustrations of the mathematical results. The math contained within the chapters should not be a problem for anyone familiar with high school geometry, algebra, trigonometry, and introductory calculus. (The reader is invited to skip over the equations if he or she doesn't like them; they are included primarily for completeness and are not essential to understanding the material.) However, more difficult mathematics is used in the Appendices, which are intended for people with a more advanced mathematical background.

Haverford, PA, USA   William Ralph Bennett, Jr.
February 2008

Acknowledgements

I am especially indebted to my wife, Frances Bennett, for her patience, encouragement, numerous helpful discussions, and editorial comments. A number of people have provided critical readings of different portions of the book. Physics professor Eugene Commins of the University of California at Berkeley checked most of the equations and provided many useful suggestions in several portions of the book. Harpsichord builder Paul Kennedy provided a critical reading of Chap. 3 devoted to the harpsichord, and Curator of the Yale Musical Instrument Collection, Richard Rephann, provided helpful information on that subject, as well as allowing me to analyze many instruments from the Yale Collection. I am especially indebted to luthier Hiroshi Iizuka for his advice and numerous comments related to my chapter on bowed strings. I also wish to acknowledge helpful discussions with violin authority Carleen Hutchins and pianist Virginia Benade Belveal regarding her late husband Arthur Benade's work on musical acoustics. Christopher Robinson provided a critical reading of the piano chapter. I received a number of suggestions from oboist William Robert Bennett about material in the book and from Nancy Bennett regarding computer graphics. Geneticist Jean Bennett provided photomicrographs of violin bow hair and helped take data on harpsichord waveforms. Maurice and Frone Eisenstadt provided a careful reading of the manuscript and a number of helpful suggestions. Mari Kimura collaborated with me in a study of the subtones she produces on the violin G-string. Dan and Dorothy Kautzman provided critical readings of my chapter on violins. Joseph Dzeda of Yale University provided helpful information on pipe organs, as did organists Paul Jordan and Charles Krigbaum. I am indebted to James Undercoffler for demonstrating the various brass instruments discussed in the chapter on horns. I also benefited from helpful comments by numerous other musicians whom I have known personally.

There are many other people whom I haven't named specifically here who have also contributed to this work. I have tried to give specific acknowledgment to them in appropriate places in the text. In addition, as acknowledged in the various figure captions, I am indebted to the people and organizations cited throughout the book for permission to reproduce various photographs and illustrations.

New Haven, CT, USA   William Ralph Bennett, Jr.

Contents

About the Author and Editor

William Ralph Bennett, Jr. (1930–2008) was a renowned physicist and professor at Yale University. Prof. Bennett is best known as co-inventor of the helium-neon laser. For over 10 years, Prof. Bennett taught a widely popular undergraduate course on the physics of music at Yale, upon which this two-volume text is based. Prof. Bennett completed his B.A. at Princeton and received his Ph.D. from Columbia University in 1957. Over the course of his long career, Prof. Bennett was the recipient of numerous awards and honors and served as master of Yale's Silliman College from 1981 to 1987.

Andrew C. H. Morrison is an Associate Professor at Joliet Junior College in Joliet, IL. He completed his B.S. at the University of Northern Iowa in 2000 and his Ph.D. at Northern Illinois University in 2005. Prof. Morrison is Chair of the Acoustical Society of America Musical Acoustics Technical Committee.

Chapter 1
Wave Motion

An editor once warned me that every equation in a book reduces the readership by a factor of two. Unfortunately, a few equations really are necessary to understand this subject. But, the reader should rest assured that there is not much math in this chapter other than a little high-school algebra and trigonometry. Throughout the book, the more difficult derivations involving calculus have been relegated to appendices designed to provide a more rigorous development of the subject. The more difficult concepts are also placed at the end of chapters. In fact, the general reader should be able to understand most of the material in this book merely by skipping the equations and by looking at the figures, many of which illustrate the mathematical results.

Many aspects of both physics and music deal with wave motion. Musical sound involves the generation of vibration in different media—strings stretched under tension in a piano, the taut membranes on a kettledrum, the wood fiber in a violin front and back plates, the reeds made from cane in woodwind instruments, and the transmission of sound through the air. The vibration is usually set up within a medium where a localized disturbance propagates from one region to another. In a gas such as air, neighboring molecules bump into each other to transmit energy throughout the medium. The result is a compression wave moving through the medium in the same direction as the vibration amplitude. In solids, the localized microscopic vibrations are also coupled to neighboring atoms through intermolecular forces. The process through which these vibrations are coupled throughout a medium generally results in wave motion of some sort.

Many people first see evidence of wave motion in the ripples on the surface of a pond, for example, after dropping a pebble into the water. That case produces a localized vertical displacement of the water surface resulting in a wave whose amplitude is transverse to the direction of wave propagation. It is important to realize that the individual molecules do not have to move very far in order to transmit waves

---

The original version of this chapter was revised: Equations on pages 16 and 25 were corrected. The correction to this chapter is available at https://doi.org/10.1007/978-3-319-92796-1_8

over a large distance and bulk motion of the material is not involved in the wave propagation. Here, as in most cases of wave motion, it is the surface disturbances and not the individual molecules that move throughout the medium. In the case of water waves, the particles undergo up and down motion. There, surface tension as well as gravity acts as a restoring force that produces wave motion.[1] The velocity with which the wave travels is normally much less than the speed of the individual atoms or molecules.

In contrast, sound waves consist of compressional variations in the medium that travels in the same direction as the initial disturbance. That is, the particles in the medium move back and forth in the direction of the wave motion and the waves are said to be "longitudinal" rather than "transverse." In most cases, the medium is uniform and isotropic (has the same properties in all directions) and the waves tend to expand at equal rates in all directions. The magnitude of the pressure fluctuation in the wave is generally quite small compared to the average background pressure—for example, the local atmospheric pressure in the case of sound waves moving in air.

1.1 Frequency, Period, and Wavelength

Musical sound very often involves the motion of periodic waves of the type illustrated in Fig. 1.1. Here, a transverse wave with amplitude y is shown traveling at velocity c in the x-direction through some material. By "periodic," we mean that the disturbance repeats itself regularly both in time (at one point in space) and in space (at one instant of time). If we stand at one point along the x-axis and watch the wave go by, we might count the number of peaks whizzing past us per second. That number is known as the *frequency* and is often designated by the letter f in acoustics and electronics. (The time between peaks is called the *period* and equals $1/f$.) Unfortunately, frequency is now measured in a nonintuitive unit called the *hertz*—abbreviated Hz, in honor of Heinrich Hertz, who died in 1894 and did early experiments on radio waves. One Hz is defined to be one cycle per second.[2] The situation is often compounded by the addition of Greek prefixes such as kilo (as in kHz for 1000 Hz) and *Mega* (abbreviated MHz for one million Hz), and so on. Fortunately, the period of the wave is still measured in units of time (seconds) but with still more Greek prefixes. (See Table 1.1.)

If one were to freeze time and walk along the wave with a meter stick, one would find that it repeated itself spatially in a distance known as the wavelength, which is often abbreviated λ (the Greek letter for L) and fortunately is still measured in standard units of length such as centimeters, meters, and feet.

---

[1] Reducing the surface tension tends to dampen out the waves—especially, the rapid ripples with short wavelengths (hence the expression "pouring oil on troubled waters").

[2] People used to measure frequency directly in "cycles per second" until some committee got a hold of the problem. One suspects that the term Hz might have been coined to keep newcomers out of the field. The situation becomes more confusing in the Russian language where there is no equivalent of "H" and people use the Russian "G" (Г) instead. Thus, "gigahertz" for a billion cycles per second becomes "gigagertz."

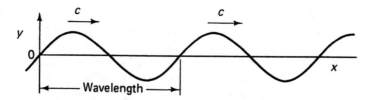

Fig. 1.1 A periodic wave traveling through some medium at velocity c

Table 1.1 Frequency and period measurement nomenclature

Frequency (f)	Written	Meaning	Period ($= 1/f$)	Written
0 hertz	DC	Direct current		
	AC	Alternating current		
1 hertz	1 Hz	1 cycle/second	1 second	1 s
1 kilohertz	1 kHz	1000 cycles/second	1 millisecond	1 ms
1 megahertz	1 MHz	1 million cycles/second	1 microsecond	
1 gigahertz	1 GHz	10^9 cycles/second	1 nanosecond	1 ns
1 terahertz	1 THz	10^{12} cycles/second	1 picosecond	1 ps
1 petahertz	1 PHz	10^{15} cycles/second	1 femtosecond	1 fs
1 exahertz	1 EHz	10^{18} cycles/second	1 attosecond	1 as

As one might suspect from the dimensions involved, the velocity c of the wave is generally related to the wavelength and frequency by the equation

$$\lambda f = c \tag{1.1}$$

That is, in the time T of one period the wave moves a distance λ; hence, $c = \lambda/T = \lambda f$. Although the human ear only responds to audio frequencies in the range from about 20 Hz to 20 kHz, some of the higher frequencies listed in Table 1.1 are used in devices related to sound transmission, recording, and reproduction. For example, sampling rates on digital recordings range from about 44 kHz to 200 kHz; the high-frequency bias on good tape recorders is typically at least 150 kHz; AM radio transmission uses carrier frequencies of about 1 MHz; FM broadcasting is done at about 100 MHz; present digital computers, TV sets, and cell phones use frequencies up to about 2 GHz; the lasers used in CD and DVD recordings typically emit light at about 0.6 PHz.

1.2 Why Sine Waves?

The reader will notice that we talk about sine functions a great deal throughout this book. The reason is that they are a fundamental building block in the physics of vibrational phenomena and, as such, are among the most important things one learns about in plane geometry or trigonometry. Indeed, one might go so far as to say that the identities one derives for the addition, subtraction, and multiplication of sine and cosine terms are among the most useful things taught in high school. The motion of a harmonic oscillator is a sine function of the time (see Appendix A), as is that of a vibrating reed at low air pressure or an oscillating pendulum or the human vocal cords at low levels of excitation. All vibrating string problems involve sine waves, from the plucked string of a harp to a bowed violin string. Solutions for the motion of a light wave through space or those for a sound wave propagating through air also involve sine waves. Of course, all these things could be expressed in terms of cosine waves just as well since the cosine is merely a sine shifted in phase by 90° ($\pi/2$ radians).

1.3 The Wave Equation

In different areas of physics, the disturbance y (which might be the amplitude of a vibrating string or the pressure fluctuation in a sound wave, or even the electric field in a light wave moving through space) obeys something known as the Wave Equation. The derivation of the wave equation involves more mathematics than we wish to include here and depends on the laws of physics holding in the particular area of interest. But, the form of the equation is much the same in these different areas.[3] In most cases in the present book, it arises merely by application of Newton's Laws of Motion. Because the form of the wave equation is essentially the same in different areas of physics, the solutions are all quite similar.

1.4 Running Waves

As shown in Appendix B, any linear function of the type $f(x \pm ct)$ can be a solution to the wave equation. However, sinusoidal solutions have a special relevance to the physics of musical sound because of their periodicity. In particular, solutions to the one-dimensional wave equation of the form

$$y = \sin 2\pi(x/\lambda - ft)(\longrightarrow)$$

and   (1.2)

$$y = \sin 2\pi(x/\lambda + ft)(\longleftarrow)$$

---

[3]The wave equation for vibrating strings is derived in Appendix B and that for sound in Appendix F.

describe running waves. The one marked ($\longrightarrow$) travels to the right ($+x$ direction) at velocity $+c$ and the one marked ($\longleftarrow$) travels to the left ($-x$ direction) at velocity $-c$. To verify the direction and speed of travel, note how x must vary with time when the argument of the sine wave is held constant. For the upper solution, $(x/\lambda - ft) =$ constant, which means that x is increasing at the rate $+\lambda f = +c$. Similarly, for the lower solution, $(x/\lambda + ft) =$ constant, which means that x is decreasing at the rate $-\lambda f = -c$.

1.5   π and Other Greek Letters

Physicists and mathematicians often use Greek letters for mathematical quantities, not to exhibit erudition, but for quite opposite reasons. They usually do not know Greek at all and want symbols that will not be mixed up with letters in their own alphabet. For the same reason, editors of papers and books like to italicize Roman letters when they are used as mathematical variables to distinguish them from the text.

Several Greek letters have taken on specific meaning through historic usage. Leading among them is the letter π (Pi), which traditionally stands for the ratio of the circumference-to-the-diameter of a circle. That quantity is of basic importance in describing wave motion. It is an irrational number given by

$$\pi = 3.14159265358979323846264338327 9 \ldots \tag{1.3}$$

a number whose digits go on and on forever. Shanks and Wrench (1962) computed the value of π to 100,000 decimal places.[4] One seldom needs to know π to that number of digits, but if some day you are stuck on a desert island and need to do a calculation in "double precision," the first 15 digits are given by a mnemonic due to Sir James Jeans:

> How I want a drink alcoholic of course after the heavy lectures involving quantum mechanics.

---

[4]The current record was set by Yasumasa Kanada of Tokyo who calculated π to over 200 billion places. (An early calculation done in 1873 by William Shanks to 707 places turned out to have a mistake in the 528th place, after which all the subsequent digits were wrong.) In 1897, the Indiana State House introduced Bill No. 246 to define π equal to the more convenient "Biblical value" of 3. Fortunately, that law was never passed by the State Senate. At the 762nd digit, π goes 999,999, but then reverts to a near random sequence. Daniel Tammet set a record by reciting π to 22,514 places in 5 h (BBC News, 3/15/2004). See "A Very Large Slice of Pi" by Tammet (2007, Chapter 10). Why do it? Because: "It's extremely beautiful...like a Mozart Symphony."

(The number of letters in each word corresponds to the successive digits in π.) Early mathematicians investigated the number to see if the digits ever started to repeat themselves. (They have not yet.) Von Neumann had hoped that the successive digits might make a good pseudo-random number series, but that turned out not to be the case, either.

Most pocket calculators or computer programming languages already have π built into the appropriate number of places.[5] Note that π is an angular measure: π radians = 180°.

1.6   Some Basic Trigonometry Relations

Figure 1.2 shows the basic definitions of the sine, cosine, and tangent functions as applied to the angle ϑ shown in a right triangle. The identity, $\cos 2\vartheta + \sin 2\vartheta = 1$, follows immediately from Pythagoras' theorem (the square of the hypotenuse equals the sum of the squares of the other two sides.) The shape of a sine wave is shown in Fig. 1.3.

Fig. 1.2 A right triangle illustrating the definitions of the sine, cosine, and tangent, plus two useful trigonometry identities

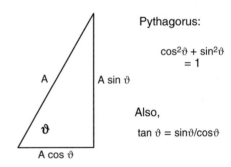

Fig. 1.3 The sine function over two periods expressed in radians and degrees

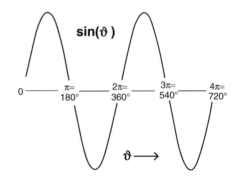

---

[5]If not, the constant can usually be entered easily by the statement $Pi = 4 * ATN(1.0)$[radians].

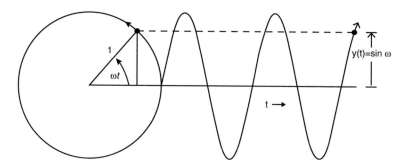

Fig. 1.4   Generation of a sine wave by a unit radius vector rotating at angular velocity ω

1.7   Angular Frequency ω

The Greek lower case omega (ω) is often used to denote angular frequency. Since there are 2π radians in a circle, the angular frequency is related to the cyclical frequency f by

$$\omega = 2\pi f. \text{ [radians per second]} \qquad (1.4)$$

This definition provides a useful shorthand notation in writing later equations. From the basic trigonometric definition of the sine (Fig. 1.2), it is easy to see how a radius vector of unit length rotating at angular velocity ω generates a sine wave. In Fig. 1.4, the radius vector started rotating at $t = 0$ when aligned with the horizontal axis. The vertical position of the tip of the vector then traces out a sine wave as a function of time. In the illustration, the sine wave is spread out with time to the right of the circle, and the vector has rotated through a little more than two complete turns.

1.8   Phase Angle ϕ

The concept of phase for a sine wave is illustrated in Fig. 1.5. The phase angle ϕ is added to the argument of the sine function and is essentially a measurement of the time delay (in units of ω) between different sine waves. The figure illustrates the relative positions of sine waves for three different phase angles spaced by 45° ($\pi/4$ radians). A phase shift of 90° ($\pi/2$ radians) would convert a sine wave into a cosine wave.

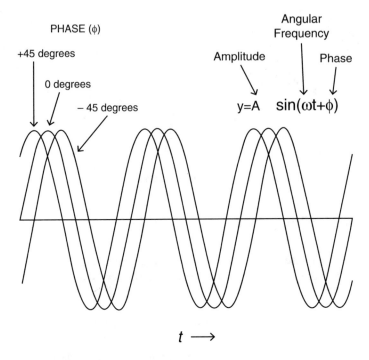

Fig. 1.5 Sine waves with different phase angles

1.9   Propagation Constant *k*

Another useful notation is that for the propagation constant k of a wave, which by definition is[6]

$$k = 2\pi/\lambda. \tag{1.5}$$

Hence, by use of these definitions, solutions to the wave equation such as those in Eq. (1.2) might be written more simply as,

$$y = \sin(kx \pm \omega t) \tag{1.6}$$

for which Eq. (1.1) becomes

$$\lambda f = \omega/k = 2\pi f/k = c. \tag{1.7}$$

---

[6]In two- and three-dimensional problems, k becomes a vector quantity that also describes the direction of propagation of the wave as well as having the magnitude, $2\pi/\lambda$. Its components play an important role in determining the allowed modes of vibration in two-dimensional resonant systems from spatial boundary conditions.

An important property of the wave equation is that it is linear; any linear combination of solutions of type (1.6) will also be a solution of the wave equation, provided that condition (1.7) is satisfied for each component.

We have defined many mathematical terms and equations, but they allow a concise description of the physical phenomena involved. If everything had to be explained in words, little progress in understanding would have been made over the centuries.

1.10 Reflections, Standing Waves, and the Vibrating String Modes

Whenever there is a discontinuity in the medium through which a wave propagates, there will be a change in the velocity of propagation and the generation of a reflected wave. For example, light is partly reflected when it hits a glass window. (The velocity of light is generally slower in the glass than in the air.) But, immersing a quartz lens in a beaker of carbon tetrachloride makes it disappear. (The refractive index, hence velocity of visible light, is the same in those two materials.) Sound waves running down a pipe are reflected at the open end. (The effective velocity of sound is reduced by reflection from the walls of the pipe compared to that in the room.) Similarly, the sound waves running through a brass instrument are reflected by the flaring bell of the horn at the open end. A pulse sent down a string under tension will be reflected by the end support. The "crack" of a bull whip is created by the reflection of a pulse at the free end of the whip, and so on. (The loud "crack" of the bull whip is created because the whip tapers toward the end so that the wave velocity speeds up and the pulse travels at supersonic velocity when it is reflected. One actually hears the tip of the whip breaking the sound barrier.)

It is useful to distinguish between "soft" and "hard" reflections. A "hard" reflection is associated with a change of phase of 180° (π radians) as, for example, occurs at the end of a fixed string where the total amplitude is forced to be zero. The amplitude of the reflected wave changes sign, and the two running waves traveling in opposite directions cancel each other out at the point of support. A "soft" reflection involves no change of phase at the discontinuity and occurs at the end of a string that is free to move. In that case (for instance, the free end of a bull whip), the reflected running wave has the same sign as the incident wave. If the wave velocity decreases in the new medium after the discontinuity, the reflection is said to be "hard," and vice versa.

Suppose we do have a medium containing two such discontinuities—for example, a vibrating string under tension fastened rigidly at the points $x = 0$ and $x = L$. Further, let us suppose that the two running waves generated by the reflections add up to the form

$$y = \sin(kx - \omega t) + \sin(kx + \omega t) \qquad (1.8)$$

where the first term represents a running wave in the $+x$ direction and the second term is a running wave in the $-x$ direction. If we make use of the identity from trigonometry that

$$\sin(A - B) + \sin(A + B) = 2 \sin A \cos B , \tag{1.9}$$

we can rewrite Eq. (1.8) as

$$y = 2 \sin kx \cos \omega t . \tag{1.10}$$

That is, we let $A = kx$ and $B = \omega t$ in Eq. (1.9) in order to get Eq. (1.10).

Equation (1.10) can only be satisfied by particular values of the propagation constant that we will denote by kn. Because we must have $y = 0$ at both ends of the string (at $x = 0$ and $x = L$) for all values of the time, the spatial boundary conditions result in special solutions for which

$$\sin k_n L = 0 ,$$

which in turn means that we must have

$$k_n L = n\pi \text{ or } \lambda_n = 2L/n , \text{ where } n = 1, 2, 3 \ldots \tag{1.11}$$

This condition defines a set of allowed modes on the string whose resonant frequencies are given by substituting the values of λn into Eq. (1.7). The corresponding frequencies for these modes are given by

$$f_n = n(c/2L) \text{ where } n = 1, 2, 3 \ldots \tag{1.12}$$

Note that the resonant frequencies of the string increase as the reciprocal of its length (i.e., as $1/L$.)

The complete solution for the resonant modes of the string is obtained by substituting Eqs. (1.11) and (1.12) in Eq. (1.10). In that way, the full solutions for the vibrating string are found to be of the form

$$y_n = \sin(n\pi x/L) \cos \left(2\pi n \frac{c}{2L} t \right) \text{ where } n = 1, 2, 3 \ldots \tag{1.13}$$

For a given mode (value of n), the spatial variation is multiplied by a time-dependent factor that oscillates between -1 and $+1$ at frequencies that are harmonics (integer multiples) of $c/2L$. Because the wave equation for which these are particular solutions is linear, any linear combination of the solutions in Eq. (1.13) is also a solution to the vibrating string problem. That property was tacitly employed in writing Eq. (1.8) in the first place. The first three resonant modes of the vibrating string are illustrated in Fig. 1.6. As shown in the figure, the values of x for which the amplitudes of the standing waves are always zero are called "nodes." (The points at $x = 0$ and $x = L$ are nodes for all resonant modes.) The maxima are often called "antinodes."

Fig. 1.6 Resonant modes of
a vibrating string

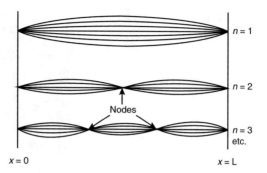

As shown in Appendix B, the velocity of the transverse waves on a vibrating string is given by

$$c = \sqrt{T/\mu},\qquad(1.14)$$

where T is the tension in the string and μ is the mass density per unit length. (The result follows by application of Newton's law of motion to a string vibrating at small amplitude.) The inverse square-root dependence on the mass is characteristic of vibrating systems ranging from the harmonic oscillator (see Appendix A) to kettle-drum heads. This general dependence on mass explains why winding copper wire around the base strings of a piano lowers their pitch for the same tension and length. It also indicates why the pitch of a wine glass, when rubbed by a wet finger on the rim, goes down as the glass is filled with more and more wine—thus, increasing the effective mass in the vibrating system.[7] (Intuitively, most people would expect the pitch to go up because the vibrating length of the glass seems to get shorter.)

1.11 Sound Waves and Open Pipes

Surprisingly, the boundary conditions in the case of the open pipe are exactly the same as those for the vibrating string. The time-varying pressure amplitude must go to zero at both ends of an open pipe because at that point the pressure is forced to equal the air pressure in the room. Consequently, the modes of an open pipe are just the same as those for the vibrating string shown in Fig. 1.6, except that the quantity plotted should be the pressure amplitude in respect to the room pressure rather than the displacement of the string. Of course, the velocity now becomes the velocity of sound (about 1130 ft/s in air), instead of that given by Eq. (1.14).

---

[7]If you do not have access to wine and a fine crystal goblet, try it with an ordinary glass of water, hitting the edge with a spoon.

Fig. 1.7 Resonant modes of
an open pipe

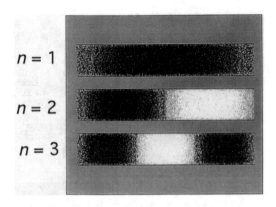

In contrast to the transverse waves propagated along a vibrating string, sound waves consist of alternate longitudinal pressure maxima ("compressions") and pressure minima ("rarefactions"). These pressure fluctuations are sometimes represented in schematic drawings by a varying density of dots. However, it is a little tricky to represent the modes accurately in that way because the rarefactions represent a negative amplitude (absence of dots) in respect to the ambient room pressure. To illustrate the problem here, I have used black dots for positive pressure and white dots for negative pressure on a gray background representing the constant ambient pressure in the room. The first three open-pipe modes are illustrated in that manner in Fig. 1.7.

Note that the black-dot density for $n = 1$ traces out half of a sine wave along the length of the pipe, whereas that for $n = 2$ represents the full period of a sine wave. Without a motion picture representation, it would be hard to show how the standing waves varied with time over one period as done for the vibrating string in Fig. 1.6. At half the period, the black and white regions would be interchanged and at the quarter-period points, the entire figure would be gray. The nodes occur at the boundaries between the black and white regions. It is easy to see that a hole drilled in the side wall of the pipe at half its length would kill the mode for $n = 1$. (The hole would force the inside pressure to equal the room pressure at that point.) If the pipe were oscillating during the drilling operation, one would hear the pitch suddenly jump up an octave as the oscillation switched from the $n = 1$ mode to the $n = 2$ mode. The latter has a node at the midpoint and would not be affected by the hole. The experiment is analogous to forcing a violin string to vibrate on the second harmonic by touching the string lightly in the middle.

1.12 Closed Pipes

The resonances in a closed pipe are trickier to derive because the boundary conditions are different at each end. At the open end (which we will take to be at $x = 0$), it is just the same as with the vibrating string, and the fluctuating pressure amplitude must be zero. However, at the closed end of the pipe ($x = L$), there will be a pressure maximum because at that point the air molecules bump against the end of the pipe rather than moving against an opening to the air in the room. (As we will discuss later, the flow of air molecules behaves in opposite sense to the pressure in such a resonance and must be zero at the closed end. In contrast, the flow will be a maximum at the open end, where the driving pressure for the resonance is nearly zero.) Hence, the closed-pipe resonances are now ones for which

$$\sin k_n L = \pm 1$$

rather than zero. (Of course, the amplitude of the pressure standing wave still goes to zero at $x = 0$.) The closed-pipe modes are then determined by requiring that $k_n L$ be an odd multiple of $\pi/2$, hence

$$k_L = (2n - 1)\pi/2 \text{ where } n = 1, 2, 3 \ldots \tag{1.15}$$

or

$$k_n L = \pi/2, 3\pi/2, 5\pi/2, \ldots$$

and the resonant frequencies for the closed-pipe modes are given from Eq. (1.7) by

$$f_n = c/4L, 3c/4L, 5c/4L, \ldots \tag{1.16}$$

The pressure standing waves at $t = 0$ for the first few resonances are shown in Fig. 1.8, using the convention of Fig. 1.7. These resonances are all odd harmonics of the fundamental closed-pipe resonance $(c/4L)$, which itself is half the fundamental frequency of an open pipe of the same length.[8] Consequently, simple closed organ

Fig. 1.8 Resonant pressure modes in a closed pipe

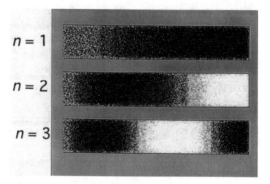

$n = 1$

$n = 2$

$n = 3$

---

[8]Organists have the quaint custom of specifying stops in terms of the length of an open pipe that produces the same pitch as the one in question. Thus, a "16-ft Gedakt" (old German for "closed pipe") on the stop list would actually be about 8-ft long and both it and the hypothetical open pipe would resonate at ≈ 35 Hz.

Fig. 1.9 Time variation of
the pressure modes in a
closed pipe

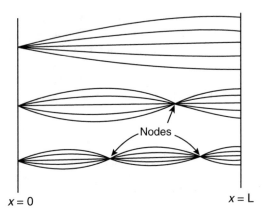

$x = 0$ $x = L$

pipes tend to sound a little like clarinets, which also mainly produce odd harmonics
because they have single reeds that vibrate in such a manner as to be closed most of
the time. (As we shall see, square-wave generators also produce only odd harmonics
and tend to sound like clarinets, too.) The time variation of these mode distributions
is illustrated in Fig. 1.9.

1.13 Determining Resonances from Phase Shifts

There is a very simple way to determine the fundamental frequencies of these
vibrating systems that is just based on phase shift analysis. The main point is that
the running waves have to close on themselves in a round-trip through the system.
That fact implies a condition on the phase of the running wave: The round-trip phase
shift has to be an integral multiple of 2π for a resonance to build up. (That is just as
true in laser oscillators as it is in vibrating strings or organ pipes.) The total phase
shift at frequency f_n in a round-trip through length $2L$ is made up of two parts: first,
the time-delay phase shift, $2\pi f_n(2L/c)$, where $(2L/c)$ is the time for the running
wave to make a round-trip during which $f_n(2L/c)$ cycles occur (each contributing
2π radians to the total phase shift); and second, the phase shifts that take place upon
reflection at the end points.

In the vibrating string and open pipe, there are two "hard" reflections (one at
each end) which each introduce a phase shift of π radians. Hence, the resonances
are given by

$$2\pi f n(2L/c) + 2\pi = n2\pi \quad \text{where } n = 1, 2, 3, \ldots$$

Dividing the equation by 2π and rearranging terms, we get

$$f_n = (n - 1)c/2L \quad \text{where } n = 1, 2, 3, \ldots \tag{1.17}$$

Here, the case $n = 1$ corresponds to the uninteresting DC solution where $f = 0$. The frequencies of interest are all uniformly spaced at $c/2L$ as we found previously. The result is the same as that in Eq. (1.12) if we replace $n - 1$ by n in Eq. (1.17). In the case of the closed pipe, there is only one hard reflection (at the open end). Hence, the resonances are given there by

$$2\pi \times f_n(2L/c) + \pi = n2\pi ,$$

or rearranging terms,

$$f_n = (2n - 1)c/4L \text{ where } n = 1, 2, 3, \ldots \tag{1.18}$$

Note that the resonant frequencies automatically turn out to be odd multiples of $c/4L$. In non-cylindrical pipes, additional terms are sometimes added to the running wave phase shift. For example, an open conical pipe has a hard reflection at both ends and one gets both even and odd harmonics and a lowest frequency of $c/2L$ rather than $c/4L$.

In general, quite a few different modes will be excited in both vibrating strings and in open and closed pipes. The relative energy distributions over these modes affect the tonal color (or *timbre*) of the sound. That energy distribution is determined by characteristics of the excitation process, many of which will be discussed later in this book. Generally, the narrower the diameter of the pipe (and the narrower the diameter of the string), the easier it is to excite higher harmonics.

In the case of both open and closed pipes, the fundamental resonant frequency increases linearly with the velocity of sound. Because that velocity varies with the media, the pitch of the pipe is directly affected by things such as the air temperature and humidity content of the air, or the nature of the gas blown through the pipe. (See Table 1.2.) For example, using helium instead of air can increase the pitch by a factor of about 2.8—or much more than an octave, and the speaker who has just inhaled helium sounds like "The Munchkins" (from The Wizard of Oz). Similarly, tightening the tension on a violin string (which increases the velocity of wave propagation) raises the pitch of the string.

The velocity of sound in a gas varies as

$$c = \sqrt{\gamma RT/M} \tag{1.19}$$

where T is the absolute temperature in degrees Kelvin, M is the molecular weight of the gas, γ is the ratio of the specific heat of the gas at constant pressure to that at constant volume, and R is the universal gas constant ($R = 8.31441$ J/K mol). As in Eq. (1.14), the velocity decreases with $1/\sqrt{M}$.

Table 1.2   Velocity of sound waves in different media

Material	Temperature (°C)	Velocity (m/s)	(ft/s)
Air: (dry)	0	331.4	1087.3
(dry)	20	343.4	1126.6
(50% humidity)	20	344.1	1128.8
(100% humidity)	20	344.7	1130.9
(at sea level)	0	340.3	1116.5
Helium	0	965	3166
Carbon dioxide	0	259	850
Water: (distilled)	25	1496.7	4910.4
(sea)[a]	25	1531	5023
Aluminum		5000	16,404
Brass		3480	11,420
Steel: (mild)		5200	17,060
(Stainless)		5000	16,400
Lead (rolled)		1210	3970
Glass (Pyrex)		5170	16,960
Quartz (fused)		5760	18,900
Wood: Ash (across rings)		1390	4560
Ash (along fiber)		4670	15,320
Beech (along fiber)		3340	10,958
Maple (along fiber)		4110	13,480

Source: CRC Handbook of Chemistry and Physics (69th Edition)

[a]According to Kuperman and Lynch (2004), the velocity of sound in shallow sea water is $c \approx 1449 + 4.6T + (1.34 - 0.01T)/(S - 35) + 0.0216z$ m/s where T is in degrees Celsius, S is the salinity in parts per thousand, and z is the depth in meters. Values for different woods used in musical instruments are given in Bucur (1988 and 2006)

1.14   Wave Propagation

The seventeenth-century physicist Christian Huygens invented a method for predicting the motion of a wave front that is useful in describing the propagation of sound waves. His principle states that for each point on the wave front at one instant in time, the position of the wave a time t later is given by constructing a series of spherical wavelets of radius $r = ct$ centered at each point along the original wave front. (The radius r, of course, is simply the distance the wavelet would expand in time t at velocity c.) This type of construction is illustrated in Fig. 1.10 for a few points on a curved wave front. Although tedious to carry out by hand, it is easy to write a computer program that will draw as many wavelets as you want. The approach is illustrated in Figs. 1.10 and 1.11.

Fig. 1.10 Illustration of a
Huygens' wavelet
construction for a wave front
moving to the right. This is an
expanding spherical wave

Fig. 1.11 Plane wave
propagating to the right

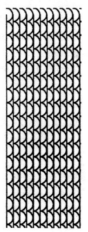

1.15   Refraction by a Continuously Varying Medium

A less familiar example of sound propagation occurs when there is a continuous variation in the velocity of sound throughout a medium. An example of that is shown in Fig. 1.12 where a spherical wave is assumed to expand from the top left into a region where the velocity of sound increases continuously with decreasing height. This type of phenomenon occurs when the ground is much warmer than the air (for example, at sunset). The air density then increases gradually with height above the ground, which from Eq. (1.19) implies that the wave velocity is continuously decreasing toward the top of the figure. As is evident from the figure, these conditions imply a continuous bending of the wave front toward the ground as it progresses to the right. This in turn can lead to focusing effects where a source that is normally inaudible such as children playing at dusk in the distance becomes unusually clear. A similar effect can occur when there is wind flowing

Fig. 1.12 Refraction of a
spherical wave by a medium
in which the velocity of
sound increases toward the
ground level

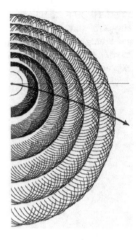

over the ground. The boundary conditions on air flow near the ground can result in
density variations and focusing effects on the downwind side. For example, one may
suddenly hear distant voices, giving the erroneous impression that they have been
blown in by the wind.

1.16   Reflection of Sound Waves from a Flat Mirror (Snell's Law)

The Huygens' method also permits illustrating the spread of a wave through a hole
in a blocking screen and permits deriving the laws for reflection and refraction
("Snell's Laws" in optics) when a plane wave encounters a discontinuity in wave
velocity between two media. The reflection of sound waves is of considerable
importance in acoustic problems.

Consider a plane wave front hitting a plane (mirror) surface as shown in Fig. 1.13.
We only have to construct a wavelet at one point (A) to see what happens. Consider
a wave coming down from the left and striking the mirror at an angle ϕ in respect
to the surface normal at point A. The initial plane wave front extends along the line
from A to B, at right angles to the initial direction of propagation. We construct a
Huygens wavelet of radius r centered at point A. In the time taken for that wavelet
to expand by the distance r, the point B on the initial wave front will have moved
through the distance r to point C on the mirror. According to Huygens' principle,
the reflected wave front will lie along the line through point C that is tangent to
the expanded wavelet at point D. By definition, that line is perpendicular to the
wavelet radius vector from point A to D, which is the direction of propagation of
the reflected wave. The two right-angle triangles (ADC and CBA) with base equal
to r and sharing a common hypotenuse along the mirror surface are congruent from
a theorem in plane geometry. Hence, the angles designated by ϑ in the two triangles
are equal. Similarly, because the sides forming the angle ACD are perpendicular to
the angle formed by the surface normal and the line AD, the latter angle (defined as

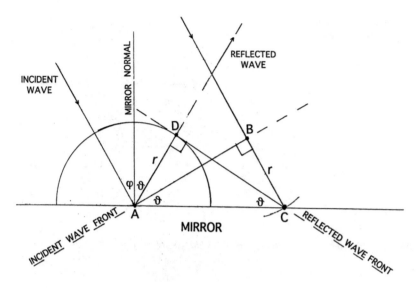

Fig. 1.13 Diagram to illustrate the derivation of Snell's Law for Reflection using a Huygens' wavelet construction

the "angle of reflectance") is also equal to ϑ. Finally, because the angle ϕ is formed by lines that are perpendicular to the mirror surface and to the initial wave front (the line AB), we know that

$$\vartheta = \phi , \tag{1.20}$$

which is Snell's Law for Reflection. That is, the angle of reflection equals the angle of incidence for plane waves hitting a mirror surface.

1.17   Focal Point of a Spherical Mirror

Armed only with Eq. (1.20) and a little plane geometry, we can determine the approximate focal length (f) of a spherical mirror. Consider a plane wave incident from the right on a spherical mirror as shown in Fig. 1.14. Consider the triangle shown in the figure whose long side R is the radius of curvature of the mirror drawn from the point O surface. From Snell's Law, the angle of incidence equals the angle of reflection, as indicated by ϑ in the drawing. The line R crosses two parallel lines (the rays from the incident plane wave). Hence, from plane geometry, the angle of incidence also equals the acute angle to the right inside the triangle. From that, we conclude that the triangle must be isosceles and that the two shorter sides must both be equal. Because the line from point O through point F to the mirror is equal to the radius of curvature, R, the two short sides of the triangle are both equal to $(R - f)$.

Fig. 1.14 Diagram used to determine the focal length of a spherical mirror

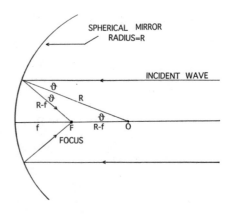

We next apply the Law of Cosines to the triangle, obtaining[9]

$$R^2 = 2R(R - f)\cos\vartheta ,$$

which reduces to

$$f \approx R/2 \tag{1.21}$$

for small values of ϑ. The approximation, $\cos\vartheta \approx 1$, is much better than one might intuitively think and holds within 10% up to angles of $\vartheta \approx 25°$—hence, for a full aperture as seen from the point O in Fig. 1.14 of 50°. Focusing by spherical surfaces is extremely important in determining the acoustic properties of architectural structures since many of them have curved ceilings or domes. A curved ceiling increases the reverberation time enormously because of focusing effects and, generally, results in very poor acoustic properties in auditoriums and even small rooms.

1.18 Reflection from a Moving Plane Mirror

Consider the case shown in Fig. 1.15 in which a plane mirror is moving normal to its surface and toward a point source of sound located at S. When the mirror surface passes through the point D, it is apparent from Snell's Law of Reflection that the image of the source heard by the observer seems to be located at point B, which is located at a perpendicular distance behind the mirror equal to the distance from the source to the mirror. That is, the $BD = DS$ in Fig. 1.15.

---

[9]The Law of Cosines (derived from fundamental theorems in plane geometry) states that $c^2 = a^2 + b^2 - 2ab\cos\vartheta$ where a and b are the two sides of a general triangle enclosing the angle ϑ and c is the side opposite to that angle. For the case in the text, $c = b = R - f$, and $a = R$. (Note that the general result reduces to the Pythagorean Theorem when $\vartheta = 90°$.)

Fig. 1.15 Image location in a
moving plane mirror

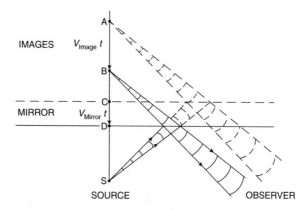

The dashed lines in the Fig. 1.15 refer to an earlier position of the mirror surface
when it passed through the point C. Again, because the image is an equal distance
behind the mirror to the distance of the mirror from the source, $AC = CS$. Hence,
during the time interval t in which the image at point A moved through the distance
$V_{\text{Image}}t$ to point B, the mirror at point C moved through the distance $V_{\text{mirror}}t$ to point
D. Therefore, we know that the distances are related by

$$AB + BC = CD + DS$$

and

$$BC + CD = DS. \tag{1.22}$$

Subtracting the second equation from the first yields

$$AB = 2CD \tag{1.23}$$

or[10]

$$V_{\text{Image}} = 2V_{\text{Mirror}}. \tag{1.24}$$

That is, the apparent velocity of the image toward the source is twice the
velocity of the mirror toward the source, something to keep in mind if you are a
policeman measuring car velocities with Doppler Radar or a physician measuring
blood velocities in the heart with Echo Cardiography.

---

[10]A much simpler derivation follows from a modest use of calculus. Let the distance from the
source normal to the mirror be y, hence the distance from the source to the image is $2y$. Then, the
velocity of the mirror is $V_{\text{Mirror}} = dy/dt$ and the velocity of the image is $V_{\text{Image}} = 2dy/dt = 2V_{\text{Mirror}}$.

1.19   The Inverse Square Law and Diffraction Effects

Although Huygens' principle is powerful in its simplicity, it is only an approximate method for handling wave propagation problems. For one thing, if you took the method literally and drew complete spheres about each point on the wave front, you would see a false backward wave arising in addition to the wave propagating in the forward direction that you expected. Without further modification, the method also does not depend explicitly on the wavelength and cannot predict interference effects between different sound waves with well-defined phase.

A more rigorous way to treat these problems is through solution of the wave equation in three dimensions. When the wave equation is written in spherical coordinates, the solutions in the radial direction for emission from a point source are actually similar to the one-dimensional plane wave solutions discussed in connection with Eq. (1.8). If a wave spreads outward spherically through a uniform isotropic medium, it is clear that the energy flow must fall off according to the inverse square law. Imagine drawing a spherical surface about the point source. The energy from the point source must flow through that surface and the flow per unit area clearly must fall off as $1/r^2$, where r is the radius of the sphere. (That is, the area of the sphere, $4\pi r^2$, increases with the square of the radius.) As shown in Appendix A, the energy in oscillatory motion is proportional to the square of the amplitude; hence, one would expect the amplitude of the spherical wave to fall off as $1/r$. As the wave expands to larger and larger size, it begins to approximate a plane wave over small regions. It is therefore not surprising to find that a rigorous solution of the wave equation for a point source at the origin in spherical coordinates shows that an outgoing spherical wave is of the form[11]

$$y = \sin(kr - \omega t)/r. \qquad (1.25)$$

Thus, an outgoing spherical wave from a point source (or "monopole" radiation) could be portrayed at one instant in time ($t = 0$) as shown in Fig. 1.16 where our previous graphic technique has been employed to illustrate regions where the fluctuating pressure amplitude is positive (black dots) and negative (white dots). As before, the ambient background pressure is shown in gray. However, the dot density in this case has been made proportional to the amplitude squared in order to illustrate the inverse square law decrease in intensity. The pattern shown could represent the radiation from the open end of a closed organ pipe.

---

[11]The singularity introduced near $r = 0$ would generally be removed when one summed the contributions over a real (nonpoint) source distribution. In the present case, the singularity does not occur because $\sin(kr)/r \rightarrow k$ at $t = 0$.

Fig. 1.16 Radiation of a
spherical wave from a
monopole source

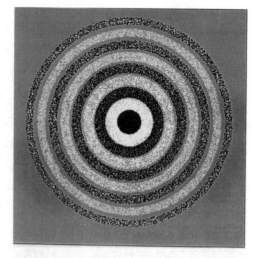

Fig. 1.17 Radiation from a
dipole source

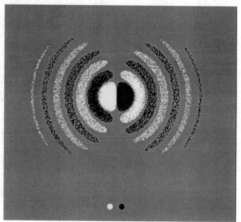

1.20   Dipole and Multipole Radiation

The radiation patterns become more interesting when we consider two or more point
sources radiating with a fixed phase relationship. One of the more well-known
cases—that of dipole radiation—occurs when two sources spaced by half of a
wavelength radiate 180° (π radians) out of phase. That case is illustrated in Fig. 1.17
where the two point sources are located on the horizontal axis. As in our previous
example, the region of positive amplitude is shown in black and that of negative
amplitude is in white, except that the dot density here has been made proportional to
the intensity (i.e., to the square of the amplitude). In this and the following figures,
the circles at the bottom indicate the relative polarity and position of the sources.
Although the intensity still falls off as $1/r^2$, the radiation is very directional and

Fig. 1.18 Radiation from a
linear multipole

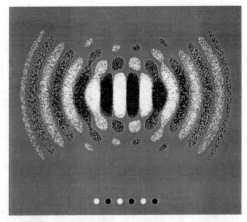

Fig. 1.19 Radiation from an
axial quadrupole

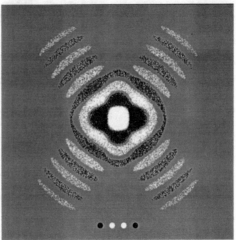

peaks along the axis of the dipole.[12] Examples of this radiation pattern occur with
open pipes (where the two open ends are separated by $\lambda/2$), in the dominant "tea
cup" mode of a kettledrum (where the dipole axis is usually toward the audience),
in many modes of vibrating plates and in instrument soundboards. As illustrated in
Fig. 1.18, adding more dipole sources in line with the first one results in a similar,
but elongated radiation pattern in the horizontal direction.

A still-more complex situation is illustrated in Figs. 1.19 and 1.20, where the
radiation patterns from "quadrupole" source distributions are shown. Again, the
configuration and relative phase relations of the sources are indicated by the circles

---

[12]People familiar with electric-dipole radiation in electromagnetic theory may find this directional
characteristic surprising. In the electric field case, the radiated amplitude is a vector quantity that
peaks in the direction perpendicular to the dipole. With sound waves, the radiated pressure is a
scalar quantity peaking in the direction of the dipole axis.

Fig. 1.20 Radiation from a
planar quadrupole

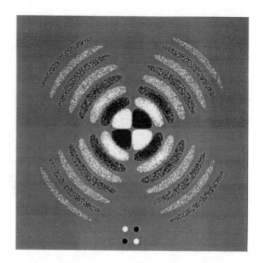

at the bottom of the figures, where black corresponds to pressure maxima and white represents pressure minima. As with the dipole, the point sources are separated by $\lambda/2$. In each case, the radiation is maximum in the plane of the quadrupole and peaks along the 45° directions in respect to the horizontal. As shown, essentially the same planar radiation patterns are obtained from these two rather different source distributions. Examples of both linear multipole and quadrupole source distributions occur in the sound boards of pianos.

1.21 The Doppler Shift

Whenever either the source of sound or the receiver is moving, there is a shift observed from the original frequency of the source. The effect was first described in 1842 by the Austrian physicist Christian Johann Doppler (1803–1853).[13] His initial treatment of the phenomenon dealt with the frequency shift in light waves. But, we will only consider sound waves here (Fig. 1.21).

First, consider the case where the source is moving at velocity V_S toward the observer while emitting sound at frequency f_S. In time T, the source emits $f_S T$ cycles that are spread over a distance in the propagating medium given by $(c - V_S)T$, where c is the velocity of sound in the medium. Consequently, the wavelength of the sound emitted in the medium is

$$\lambda_M = (c - V_S)T/(f_S T) = (c - V_S)/f_S. \qquad (1.26)$$

---

[13]The Austrian physicist Christian Johann Doppler should not be confused with the several contemporary Hungarian musicians having the same last name.

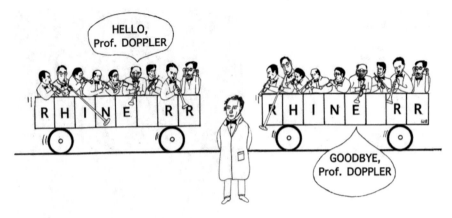

Fig. 1.21 The demonstration at Maarsen

Because the frequency and wavelength in the medium must be related by Eq. (1.7), the frequency measured by a stationary observer in the medium is given by

$$f_M = \frac{c}{\lambda_M} = \frac{cf_S}{(c - V_S)} = \frac{f_S}{(1 - V_S/c)}.$$

If V_S is positive, the observed frequency is higher than the source frequency because the number of cycles emitted in time T has been crowded into a shorter space in the medium than would occur if the source were not moving. If V_S is negative (source moving away from the observer), the observed frequency is lower. In the days of railroad trains, the effect was commonly observed on station platforms. The horn or bell on an approaching locomotive would sound higher in pitch as the train approached than it did as the train receded away from people standing on the platform. The first public demonstration of this effect was conducted by Dr. Buijs Ballot in June of 1845, using a group of trained musicians at stations along the tracks of the Rhine Railroad between the Dutch towns of Maarsen and Utrecht. They listened to the change in pitch as a group of trumpet players riding on a railroad car went by and found that the pitch change agreed well with Doppler's equations. (See Ballot 1845.)

The size of the effect differs in detail when the observer is also moving and is actually more complex with sound than in the case of light.[14] Consider the case where the observer is moving in the same direction as the source, but with velocity V_O in respect to the medium. If V_O is positive (observer moving away from source), the observer now hears a frequency decreased by $V_O T/\lambda M$ (the number of cycles in

---

[14]One conceptual difference in the analysis between sound and light waves occurs because there is no fixed material medium (or "ether") supporting the motion of the light wave. Light waves, of course, can travel through vacuum, whereas sound waves cannot.

the medium the observer passes through in time T). Hence, with both the source and the observer each moving in the same direction, the observer hears the frequency

$$f_O = f_M - \frac{V_O}{\lambda_M} = f_s \frac{(1 - V_O/c)}{(1 - V_S/c)}. \tag{1.27}$$

For $V_S \ll c$ (hence, $\lambda_M \approx c/f_S = \lambda$), one can expand the denominator of Eq. (1.26) giving the total frequency shift $\Delta f = f_S - f_O$ to be

$$\Delta f \approx f_S(V_S - V_O)/c \approx (V_S - V_O)/\lambda_M \approx V_{Rel}/\lambda, \tag{1.28}$$

where V_{Rel} is the relative velocity between the source and observer. When both source and observer are moving toward each other, V_{Rel} and the observed frequency have their maximum values, and vice versa. If the relative velocity is not in the direction of the line between the source and the observer, one needs to use the component of V_{Rel} in that direction in these formulas. The inclusion of a $\cos \vartheta$ term (where ϑ is the angle between the relative velocity vector and the direction from the source to the observer) causes the Doppler shift to decrease and then vanish as $\vartheta \to 90°$.[15]

Problems

1.1 (1) Benjamin Franklin enjoyed walking to the edge of a pond where he would raise his arms in a Moses-like gesture while slyly dropping a little oil on the water. The surface ripples would immediately die out, greatly impressing local bystanders. Explain.[16]

1.2 (a) Radio station WQXR-FM has broadcast classical music in the New York City area for well-over 50 years. It currently uses a carrier frequency of 96.3 MHz. What is the corresponding wavelength? (b) A surgeon employs a green argon ion laser with a wavelength of 488 nm (1 nanometer = 10^{-9} m) to weld detached retinas back in place. What is the corresponding frequency? (Note: both radio waves and visible light travel at about 3×10^8 m/s.)

---

[15] The Doppler shift for light waves always depends on the relative velocity between the source and the receiver because there is no propagation medium (or "ether"). But, there is an additional relativistic Doppler shift in the optical case from "time dilation" when $(v/c)^2$ is not negligible compared to unity that occurs even when the relative velocity is perpendicular to the distance between the source and the receiver. But, the first-order optical shift is also just $\approx V_{Rel}/\lambda$.

[16] Possibly inspired by Franklin, a chemistry professor at Oregon State College in the 1920s enjoyed impressing the students by spitting into puddles on the campus while secretly dropping a small piece of sodium in the water. The resulting reaction would produce a small (and rather dangerous) explosion with yellow flames shooting 6 ft up into the air.

1.3 (3) According to the New York Times (August 3, 2004, p. D1), a giant black hole in the Perseus cluster emits pressure waves through the thin hot gases at "B-flat, 57 octaves below middle C." What is that frequency? What is the corresponding period?

1.4 (a) A physicist laid out a house on the side of a hill using a red helium–neon laser oscillating at 474 THz. What was the laser wavelength? (b) In order to minimize standing wave resonances, the same physicist designed a living room that would be 10 ft longer than the wavelength of the lowest note ($C = 32.7\,\text{Hz}$) on his pipe organ. How long did he make the room? (Take the velocity of sound to be 1100 ft/s.)

1.5 A physics professor hurriedly grabbed a tank of hydrogen rather than helium on his way to give a demonstration of the "Munchkin effect" for his class in acoustics. (Fortunately, there were no open flames in the lecture hall, or the demonstration could have been really spectacular.) (a) Noting that the molecular weight of $H_2 = 2$ and that for air is about 29, by about what factor did the resonant frequencies in his vocal tract go up when he inhaled a lung full of hydrogen? (b) By how much would they have gone up if he had used helium (molecular weight = 4) as originally planned?

1.6 A string has a kink in it at two thirds of its length. If the full length of the string is L, what frequencies would you expect from the string?

1.7 An auditorium has a microphone 10 ft from a loudspeaker in its Public Address system. (a) At what frequencies would you expect the system to oscillate ("sing") from regenerative feedback? (b) Suppose it did not oscillate where you expected. What might cause a shift from the expected frequencies?

1.8 An elegant apartment owned by a French psychiatrist living in Paris near the Madeleine has a water closet containing a nineteenth-century flush toilet with a large water tank mounted on the wall 6 ft above the bowl. When the chain is pulled, a flap valve opens at the bottom of the tank causing water to pour down a narrow pipe into the bowl. The sound produced is a beautiful clarinet-like glissando rising from a low note through a two and a half octave range (See footnote 17). How does this work and what is the lowest frequency? (Take the velocity of sound to be about 1100 ft/s.)

1.9 The whistle on a steam engine consists of four closed pipes with lengths of 6.31 in, 5.30 in, 4.46 in, and 3.54 in. A composer wants to imitate the sound on a piano. Assuming that only the first two harmonics are important for each pipe, what would the chord look like if the pipe were blown with air? (Optional for music majors: How would the lower chord be resolved?)

1.10 The monks' cells in the San Marco monastery in Florence run by Girolamo Savonarola (before he was hanged and burned in 1498) have an unusual design. The ceilings are spherically concave and about 10 ft above the floor, thus producing strong resonances in the audio range for sound propagating vertically. For the purpose of this problem, assume that the radius of curvature of the ceiling is 20 ft, hence that the ceiling acts like a mirror with a focal point at the floor. One monk standing in the middle of these resonant modes mumbling prayers sounds like an entire army of monks to the uninitiated. What would the first few frequencies be? (Hint: Draw the ray diagram for sound waves from a source on the floor that are focused by the ceiling. How many trips up and down are required for the wave front to repeat itself?)

1.11 It is often noted by organists that the sound from free-standing large closed pipes "carries" much further than that from large open pipes. Why should that be the case?

1.12 How could a person with perfect pitch determine the speed of a train in which he or she were riding?

1.13 (a) About how fast would the car carrying the trumpet players have to move for the musically trained observers near the track at Maarsen to hear a shift of one-half step on the Well-Tempered Scale for the approaching train? (b) About how large an interval on the well-tempered scale would the total shift (coming plus going) be at that speed? Assume the velocity of sound is about 1100 ft/s $\approx$ 680 mph. (The interval between successive half-steps on the WTS is $\sqrt[12]{2} \approx 1.0595$.)

1.14 A clarinetist who abhors vibrato is playing in a large room with an overhead fan. He is distressed to realize that his tone has developed a pronounced vibrato. Explain.

1.15 The TGV (Très Grande Vitesse) runs at 322 km/hr (200 mph) between Paris and Lucerne. (a) Suppose the engineer blows a 500 Hz whistle as a warning to a farmer standing near the track. What pitch does the farmer hear? (b) Suppose a French driver is cruising along at 200 km/h (about 124 mph) on a road parallel to the track, but heading toward the train. What pitch does he hear? (c) What would be the approximate answer for part b) using Eq. (1.27)? (Take the velocity of sound to be 680 mph.)

1.16 A traffic officer is checking the speed of motorists with a Doppler Radar device. Suppose he stands at the edge of the road and uses a microwave source with a wavelength of 10 cm while aiming his radar gun at an on-coming car going 80 mph. Treating the car as a plane mirror, what is the frequency shift he would measure in the wave reflected from the car? (There are 5280 ft per mile and 2.54 cm per inch.)[17]

---

[17]During the summer of 1940 when the first portable radar gun developed at Loomis Research Laboratories in Tuxedo Park 35 miles north of New York City as part of the war research program was tested on unsuspecting motorists on a nearby highway, one research worker warned, "For Lord's sake, don't let the cops know about this!" (Herken 2002, p. 36).

1.17 An adult giraffe has a neck 6.9 ft long. If it acted like a closed pipe, what would the resonant frequency be?

1.18 (18) The dunes at Nevada's Sand Mountain "sing" at C, two octaves below middle C.[18] What is the wavelength?

---

[18]Kenneth Chang, "The Secrets of the Singing Sand Dunes," The New York Times, July 25, 2006.

Chapter 2
Spectral Analysis and Fourier Series

2.1 Musical Sounds

As discussed in Chap. 1 in the case of vibrating strings and organ pipes, there are generally many different modes in which a resonant system may vibrate. (See Fig. 1.6 and related discussion.) Generally, more than one of these modes are excited simultaneously in the sounding of a musical instrument. Indeed, their presence or absence is what determines the beauty of a particular tone as well as the difference in sound from one instrument to another. Which modes are excited is not only determined by the characteristics of the resonant system but also by the way in which it is excited. For example, the slipping of the violin string on the bow, or the vibration of the reed in an oboe or a bassoon excites a particular set of modes in those instruments.

Musicians often refer to the extra sounds produced above the pitch of a note as overtones. In most cases, these overtones are harmonically related to the fundamental frequency in that their frequencies are integral multiples of the fundamental. Unfortunately, that fact sometimes produces confusion between the meaning of the musician and that of a scientist analyzing the tone. For example, the first overtone of a vibrating string is actually the second harmonic of the fundamental pitch (its frequency is twice the fundamental frequency), the second overtone is the third harmonic (or three times the normal pitch), and so on.

Life is further complicated by the fact that some instruments produce overtones that are *not* harmonically related to the fundamental. Examples are the kettledrum (or timpani), bells, bars, wood blocks, and so on. Even more surprisingly, the plucked string turns out to have overtones that are not precisely harmonics of the fundamental, whereas the bowed string does.

The original version of this chapter was revised: Equations on pages 40 and 41 were corrected. The correction to this chapter is available at https://doi.org/10.1007/978-3-319-92796-1_8

© Springer Nature Switzerland AG 2018
W. R. Bennett, Jr., *The Science of Musical Sound*,
https://doi.org/10.1007/978-3-319-92796-1_2

In most instruments characterized by harmonically related overtones, the excitation mechanism itself produces a locking effect causing the various harmonics to be in phase with the fundamental. That results from the action of the vibrating reed in woodwinds, the vibrating lips in the case of brass instruments, and the stick-slip motion of the string against the bow in instruments of the violin family.

Because the relative distribution of overtones plays such a key role in defining the characteristics of musical sound, it will help to review some of the methods used to determine spectral distributions. By the term "spectral distribution," we mean the variation of amplitude or intensity of a waveform with frequency. (This chapter gives a qualitative discussion of several important methods of spectral analysis, whose mathematical bases are derived in the appendices.)

2.2 Early Methods of Spectral Analysis

The early pioneer, Helmholtz, did much of his experimental research in acoustics using volume resonators for which the design bears his name. His basic idea was to have a large spherical volume of air resonant at a given frequency that could be driven by sound waves entering through a small aperture. (See Fig. 2.1.) A much smaller tube at the other side of the sphere was designed to fit snugly through a wax seal into his ear so as to block off external sounds. Helmholtz had a set of such matched resonators made that were tuned to different frequencies and managed to accomplish an amazing amount of research with this relatively crude type of apparatus.

Michelson (1903) designed another kind of spectrum analyzer to study the fringes produced in optical interferometry. His analyzer (Fig. 2.2) consisted of a large number of vertical rods tuned to different frequencies. These would vibrate at sympathetic resonances in the audio range when a horizontal lever was made to trace out a particular waveform. The extent of vibration of each rod was recorded on paper and thus provided a measure of the spectral amplitudes.

Fig. 2.1 An original Helmholtz resonator. Sound entered the resonant volume at *a* and was monitored through the narrow tube at *b*, which was covered with wax molded to fit the experimenter's ear. From Helmholtz (1885, pp. 43, 373). The resonant frequency is derived in Appendix A. See Eq. (A.68)

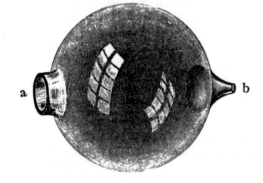

Fig. 2.2 Michelson's
spectrum analyzer consisting
of vibrating rods tuned to
different frequencies. From
Michelson (1903, p. 67)

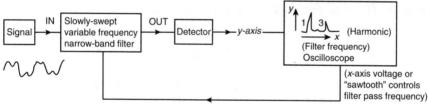

Fig. 2.3 Schematic diagram of an analog electronic spectrum analyzer. The illustration shows the
waveform from a closed organ pipe being broken into its spectral components—principally, the
first and third harmonics (in practice, such devices often use one very good narrowband filter at a
fixed high frequency which looks at the difference frequency produced when the audio input signal
is multiplied by a sine wave from a swept, high-frequency oscillator)

A more-sophisticated electronic approach was developed at the Bell Laboratories
during the 1930s to study sound in which a signal could be recorded on a magnetic
disc that was repeatedly scanned while a narrowband filter swept slowly through
the audio frequency range. The spectra were shown by using the rectified output of
the filter to darken a piece of paper. Apart from the time required to observe the
spectrum, the recording medium had very limited dynamic range.

A more recent version of that approach having a much wider dynamic range
is shown in Fig. 2.3. The main difficulty with such analyzers is that you need a
sustained source of sound (or a tape loop) for analysis (Fig. 2.4).

Fig. 2.4 Odd-harmonic
spectrum from a square wave
determined with this analyzer

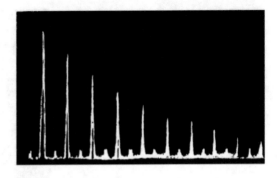

2.3   The Decibel (dB)

Relative spectral amplitudes are often described in terms of "dB," or decibels. The
"decibel" was originally called the "transmission unit" and referred to the loss in
a standard length of telephone wire (Martin 1924). It was subsequently renamed
in honor of Alexander Graham Bell, but with his name misspelled and entered in
lower case. The more recent abbreviation for the unit (the "dB") at least capitalizes
the "B." The most important thing to remember about the unit is that it represents
a logarithmic measure of the *ratio* of two intensity (or power) levels. Specifically,
the ratio of the intensity levels I_2 to I_1 is defined in dB as

$$10 \log_{10} \left(\frac{I_2}{I_1} \right) . \tag{2.1}$$

Because the intensity is generally proportional to the square of an amplitude (for
example, $I_2 \propto A_2^2$ and $I_1 \propto A_1^2$), the intensity ratios in dB may also be written as[1]

$$10 \log_{10} \left((A_2^2)/(A_1^2) \right) = 10 \log_{10} (A_2/A_1)^2 = 20 \log_{10} (A_2/A_1) \tag{2.2}$$

where A_2 to A_1 is the amplitude ratio. If the wave is attenuated in passing through
a medium, the result is a negative number of dB, and vice versa. Some useful
benchmarks to keep in mind: 10 dB corresponds to an intensity ratio of 10:1 whereas
doubling the intensity only amounts to about 3 dB. On the other hand, doubling the
amplitude results in a gain of about 6 dB. Conductor Leopold Stokowski became
enamored with decibels during the 1930s. People at the Bell Laboratories gave him
a dB meter hooked to a microphone which he used on the podium of the Philadelphia
Orchestra. One can imagine comments during rehearsals such as, "Mr. Tabuteau, I'd
like 6 dB more in the crescendo at letter *A*."

---

[1]Note by definition,

$a = \log_{10} b$ means that $b = 10^a$.

Therefore,

$\log_{10} b^2 = \log_{10} 10^{2a} = 2a = 2 \log_{10} b$

Table 2.1 Sound pressure levels (SPLs) referred to 2×10^{-4} dynes/cm^2 (100 dB $= 1\,\mu$W/cm^2)

SPL (in dB)	Source
200	16 in. naval gun at 12 ft
140	Jet taking off
130	Jet taxiing
125	Student rock concert at Yale university at 50 ft
120	Threshold of pain
110	Construction site (pneumatic drills at 100 ft)
100	Lawn mower
95	Good Stradivari violin at 3 ft; most car interiors; times square traffic.
90	San Francisco symphony in Carnegie hall playing Mahler (100 ft)
85	Kirov orchestra in Verizon hall playing Wagner (100 ft)
80	City street; alarm clock (at 2 ft)
75	Frappuccino maker at 3 ft
70	Shouting at 4 ft
60	Normal conversation at 3 ft; busy office
50	Quiet office or classroom
40	Living room
30	Bedroom at night
20	Recording studio
10	Yale maintenance workers
0	Snowflake hitting ground

Note: These are total SPLs and not "A weightings." Extended listening to sound levels above 90 dB is thought by OSHA to be damaging to the ear. The Bavarian Radio Symphony Orchestra was cited for violating a new law on noise level while rehearsing the "State of Siege" by Dror Feiler, a piece containing sustained sound levels of 97 dB (N.Y. Times, April 20, 2008, p. 1)

Confusion may be introduced by people who refer to absolute sound levels in decibels. What they usually mean by that terminology is that the sound intensity ratio is in respect to a standard reference level where 0 dB corresponds to 2×10^{-4} dynes/cm^2. That value is approximately the threshold of hearing at 2 kHz. On that same scale, 120 dB is about the threshold of pain. By coincidence, an increment of 1 dB is about the smallest change in intensity ratio that the average human ear can detect, although the value varies somewhat with individuals, with frequency, and with sound level. (See Riesz 1932.) Using such SPL ("Sound Pressure Level") meters, the various peak absolute sound levels shown in Table 2.1 were obtained.

Recently, circuits containing multiple frequency-band transmission filters to cover the audio spectrum have flooded the "Hi-Fi" market and are generally calibrated in decibels. The frequency bands are often spaced at octave, or even one-third octave, intervals and use light-emitting diodes to indicate the relative sound intensity levels within the different bands. Although they have the advantage of rapid response and can provide a rough portrayal of spectra in "real time," the resolution is limited by the number of filters one can crowd into a small circuit. A nice application of this display has been incorporated in the sound level meter

Fig. 2.5 A portable real-time spectrum analyzer and sound level meter

made by the AudioSource company. As illustrated in Fig. 2.5, the meter is portable and provides a real-time spectral display of absolute sound levels in dB detected from a calibrated condenser microphone.

With the computer methods discussed below, one can increase the resolution merely by increasing the dimension of a column array. It used to be that machine "running times" were an impediment to mathematical analysis. But, now that we have computer speeds in the GHz domain and nearly unlimited random access memory storage capability, Fourier analysis can be done throughout the entire audio spectrum in real time with good resolution.

2.4 Fourier Analysis

Although the mathematical techniques involved in Fourier analysis have been known since the 1820s, the ability to use this method rapidly in real-time analysis required innovations in computer technology that did not arise until the 1970s. One major advantage is that you do not need a very long sample of a waveform to determine its spectral distribution. Indeed, if you know that the waveform is truly periodic, you only need one period for analysis. Hence, in many cases, the spectra can be captured "on the fly."

2.5   A Brief Historical Background of Fourier Series

Fourier began the mathematical work which led to his formulation of what we now call "Fourier Series" in a theoretical study of heat flow in 1807, stimulated by engineering problems encountered in the boring of Napoleon's cannons.[2] Fourier solved the heat-flow equation for sinusoidal distributions of temperature. But, he needed an infinite series of such solutions to describe the results of arbitrary temperature distributions on the walls of the material. Fourier's initial paper on this subject was highly controversial. Several outstanding mathematicians did not believe what he was saying and urged the paper's rejection. In fact, his paper was refused publication until Fourier himself was elected President of the French Mathematical Society in 1822.

According to Whittaker and Watson (1920), the major background developments were as follows:

1. *D'Alembert* had solved the wave equation for the vibrating string problem and obtained a solution of the form

$$y(x, t) = 1/2[f(x + ct) + f(x - ct)]. \tag{2.3}$$

   Note that $y = f(x)$ is the shape of the string at time $t = 0$.
2. *Daniel Bernoulli* next showed that a formal solution to the problem was also given by a sum of solutions of the type summarized in Chap. 1 by Eq. (1.13):

$$y(x, t) = \sum_{n=1}^{\infty} (A_n \sin(n\pi x/L) \cos(n\pi ct/L)) \tag{2.4}$$

   where the A_n are adjustable constants.[3] Bernoulli went on further to claim that this result was the most general solution to the problem possible. (Although the claim sounded like a Madison Avenue advertising slogan, it turned out to be right.)
3. Neither *d'Alembert* nor *Euler* believed Bernoulli and protested that such a series could not possibly converge to a function such as $f(x) = x(L - x)$ at

---

[2]Baron Jean-Baptiste-Joseph von Fourier (1768–1830) accompanied Napoleon in 1798 on his expedition to Egypt, where he served as Secretary for Napoleon's newly formed Institut d'Egypt. In Cairo, he did extensive research on Egyptian antiquities and gave advice on engineering matters. He returned to France in 1801, about the same time that the Rosetta Stone and other major ancient Egyptian relics were surrendered to the British. Back in France, he was charged with the publication of an enormous mass of Egyptian material which became known as Description de l'Egypt (in 21 volumes from 1808 to 1825). He was also the first to describe the atmosphere's trapping of heat as "The Greenhouse Effect" in the 1820s. (See Segré 2002, p. 119.)

[3]The notation $\sum_{n=1}^{\infty} (A_n \sin(n\pi x/L) \cos(n\pi ct/L))$ means that you sum the expression for the values of $n = 1, 2, 3, \ldots$ to infinity.

$t = 0$, or even worse, the boundary conditions at $t = 0$ on a plucked harpsichord string.

4. *Fourier* (1822) proved for the first time that such a series did indeed converge in a large number of specific cases while discussing his analytic theory of heat flow.

5. Others (*Poisson*, *Cauchy*, *Dirichlet*, and *Bonnet*) went on to attempt more general proofs (some of them wrong). According to Whittaker and Watson (1902), the first correct proof of convergence was given by Dirichlet.

2.6   A Note on the Convergence of Infinite Series

The concept of convergence of a sum such as that in Eq. (2.4) at $t = 0$ is of fundamental importance in establishing the usefulness of Fourier series. For a rigorous discussion of convergence, the reader should consult a treatise on mathematical analysis such as that by Whittaker and Watson (1902). What follows here is a more pragmatic approach to the problem.

Suppose we have a sum of numbers of the form

$$S = a_1 + a_2 + a_3 + \cdots + a_n \cdots$$

where the nth term is a known function of n. For the sum to converge to a limiting value, a_n clearly must go to zero as $n \to \infty$. Although that is a necessary condition for convergence, it is not a sufficient one. For example, the well-known series

$$S = 1 + 1/2 + 1/3 + \cdots + 1/n + \cdots$$

does not converge, but obviously satisfies that "necessary" condition. Convergence does occur when $a_n + 1/a_n$ goes to zero in the limit that $n \to \infty$. (The divergent case quoted above obviously does not satisfy that requirement.)

In the present computer age, it is often adequate to run off the sum of the series to a few dozen terms to see what actually happens. In that approach, if you stop calculating the sum after $|a_n| < 10^{-7}|S|$, you will usually have reached the convergence limit within the accuracy of the computer. That is, "single-precision" computer calculations in which the mantissa is evaluated to 24-bit accuracy are typically good to only about one part in 10^7. (Of course, convergent series can always be computed in extended precision using more bits for the calculation.)

A numerical example will help for clarification. The infinite series for e^x is given by

$$S = 1 + x + x^2/2 + x^2/3! + \cdots + x^n/n + \cdots \tag{2.5}$$

It is useful to note that the nth term of the series is easily related to the $(n - 1)$th term by

Fig. 2.6 The first 30 terms
for the series in Eq. (2.5) for
$x = 10$

n	a (n)	s
1	0	1
2	10	11
3	50	6.1
4	166.667	22.77
5	416.667	64.44
6	833.334	147.77
7	1388.889	286.66
8	1984.127	485.07
9	2480.159	733.09
10	2755.732	1008.66
11	2755.732	1284.24
12	2505.211	1534.76
13	2087.676	1743.52
14	1605.905	1904.11
15	1147.075	2018.82
16	764.717	2095.29
17	477.948	2143.09
18	281.146	2171.2
19	156.193	2186.82
20	82.207	2195.04
21	41.104	2199.15
22	19.573	2201.11
23	8.897	2202
24	3.869	2202.39
25	1.612	2202.55
26	0.645	2202.61
27	0.248	2202.64
28	0.092	2202.65
29	0.033	2202.65
30	0.012	2202.65

$$a_n = a_{n-1}x/n. \tag{2.6}$$

The series will converge for any finite value of x because

$$a_n/a_{n-1} \to 0 \text{ as } n \to \infty.$$

The first 30 terms for the series are illustrated in Fig. 2.6 for the case $x = 10$. As can be seen from the figure, the increment a_n rapidly builds up for the first few powers of x but goes through a maximum value at about the 11th term. After that, the $n!$ in the denominator rapidly reduces the increment to zero and the series converges to 1 part per million by the 30th term to $S = 2202.65$. To get the numerical value for e (= 2.718282..., the base of the Naperian logarithms), one merely lets $x = 1$ in the series Eq. (2.5). The number π is also the result of a convergent infinite series, as are all the transcendental trigonometric functions.[4]

---

[4]Ramanujan (1914) gave the most rapidly convergent series for $1/\pi$ ever discovered: $\frac{1}{\pi} = \frac{1}{4}\left[\frac{1123}{882} - \frac{22,583}{882^3} \cdot \frac{1}{2} \cdot \frac{1\cdot3}{4^2} + \frac{44,043}{882^5} \cdot \frac{1\cdot3}{2\cdot4} \cdot \frac{1\cdot3\cdot5\cdot7}{4^2\cdot8^2} - \cdots\right]$. Amazingly, the first term by itself gives $\pi = 3.141585041\ldots$ (Ramanujan liked to entertain his friends by reciting the endless digits of π at parties.

2.7   Specific Examples of Convergence for Periodic Series

The following three examples involve convergence of an infinite series for each value of x over the domain $0 \le x \le 4\pi$. All three represent periodic functions that repeat themselves over the range from 0 to 2π. (The range for x from 0 to 4π was chosen to illustrate two periods of the function in each case.) Here, we have used a computer to demonstrate convergence by adding up the terms for different values of n at each value of x. For each of the three cases listed below, a superposition of the first ten terms is shown at the left in Fig. 2.7, and the limit of the series after 100 terms is shown at the right. Although the three cases look superficially similar, the results converge in each case to very different, highly non-sinusoidal functions.

Case 1: "Sawtooth":

$$y = \sin x + \frac{1}{2}\sin 2x + \frac{1}{3}\sin 3x + \frac{1}{4}\sin 4x + \cdots + \frac{1}{n}\sin nx \tag{2.7}$$

Case 2: "Square Wave":

$$y = \sin x + \frac{1}{3}\sin 3x + \frac{1}{5}\sin 5x + \cdots + \frac{1}{n}\sin nx\ [n\ \text{odd}] \tag{2.8}$$

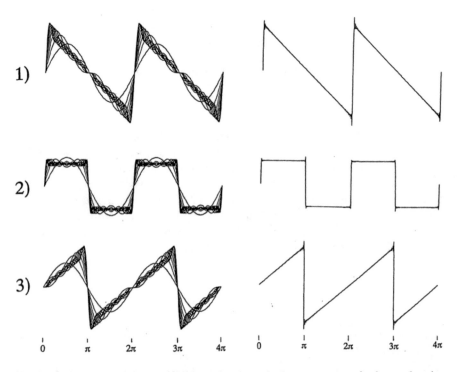

Fig. 2.7 Convergence of the three infinite series shown in the text over two fundamental cycles ($0 \le x \le 4\pi$.) Case (1) Sawtooth waveform; (2) Square wave; and (3) the Gibbs Zigzag. The column on the left shows the superposition of the buildup of the series through the first 10 terms. The column on the right shows the series after 100 terms were added in each case

Case 3: Gibbs "Zigzag":

$$y = \sin x - \frac{1}{2} \sin 2x + \frac{1}{3} \sin 3x - \frac{1}{4} \sin 4x + \cdots + \frac{(-1)^n}{n} \sin nx \qquad (2.9)$$

2.8   The "Gibbs Phenomenon" or Wilbraham Effect

If you look in the vicinity of the vertical discontinuities in the development of the infinite series shown in Fig. 2.7, you will notice a small "horn" sticking up above the waveform. That effect was first discovered by the Scottish mathematician Henry Wilbraham in 1848. It was rediscovered some 50 years later by Gibbs (1899) and has since come to be known as the "Gibbs Phenomenon" in Fourier Series. The width of the horn gets narrower and narrower as the number of terms added to the series increases, but it never disappears. It arises because the convergent limit of the series at the discontinuity differs by about 14% from the limit a small distance away on the curve. Since the waveforms of interest in the present study are not characterized by vertical discontinuities, the effect does not show up in musical instrument waveforms and is merely of historical interest here.

2.9   Basic Aspects of Fourier Series

In what follows here, we will restrict ourselves to periodic functions that are "well-behaved" in the sense that they are continuous and their slopes are finite. By a periodic function $V(\theta)$ such as shown in Fig. 2.8, we mean that

$$V(\theta + 2\pi) = V(\theta) . \qquad (2.10)$$

Many musical instrument waveforms are periodic in the time, or at least quasi-periodic after an initial excitation transient has died down. For example, the sound pressure wave produced by a closed organ pipe is shown in Fig. 2.9, where the pipe was turned on at the start of the oscillogram. As can readily be seen by eye, the waveform settles down to a periodic one after about ten cycles of the fundamental pipe resonance.

Fig. 2.8   A hypothetical periodic waveform

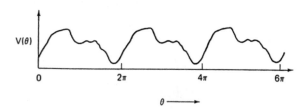

Fig. 2.9 Oscilloscope
display of the waveform from
a quintadena (closed organ
pipe of circular cross-section)

TIME ⟶

As Fourier showed, any such periodic function can be represented by an infinite
series of harmonics of sine and cosine functions over the fundamental period. Thus,
$V(\theta)$ in Eq. (2.10) could be written

$$V(\theta) = C_0 + A_1 \sin 1\theta + A_2 \sin 2\theta + A_3 \sin 3\theta \tag{2.11}$$
$$+ \cdots + B_1 \cos 1\theta + B_2 \cos 2\theta + B_3 \cos 3\theta + \cdots$$

or

$$V(\theta) = C_0 + \sum_{n=1}^{\infty} A_n \sin n\theta + \sum_{n=1}^{\infty} B_n \cos n\theta \tag{2.12}$$

Here, the constant C_0 allows for a net DC ("Direct Current") offset of the waveform from the horizontal axis. The terms involving $\sin n\theta$ and $\cos n\theta$ are called the nth harmonic terms and the coefficients A_n and B_n are called harmonic amplitudes.

2.10 Calculating the Fourier Coefficients for $V(\theta)$

We could just postulate different values for the coefficients C_0, A_n, and B_n, and evaluate the series in Eq. (2.11) with a computer in the same way that we computed those for Fig. 2.7. We might even try to narrow in on a set of coefficients that would match a particular waveform. However, that would be an extremely tedious and inefficient approach. Fortunately, Fourier worked out a systematic method to compute the coefficients from the waveform directly. The method involves integral calculus and is described in detail in Appendix C. Essentially, the different coefficients are determined by finding the areas under various curves related to the initial waveform over one fundamental period. The DC constant C_0 is the average value and is determined from the area under the curve for $V(\theta)$ itself, whereas the coefficients A_n are determined from the area under the curve, $V(\theta)\sin(n\theta)$, and those for B_n from the curve, $V(\theta)\cos(n\theta)$. For musical instruments, the waveforms can be measured numerically using an A-to-D converter, a circuit that converts Analog microphone voltages to Digital output values to be read by a computer. (Microphone voltages are usually proportional to the sound wave pressure.)

Once numerical values have been determined for the sine and cosine terms (A_n and B_n) in the series, it is desirable to express the results in terms of one net coefficient and phase for each harmonic (value of n). That process just involves a little trigonometry. We rewrite the original Fourier series in Eq. (2.11)

$$V(\theta) = C_0 + \sum_{n=1}^{\infty} A_n \sin n\theta + \sum_{n=1}^{\infty} B_n \cos n\theta$$

as an equivalent series involving one sine and a phase angle for each harmonic:

$$V(\theta) = C_0 + \sum_{n=1}^{\infty} C_n \sin n\theta + \phi_n . \tag{2.13}$$

One then evaluates the coefficients C_n and the phases ϕ_n in terms of A_n and B_n by comparing like terms in the two different expressions for the infinite series.

Thus,[5]

$$C_n = \sqrt{A_n^2 + B_n^2} \text{ for } n \geq 1 \tag{2.14}$$

and

$$\phi_n = \arctan(B_n/A_n).$$

Often, one is primarily interested in the relative distribution of the net harmonic amplitudes C_n because they correspond roughly to the psychological impression the sound makes on the human ear. As shown in Appendix C, the relative energy distribution in the harmonics of a Fourier series goes as the square of the amplitudes. Some people prefer to convert that number into decibels because the ear responds logarithmically to the harmonic intensity.

Although it is straightforward (but tedious) to do a Fourier analysis by hand, the calculation is a simple matter with a high-speed computer. A program for doing that is given in Appendix C, together with a derivation of the mathematical quantities involved. Not only is that approach to the problem much faster than the older methods of spectral analysis, you only need one period of the waveform in order to determine the harmonic structure. Thus, you can catch the spectral distribution in the time of one period of the waveform rather than, for example, spending a long time scanning the output of a tape loop (as in Fig. 2.3) while a narrow frequency filter is slowly swept through the spectrum.

2.11   An Example of Discrete Fourier Analysis

In order to do the computations involved in Fourier analysis, one needs to sample at least one period of the waveform digitally and obtain lots of points. In doing that sort of analysis myself, I used a high-quality condenser microphone to pick up the sound and fed its output into a high-speed A-to-D ("Analog-to-Digital") converter controlled by a computer. The computer recorded the data, showed the waveform, did a Fourier analysis, and then displayed the relative amplitudes of the harmonic coefficients. The photograph in Fig. 2.10 was taken during a lecture I once gave at Yale in which a student (William C. Campbell) blew a note on a 50-ft garden hose. The hose behaved like a narrow-scale open pipe with modes spaced at about 11 Hz. Campbell was able to phase-lock a large number of those modes in the mid-audio range (at a fundamental frequency of about 307.7 Hz) and produce a waveform with a sharp, periodic pulse that sounded much like a braying elephant. (The waveform

---

[5]These results are obtained by applying the trigonometry identities $C_n \sin(n\theta + \phi_n) = C_n \sin n\theta \cos \phi_n + C_n \cos n\theta \sin \phi_n$
and
$\tan \phi_n = \frac{\sin \phi_n}{\cos \phi_n}$ together with $\cos^2 \phi_n + \sin^2 \phi_n = 1$.

Fig. 2.10 Photograph taken in Davies auditorium at Yale University during a lecture by the author on "Live Fourier Analysis" in the 1970s. The garden hose was played by Yale student, William C. Campbell

and amplitude spectrum are shown on the oscilloscope in Fig. 2.10, together with the HP-2116B computer used.)

The Campbell waveform provides a nice example of the way in which a sum of sine waves can add up to produce a sharp pulse. At the same time, it provides a useful example to illustrate the convergence of a Fourier series. The amplitude coefficients and phases shown in Fig. 2.11 were computed from the digitized waveform using the program described in Appendix C. A histogram of the Fourier coefficients $C(n)$ is shown as a function of harmonic number starting from the left with $n = 1$ in Fig. 2.12, together with the waveform over one cycle. (The DC offset, $C(0)$, probably resulted from air coming out of the hose near the microphone and is not included in the histogram.)

The original waveform can, of course, be reconstructed by putting the amplitudes and phases from Fig. 2.11 back into Eq. (2.12). That process illustrates the convergence of the Fourier series with increasing number of harmonics, as has been done in Fig. 2.13 where the numbers represent the maximum number of harmonics used in the reconstruction. The values of the phase are very important in determining the visual shape of the waveform, whereas the harmonic amplitudes are more related to the sound heard by ear (Fig. 2.14).

The oscillogram of the closed pipe waveform in Fig. 2.9 provides an example of the pitfalls involved in Fourier analysis. If you had started analyzing that data when the organ pipe was initially turned on, you probably would not even have been able to determine the fundamental period. There was an initial transient during which only higher modes of the pipe were excited. Then, as time went on, the fundamental mode slowly built up and became the dominant source of sound in the spectrum. At the extreme right end of the oscillogram, the waveform has become strongly

Fig. 2.11 Relative amplitude
coefficients $C(N)$ and phases
$P(N)$ computed from one
cycle of the garden hose
waveform for the first 20
harmonics

N	C(N)	P(N)
0	-5.1666666667	---
1	200.18752265	1.1031844815
2	187.226833595	0.2328547785
3	133.673566323	-0.4481088423
4	99.610983131	-1.0783248202
5	78.33444076	4.5807732471
6	63.36763511	4.0453575124
7	49.62334311	3.4573178766
8	42.773894111	2.9106885433
9	33.244534907	2.241522386
10	27.3404696372	1.7578400142
11	21.7070327263	1.2408095503
12	18.0638132271	0.7281888317
13	15.7394445738	0.1343847924
14	11.7314319947	-0.3948992557
15	10.5061603374	-0.9722905303
16	8.7580766154	-1.5573037672
17	7.1367499029	4.1680712601
18	6.0349917008	3.5319263251
19	4.8830470689	2.8393246777
20	3.2039572595	2.2241875619

Fig. 2.12 Waveform and
histogram of the Fourier
coefficients $C(n)$ for the
garden hose waveform

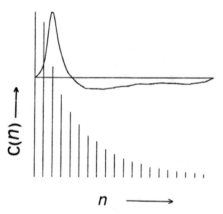

periodic and one can easily pick out the period. Just by looking at it, you can see
that only odd harmonics are of importance in the steady-state waveform and not
much more than the first three are significant. If you were to use a slow computer,
it would help to estimate just how many harmonics you need to analyze in advance,
for the running time in the computation of the discrete Fourier analysis goes up as
the square of the number of harmonics (and of the number of points). Once you
know the fundamental period, you could, of course, go back to the beginning and
Fourier analyze the entire spectrum, period by period. That process would show how
the harmonics changed during the transient.

   We can use the reconstruction of the waveform from the harmonic coefficients to
show how the Fourier series itself converges. That has been illustrated in Fig. 2.13
for the garden hose waveform in Fig. 2.12.

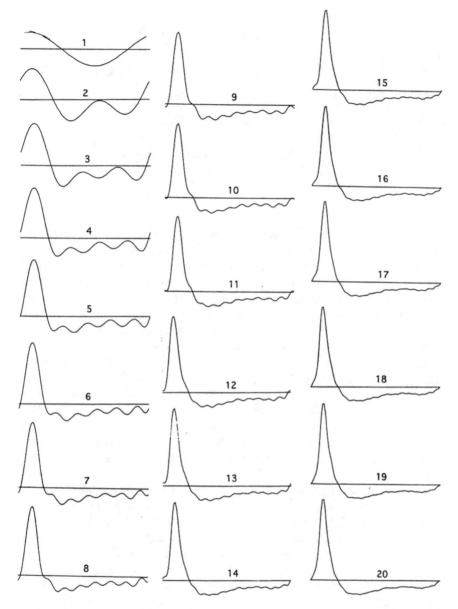

Fig. 2.13 Reconstruction of the waveform in Fig. 2.12 from the Fourier coefficients. The numbers represent the maximum number of harmonics used in each case to reconstruct the waveform

Fig. 2.14 Top: Waveform from a tuba at 33 Hz reconstructed from the original 50 amplitudes and phases obtained from Fourier analysis. Bottom: Waveform reconstructed from exactly the same set of 50 amplitude coefficients but with randomly selected phases for the different harmonics

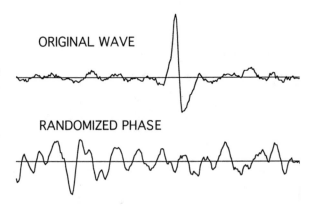

The cochlea in the human ear acts somewhat like a spectrum analyzer in that thousands of different channels respond to sound waves of different frequency and transmit pulses to the brain such that their rate increases with the loudness detected by each channel. To a large extent, the apparent tonal color of the sound is determined by that distribution—hence by the energy content in each harmonic component. The harmonic distribution thus gives the listener the main perception of tonal color. However, that is not the entire story. The ear is also somewhat sensitive to the actual shape of the waveform. Hence, a waveform consisting of periodic sharp spikes sounds somewhat different from that produced by a periodic waveform with the same relative harmonic amplitudes but different phases. The relative phases in a musical instrument waveform are usually determined by the excitation process— for example, the stick-slip mechanism in the bowed violin string, the vibration of the reed in a woodwind, or that in the lips of a brass instrument player. Generally, these processes produce phase locking of the different harmonics in respect to the fundamental period of the instrument so that the phases do not just wander around randomly.

A question naturally arises regarding the number of points needed for analysis of the waveform. Most instruments seldom have more than 10–30 important harmonics. (There are exceptions such as the low notes on a krummhorn, or tuba.) It is, of course, the relative distribution of the stronger harmonics that mainly determines the tonal color (not to mention their variation with time, as in the case of vibrato). Surprisingly, a criterion developed many years ago by Harry Nyquist (1924) for the transmission of telegraph pulses is relevant. He showed in general that in order to transmit signals digitally, one needs to sample the original analog signal at more than twice the maximum frequency you want to transmit. His criterion (following from something called the "Nyquist Sampling Theorem" and which plays a key role in the CD-recording industry) also works backwards. For example, if you have a waveform whose fundamental frequency is 200 Hz and want to examine 20 harmonics (i.e., up to a frequency of 4 kHz), you need to have more than 8000 samples per second, which means more than 40 points over one period of the waveform. In practice, you would want to exceed that minimum requirement by at least a factor of 2 or 3, which means that you would probably want at least 100 points over the fundamental cycle. (The minimum number required by the Nyquist criterion would just give two samples over the period of the highest frequency component.) However, for similar reasons, it is also pointless to analyze the waveform for a number of harmonics larger than half the number of points measured over one period. (If you do so, you just get the same spectral information back again in the higher harmonics, but in reverse order.) It is sometimes implied that there are only slight differences in the waveforms between different instruments and that a relatively small fraction of the sound intensity falls in the overtones. Nothing could be farther from the truth. The differences in the harmonic structure can be enormous, even between instruments of the same species. We will illustrate that fact with a variety of different examples throughout this book.

2.12   The Fourier Transform

The method of Fourier series discussed above works well when one has a precisely periodic waveform. However, in practice one often encounters situations where the waveform may be quasi-periodic, but varies significantly over the time of observation. Such cases might include the sound from an instrument played with vibrato, or the sound from an instrument such as a harpsichord or piano that is inherently transient in character. Finally, there are some instruments (e.g., tympani and bells) where the waveforms are not even approximately periodic and for which the overtones are not harmonically related. Here, there is a useful computational method based on Fourier analysis that goes under the heading Fourier Transform. In that approach, we observe the wave over a very long time T that is not the period of the vibration as discussed before. We then pretend that the waveform is periodic over that long time interval T (which generally includes many cycles of oscillation in the frequency range of interest). Now, even though it is just a mathematical fiction, we can apply our previous results for periodic waveforms to compute the harmonics present with fundamental frequency $1/T$. But, of course, the results will not apply outside the region $0 \leq t \leq T$. Most of the spectral components will be of no physical interest. However, the components we do care about will be contained in the computed frequency range and appear as harmonics of $1/T$.

It will help to illustrate with a particular example. Consider a decaying waveform of the type

$$y(t) = \exp(-\gamma t/2) \sin(2\pi F_0 t) \text{ for } 0 \leq t \leq T \qquad (2.15)$$

which might represent the sound amplitude from the fundamental mode of a string plucked at $t = 0$. (See Fig. 2.12.) Here, $\gamma/2$ is the amplitude decay rate which might result from energy being coupled to a sounding board. Because the energy in the wave motion varies as the square of the amplitude, the energy decay rate in this situation is simply γ, or twice that for the amplitude decay rate. We assume here that

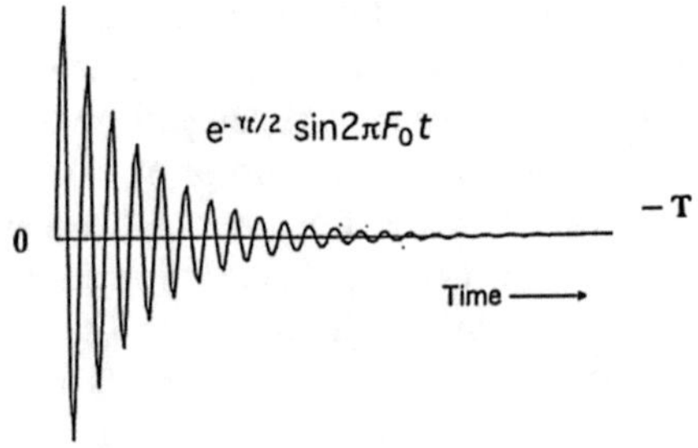

Fig. 2.15 Decay of a damped
waveform given by Eq. (2.14)

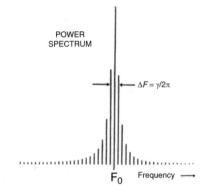

$e^{-\pi t/2} \sin 2\pi F_0 t$

$-T$

0

Time $\longrightarrow$

Fig. 2.16 Energy spectrum
for the waveform in
Eq. (1.14) computed with a
Discrete Fourier Transform

POWER
SPECTRUM

$\Delta F = \gamma/2\pi$

F_0 Frequency $\longrightarrow$

$T \gg 1/F_0$, or equivalently, that Eq. (2.14) describes the oscillation of the string over many fundamental periods of the string oscillation frequency. In practice, there might be a thousand or more digital samples taken of the string amplitude during the long time interval T.

Fourier analysis of the waveform described by Eq. (2.14) and Fig. 2.15 results in the spectrum shown in Fig. 2.16, where the square of the net Fourier amplitudes is plotted as a histogram as a function of the harmonic number. Figure 2.15 represents the energy distribution in the spectrum. Note that the spectrum peaks at F_0, which itself is a high harmonic of $1/T$, where T again is the long observation time. As shown in Appendix A, the energy or power distribution in this case has a so-called "Lorentzian shape" with a full width at half-maximum intensity of $\Delta F = \gamma/2\pi$. That is, the resonance is not perfectly sharp but is spread over a range of frequencies ΔF centered about F_0. The spread arises because the initial signal is not a constant single-frequency wave persisting for an infinite length of time. The result is actually an example of the Uncertainty Principle—something known to electrical engineers long before Heisenberg made his famous pronouncement as applied to quantum physics. As we have shown here, it is a consequence of Fourier analysis that

$$\Delta F \Delta t \approx \frac{1}{2\pi} \qquad (2.16)$$

where $\Delta t \approx 1/\gamma$. To put it in different words, the limiting uncertainty in the frequency measurement (ΔF) varies approximately as the inverse of the signal duration (Δt).[6]

2.13   Window Functions

Although the Discrete Fourier Transform worked perfectly well with the waveform shown in Eq. (2.14), there are some pitfalls in the method. The waveform in Eq. (2.14) was carefully chosen to go to zero at $t = 0$ and to become very small as $t \to T$. However, an arbitrarily chosen wave shape, $y(t)$, might be nonzero at both $t = 0$ and $t = T$ and result in a function that could not conceivably be periodic in the large time interval without having major discontinuities. They, in turn, would result in spurious frequency components during Fourier analysis. To avoid that difficulty, it has become a standard practice to multiply the data obtained in the large time interval by a Window Function which we will call $W(t)$ that goes smoothly to zero at both $t = 0$ and $t = T$. Although the process tends to broaden the computed spectral widths and produces minor distortion of resonant line shapes, it does not interfere with the determination of the resonant frequencies and, most important of all, it does not introduce spurious spectral components. There are almost as many window functions as people who have worked in this field. The most commonly used one is that proposed initially by the Austrian mathematician, Julius von Hann, which for some strange reason is now called the "Hanning Window."[7] It multiplies the data by the function

$$W(t) = 0.5[1 - \cos 2\pi t/T] \text{ for } 0 \leq t \leq T. \qquad (2.17)$$

This "Hanning Window" has been adopted as a standard by the IEEE ("Institute for Electrical and Electronics Engineers") and is built into a number of commercial electronic spectrum analyzers, including the one used by the author to take much of the data presented in this book.

---

[6]In Heisenberg's formulation, the energy of the electron (or other particle) is given by $\mathcal{E} = h\nu$ where ν is a frequency corresponding to the de Broglie wavelength and h is Planck's constant. Hence, in Heisenberg's formulation of the Uncertainty Principle, $\Delta \mathcal{E} \Delta t \approx h/2\pi$.

[7]My personal suspicion is that the peculiar nomenclature arose as a typographical error, compounded by the fact that there actually was an electrical engineer at the Bell Laboratories named Richard W. Hamming who developed his own window function for numerical analysis that is called the "Hamming Window" in the literature.

2.14 The Fast Fourier Transform

The main problem in applying the straightforward Discrete Fourier Transform to the analysis of data is that the running time for the calculation increases as n^2 where n is the number of data points to be analyzed. Because one often wants to analyze waveforms consisting of 1000 points or more (for example, one convenient block size is $2^{10} = 1024$ points), running time is of major importance. Methods to reduce the running time by making use of the redundancy contained in the sine function date at least to the early work of Runge (1903). Most current processors use something known as the fast Fourier transform (FFT) algorithm devised by Cooley and Tukey in 1965. The Cooley–Tukey algorithm reduces the running time from an n^2-dependence on the number of data points to one that goes up as $n \log 2n$. For $n = 2^{10}$, that saving can reduce the running time for computer analysis by a factor of 100. (See Brigham and Murrow 1967.) With the advent of high-speed, hardwired FFT processors, it is now possible to do spectral analysis over the entire audio band in real time.

Within the limits imposed by Eq. (2.15), one can use the FFT to study spectra as a function of time. That not only has broad applicability to the study of musical instrument sound generation, but to numerous other areas of science—especially, to medical diagnostics. For example, the FFT is an essential tool for unfolding the data in magnetic resonance imaging (or MRI.) It also can be applied to acoustic diagnostics in medicine. As an example, the variation of the acoustic spectrum of heart sounds with time can be used to diagnose and categorize heart murmurs. (See Bennett and Bennett 1990.)

Figure 2.17 illustrates this technique using the sound monitored by a high-quality condenser microphone placed on the chest at the apex of the heart. The top figure is for a normal 28-year-old male where the spectrum is concentrated below 200 Hz and shown in yellow. The lower figure is for a 54-year-old patient with prolapse of the mitral valve.[8] The data are presented here as a three-dimensional surface in which frequency runs horizontally from near DC to 1000 Hz (left to right) and time advances diagonally from the upper right to the lower left in increments of 0.1 s. The amplitudes of the Fourier components are plotted vertically. Before analysis, the signal was run through an "A-Weighting" Filter that fell off at the low-frequency end so as to mimic the response of the human ear. Hence, what one sees in the figure corresponds to what one would hear through a stethoscope, except that the electronic technique is far more sensitive. Four heart beats are shown in the figure. The signal running from about 200 to 1000 Hz in between the "first" (S_1) and "second" (S_2) heart sounds from each beat and shown in red is due to the murmur. (The signal below 200 Hz was fairly normal.)

---

[8]Mitral prolapse (verified in the present case by an echocardiogram and open-heart surgery) is a common condition in which the mitral valve (so-called because it is shaped like a Bishop's mitre) is pushed backward toward the left atrium when the left ventricle contracts. Some blood from the ventricle is then forced back through the leaky mitral valve into the atrium in turbulent flow, instead of going out through the aortic valve in laminar flow, as in the normal case.

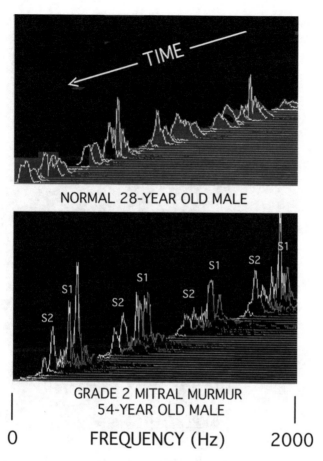

Fig. 2.17 Spectral surfaces of stethoscopic heart sounds. Upper figure: normal heart sounds from a 28-year-old male. Lower figure: heart murmur arising from mid-systolic mitral prolapse in a 54-year-old male. The RMS (root-mean-square) amplitude is shown vertically and the frequency scale runs from near 0 to 2000 Hz. Time advances diagonally in the plot in increments of 0.1 s. Source: Bennett (1990). The author is indebted to Dr. Lawrence Cohen for helpful discussions

As can be seen from Fig. 2.17, the murmur peaks in the middle of systole (during contraction of the heart) and has its strongest components at about 600 Hz. The murmur was generated by turbulent flow of blood back through the mitral valve into the left atrium when the heart contracted. In contrast to most musical instruments, the relative phases of the spectral components are random. The sound of the murmur is actually very similar to that produced by an African percussion instrument called "The Lion's Roar." (It is also akin to the noise made by a crosscut hand saw going through a piece of wood.)

One shortcoming of single FFT-based analysis is that the use of a window function precludes the possibility of reconstructing the original waveform exactly, since part of the information contained in the original waveform is discarded. That

Fig. 2.18 A real-time FFT
(fast Fourier transform)
analyzer

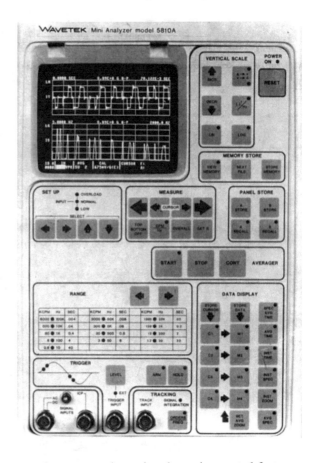

is not a problem when one merely wants to determine the main spectral features. However, in cases where one might want to manipulate the data in the frequency domain and then reconstruct a signal in the time domain, that limitation can be a problem. Although one can get around that difficulty by using two FFT processors having identical time windows staggered by half their common duration, the method is cumbersome. A much-touted recent development called Wavelet Analysis appears to offer a more mathematically elegant solution. There, one devises a complete set of wavelet functions that look somewhat like windowed Fourier transform integrands. Since a complete set is involved, the original signal can be reconstructed. (See Rioul and Vetterli 1991.) However, there is a disadvantage in the wavelet analysis method for our present purposes in that the measured frequency intervals increase in powers of two. Although that tends to mimic the logarithmic frequency response of the human ear, such a logarithmic display makes it much harder to pick out the fundamental frequency visually from the spectra of periodic waveforms. In a linear display based on FFT analysis, the harmonic terms are separated by a constant which is usually equal to the fundamental frequency. Much of the data presented in the present book were taken with a real-time, hard-wired FFT analyzer using a "Hanning Window." (See Fig. 2.18.)

Problems

2.1 Suppose the amplitude of the rod vibration in Michelson's spectrum analyzer decayed to $1/e$ of its initial value in about 5 s. What would the minimum frequency width be that the analyzer could resolve?

2.2 (a) If the narrowband filter in Fig. 2.3 were 50 Hz wide, what would be the least time you would need to scan through the spectrum from 0 to 10 kHz without distorting the data? (b) What would the frequency scanning rate be in Hz/sec?

2.3 Draw the amplitude spectrum for the first 15 harmonics of a sawtooth and of a squarewave.

2.4 If the paper used on an early spectrum analyzer can only be darkened in intensity by a factor of ten, what is its maximum dynamic range?

2.5 Citizens of Leyden, Massachusetts reported hearing the cannons at Bunker Hill some 80 miles away. Suppose that the sound level in Leyden was about 60 dB ("normal conversation at 3 ft" from Table 2.1). What would the sound level have been 10 ft from a cannon? (Assume the sound pressure falls off as the inverse square law.) [Reference: Arms 1959.]

2.6 When the first atomic bomb was exploded at the "Trinity" test site in New Mexico, Robert Serber (Oppenheimer's assistant at Los Alamos) heard the sound of the explosion about a minute and a quarter after the flash. How far was he from the explosion? Noting that the blast was heard at Los Alamos 20 min after the flash, how far was Los Alamos from the test site? Use the value for the velocity of sound at $0°$ in dry air from Table 1.2. About how many dB louder would the blast have been at Serber's location than at Los Alamos? (Data from Serber and Crease 1998, pp. 91, 93).

2.7 The loudest natural noise in recorded history is said to have occurred on August 27, 1883 when the volcanic island Krakatoa blew up. The sound was heard on Rodriguez Island 3000 miles away. If the level there were 60 dB ("normal conversation" at 3 ft), what would it have been one mile away on the island of Verlaten? (Assume the inverse square law.[9]) [Reference: Winchester 2003.]

2.8 An A-to-D converter samples a microphone voltage with 10-bit accuracy. What is its limiting dynamic range in dB? (Note: $2^{10} = 1024$.)

2.9 A CD recording uses 16-bit samples. About how big a dynamic range in dB would it provide?

2.10 A large gymnasium is to be constructed in the middle of a residential area in Bryn Mawr, PA with an air conditioning unit installed on the roof that will produce

---

[9]The assumption of an inverse square law is not terribly good here because the source of noise extended over a distance larger than the island of Verlaten. Also, the radiation pattern appears to have a strong dipole component. (See Winchester 2003, p. 271.)

a sound level of about 80 dB at a distance of 50 ft. Assuming an inverse square law loss, what intensity will this sound have at a neighbor's house 300 ft away?

2.11 Suppose four identical air conditioning units were to be placed on the roof of the gymnasium in Problem 2.7). (a) What would the increase in sound level be if the four sources were in phase? (b) What would it be on the average if the four phases were randomly related?

2.12 The air conditioning unit in the previous problem produced the following spectrum, as measured by the instrument in Fig. 2.4: What was the amplitude spectrum?

Frequency (Hz)	31.5	63	125	250	500	1000	2000
Signal (dB)	65	78	88	80	70	67	50

2.13 The emergence of 17-year cicadas on the weekend of May 22, 2004 resulted in the following spectrum at noon measured in the author's backyard at Haverford, PA: Draw the amplitude spectrum.

Frequency (Hz)	31.5	63	125	250	500	1000	2000	4000	8000
Signal (dB)	50	52.5	52.5	55	60	67.5	60	57.5	52.5

2.14 We know the harmonics of a square wave decrease in amplitude as $1/n$ (with n odd). Draw a spectrum through $n = 11$ of a square wave in dB referred to the value at $n = 1$.

2.15 Sketch waves proportional to $\sin \theta$ and $0.3 \sin 3\theta$ over the range $0 \le \theta \le 2\pi$. Then, add the two together and sketch the resultant waveform. What might produce that waveform?

2.16 Suppose the wind supply to an organ pipe were modulated in amplitude sinusoidally at 6 Hz. What would the effect be on the sound spectrum? (Hint: Note the trig identity $\cos(A \pm B) = \cos A \cos B \mp \sin A \sin B$ and take into account that each harmonic component from the sound wave is multiplied by a sinusoidal term at 6 Hz.)

Chapter 3
Plucked Strings

3.1 The Plucked String

A multitude of stringed instruments make use of plucking as a source of excitation. These range from the harp of Egyptian antiquity to the oriental Koto. Some feel that the most elegant of all plucked instruments is really the harpsichord. (See later illustrations.)

All acoustic stringed instruments (as opposed to electronic ones) are characterized by a method of producing transverse vibratory motion in highly resonant strings whose energy is coupled mechanically through a "bridge" to a soundboard of large area compared to the cross-sectional area of an individual string. This loss of energy is small during one period of vibration, and one can talk about a well-defined frequency with at least quasiperiodic behavior. The soundboard itself typically has many broad resonant modes that are efficiently excited by the bridge and which in turn couple a small fraction of the original vibrational energy to the air in the form of sound waves during each cycle. The soundboard radiates most of the sound heard by the listener and determines the directional radiation characteristics of the instrument. Electrical and acoustic engineers like to think of this coupling process as one of "impedance matching" similar to that encountered in transferring energy from one electrical circuit to another. But, the basic point is that one wants to convert relatively large-amplitude vibrations over the small cross-sectional area of the string to small vibrations over a large area in the soundboard—and, hence, to small sound pressure fluctuations over a large cross-sectional area in a room. If one merely excites a string that is supported under tension between two posts, the sound produced is nearly inaudible. The cross-sectional area of the string is too small to produce sound waves moving through the air with much efficiency. The problem is most severe at low frequencies where the wavelength of the sound is long. The coupling becomes most efficient when the dimensions of the radiating surface are comparable to the wavelength involved. However, one does not want the coupling process to be 100% efficient, for then all of the energy would be removed from the string so

© Springer Nature Switzerland AG 2018
W. R. Bennett, Jr., *The Science of Musical Sound*,
https://doi.org/10.1007/978-3-319-92796-1_3

Fig. 3.1 Egyptian harp.
After a drawing by Ippolito
Rosellini of a fresco in the
mortuary temple of Ramses
III (circa 1194 BC) at Thebes

rapidly that there would be no musical sustaining power within the instrument. In addition to soundboards, some instruments (notably harpsichords, harps, guitars, and violins) have enclosed air-resonant cavities below the soundboard. Generally, the soundboard makes up at least one wall of that cavity. The resonances produced in these cavities enhance the radiated sound at certain frequencies. In some instruments (especially violins and even pianos), wood resonances in the walls supporting the soundboard are also involved. Because of this intermediate coupling process, the sound heard in the room is changed quite substantially from the initial spectrum of the vibrating string (Figs. 3.1 and 3.2).

There are three primary means by which the strings in musical instruments are excited: plucking, striking, and bowing. The sound produced when a string is excited is greatly affected by the excitation point, as well as the mass and tension of the string, and, especially, the soundboard or resonant cavity to which the vibrating string is attached. The inherent spectrum of the plucked string is quite different from that obtained when the string is bowed or hit by a hammer and the physics involved in these three cases will be treated in separate chapters. However, there is one common characteristic in all of these acoustic, stringed instruments. Single, isolated normal modes of the vibrating string such as those shown in Fig. 1.6 are almost never produced by themselves. People who are only used to the simple sine-wave envelopes characteristic of the normal isolated modes of a vibrating string will be quite surprised to see what actually happens when a real string is plucked, or struck. Instead of the simple envelope of a vibrating sine wave like that shown in Fig. 1.6 for $n = 1$, one gets complex structures that run back and forth along the length of the string. They, of course, move too fast for the eye to follow under normal illumination since the round-trip frequency is that of the pitch to which the

Fig. 3.2 The Japanese Koto
played by Maya Masaoko
(reproduced with permission)

string is tuned. But, one can see these running pulses with stroboscopic illumination at a frequency slightly detuned from the string resonance. They arise because the string is never excited in a pure resonant mode. For example, to excite the string in only its normal lowest mode, one would have to use a plectrum or a hammer-shaped like half a sine wave over the entire length of the string.

It is easy to see why one gets a wave that oscillates back and forth between the two halves of the string at the fundamental frequency. Consider a hypothetical situation in which only the first and second harmonics of Fig. 1.6 are excited and with equal amplitudes. In that situation, the total initial waveform would be made up of two terms, $\sin(\pi x/L)$ and $\sin(2\pi x/L)$, added together over the length (L) of the string. The first term oscillates at the fundamental string resonance F_0, and the second term oscillates at twice the fundamental. The first term has even symmetry about the middle of the string, whereas the second has odd symmetry about that point. (See Fig. 3.3.) Assuming that they both start oscillating in phase at $t = 0$, the two solutions will tend initially to cancel each other out on the right half of the string and reinforce each other on the left-hand side. (The top situation in Fig. 3.3.) Because the second harmonic oscillates twice as fast as the first, the two solutions will tend to cancel on the left side of the string and add (negatively) on the right-hand side at half the fundamental period ($t = T/2$ in the figure). Finally, after one complete period, the string will be back to its original shape. The net result is that the string oscillates back and forth at the fundamental frequency F_0 from a positive bump at the left to a negative bump at the right. Qualitatively, the case with the real string just involves adding more harmonics and keeping track of their initial relative amplitudes and phases.

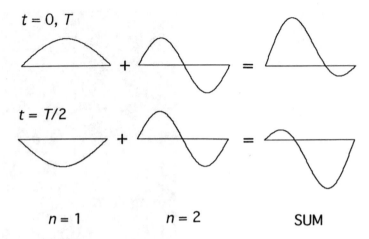

$t = 0, T$

$t = T/2$

$n = 1$ $n = 2$ SUM

Fig. 3.3 String motion for a superposition of the first two modes shown at $t = 0$ and $t = T/2$. The resultant bump oscillates back and forth at the resonant frequency F_0 between the left side of the string (where it is positive) and the right side (where it is negative.) As indicated in the figure, the bump is back at the starting point again at $t = T$

3.2 Motion of a Plucked String[1]

As discussed in Chap. 2, d'Alembert solved the wave equation for the vibrating string and obtained a solution of the form

$$\longleftarrow \longrightarrow \qquad\qquad (3.1)$$

$$y(x, t) = \frac{1}{2} f(x + ct) + \frac{1}{2} f(x - ct).$$

As indicated by the arrows, the first term represents a running wave going to the left ($-x$ direction) and the second term represents one going to the right ($+x$ direction), where c is the velocity of the wave. One important aspect of this solution is that the shape of the string for all future times is given in terms of the initial shape, $f(x)$, at $t = 0$. The situation is easiest to understand graphically when the plucking point is at the middle of the string. Here, the initial shape of $f(x)$ is a simple triangle peaked at the midpoint, as illustrated by the curve for $t = 0$ in Fig. 3.4. That triangle can be broken up into two identical triangles of half the initial amplitude, as in the equation by d'Alembert. As time increases, these two triangles move in opposite directions.

---

[1] See Appendix B for a derivation of the wave equation and various solutions to the vibrating string problem. Note that the wave equation itself is an approximation valid for very small amplitudes compared to the length of the string. The drawings given here exaggerate the size of the deflection for the sake of illustration. Typically, the deflection is only a few millimeters in a string 4-ft long.

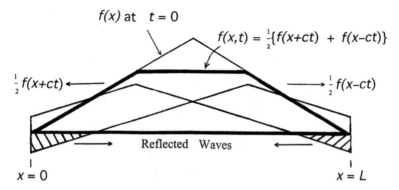

$$f(x) \text{ at } t = 0$$

$$f(x,t) = \frac{1}{2}\{f(x+ct) + f(x-ct)\}$$

$$\frac{1}{2}f(x+ct) \leftarrow$$

$$\rightarrow \frac{1}{2}f(x-ct)$$

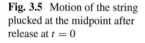

Reflected   Waves

$$x = 0$$

$$x = L$$

Fig. 3.4 Graphical illustration of d'Alembert's solution to the vibrating string problem. Here, the string is plucked in the middle at $t = 0$

Fig. 3.5 Motion of the string plucked at the midpoint after release at $t = 0$

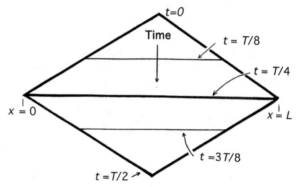

As they pull apart from each other, the two triangular waves add up to a constant plateau in the region between the two running-wave peaks. The leading edge of the left running wave undergoes a "hard" reflection[2] immediately at $x = 0$ (the shaded portion to the left of the diagram) and travels back in the $+x$ direction with negative amplitude. Similarly, the running wave initially going to the right undergoes a hard reflection at $x = L$ and then heads back in the $-x$ direction (shaded portion to the right). When the two (shaded) "Reflected Waves" in the figure (which are now negative due to the "hard" reflections at each end) are added to the two displaced, positive triangular running waves, the result is the trapezoid outlined by the thick bold lines in Fig. 3.4.

As time progresses, the trapezoid broadens out more and more and decreases in amplitude until it vanishes altogether. As shown in Fig. 3.5, that happens one quarter of the way through the full period of oscillation where the string is coincident with the horizontal axis. Only the behavior over the first half of the cycle (up to $t = T/2$)

---

[2]As discussed in Chap. 1, a "hard" reflection produces a change in sign of the running-wave amplitude.

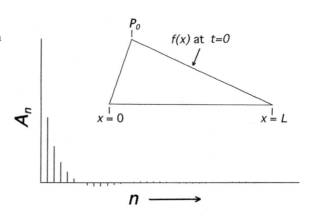

Fig. 3.6 Relative harmonic amplitudes A_n excited when a string is plucked at the seventh harmonic node for which $P_0 = L/7$ (see inset)

is shown. After that, the motion reverses and the plateau moves upward until at $t = T$, the trapezoid becomes coincident with the initial shape the string had at $t = 0$. As a result, the motion of the string is given by the simple sequence of shapes shown in Fig. 3.6 in time increments of $T/8$.

One can construct similar solutions for the variation of the string with time when the string is plucked at different places along its length, but they are harder to interpret graphically. Instead, for the general case we will use a different, more powerful method of analysis based on Daniel Bernoulli's solution to the problem that was also summarized in Chap. 2 and is derived in Appendix B.

Bernoulli noted that the solution for the motion of the string may also be written

$$y(x, t) = \sum_{n=1}^{\infty} A_n \sin(n\pi x/L) \cos(2\pi n F_0 t) \text{ where } F_0 = c/2L . \qquad (3.2)$$

This means that the shape $f(x)$ of the string at $t = 0$ is now given by a sum of sine waves,

$$y(x, t) = \sum_{n=1}^{\infty} A_n \sin(n\pi x/L) , \qquad (3.3)$$

an assertion that upset a number of famous mathematicians of the time.[3]

---

[3]Using the trigonometric identities $\sin(A \pm B) = \sin A \cos B \pm \cos A \sin B$, Bernoulli's solution may also be written as a sum of oppositely directed running waves:

$$y(x, t) = \sum_{n=1}^{\infty} A_n \left[\frac{1}{2} \sin \left(\frac{n\pi x}{L} + 2\pi n F_0 t \right) + \frac{1}{2} \sin \left(\frac{n\pi x}{L} - 2\pi n F_0 t \right) \right] . \qquad (3.4)$$

As shown in Appendix C, we can use the orthogonal properties of the sine functions to obtain the values of the coefficients A_n, yielding

$$A_n = \frac{2}{L} \int_{x=0}^{L} f(x) \sin(n\pi x/L) dx .\qquad(3.5)$$

Equation (3.4) tells us that the harmonic spectrum for the plucked string is determined entirely by the initial shape of the string when it is released. We have to do the integral to see how the actual relative spectrum appears in the general case. As shown in Appendix B, the integral in Eq. (3.4) can be evaluated in closed form giving[4]

$$A_n = \frac{2AM^2}{(M-1)n^2\pi} \sin(n\pi/M)\qquad(3.6)$$

where $M = L/P_0$ depends on the plucking point $x = P_0$, A is the initial amplitude, and it is assumed that the harmonic number n is an integer. However, M is not necessarily an integer, since the string could be plucked at any arbitrary point along its length. As can be seen from a symmetry argument, one gets only odd harmonics of F_0 for $M = 2$. [When the string is plucked at the middle, the areas in the integrand for even harmonics cancel on opposite sides of the midpoint.] Under that condition, one gets a clarinet-like sound from the instrument since the clarinet also has predominantly odd harmonics. The harp (Fig. 3.1) is one of the few stringed instruments that is often plucked at the middle, and the odd harmonics so produced give that instrument its characteristic sound. Hubbard (1965, p. 24) mentions that the notes on some virginals were plucked very close to the midpoint (e.g., at 47.2% of the length), but for practical reasons that is not the case with most harpsichords and pianos.[5] One also sees from Eq. (3.5) that $A = 0$ for $M = n$. (That is, plucking at the node for the nth harmonic kills the nth harmonic.) Note that the amplitudes in Eq. (3.5) may be both positive and negative and the angle $(n\pi/M)$ in the sine function is in radians.

Figure 3.6 illustrates the initial shape and computed spectrum when the string is plucked at the node for the seventh harmonic—i.e., at 1/7 of the length of the string. Note that harmonics at multiples of the 7th (e.g., at $n = 14$ and 21) are also eliminated. Some feel that the seventh harmonic really should be removed because it clashes with the 8th and hence sounds "unmusical." However, that notion is very subjective. Many people (among them, Beethoven) were fond of that dissonant effect. The basic fact of life is that you have to pluck the string somewhere, and

---

[4]It is also a simple matter to integrate the expression in Eq. (3.4) numerically using the sine portion of the Fourier analysis program in Appendix C.

[5]Some upright pianos may be exceptions. The geometry of an upright piano (and possibly of a square piano) could make striking the string at the midpoint feasible. With a concert grand, it would be out of the question since the hammer shanks would have to be about 4-ft long in the extreme bass.

the harmonic with a node nearest to that point will be suppressed. It would be hard to design a keyboard instrument covering any substantial range for which the plucking points remained a constant fraction of the length for each string. Because the location of the plucking point changes the tone quality, one gets a pleasing effect on such instruments in that the tonal color varies continuously over the range of the instrument. That property is especially nice for playing fugues on a single-manual instrument because the different voices seem to enter with different tonal colors.

All acoustic stringed instruments produce substantial changes in the radiated sound from the vibrating string waveform due to frequency-dependent coupling of energy through a bridge to a soundboard. Pulses reflected at the bridge result in momentum being imparted to a large area soundboard, which in turn radiates acoustic energy efficiently. Part of the challenge to the craftsman is in tailoring the components to provide desirable overall tone quality.

3.3   Motion of the Computed Plucked String

Putting the computed values for the relative amplitudes back into Eq. (3.2) permits determining the shape of the string as a function of time after it is plucked. As shown in Fig. 3.7, a string initially plucked at $x = P_0$ near the left end will produce a triangular wave running to the right. The wave bounces off the support at $x = L$ at time $T/2$ (half the period) and then heads back toward its starting point. It is reflected back and forth, over and over, at the fundamental frequency, $F_0 = c/2L = 1/T$. In a real string, the energy in the motion would decay with time due to coupling energy to a soundboard and frictional loss from air resistance. However, that process takes place over many fundamental periods of oscillation and has not been included

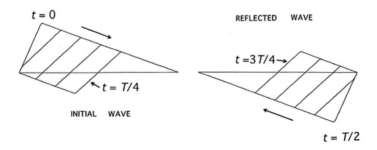

Fig. 3.7 Left: When the string is released (at $t = 0$), a triangular-shaped initial wave is launched along the string toward the right. The peak appears to slide down the top side of the initial triangle, as indicated by the arrow. Right: After the wave reaches the right-hand end of the string (at $t = T/2$), it is reflected back toward its starting point. Here, the (negative) peak slides up the bottom side of the triangular shape of the reflected wave, as indicated by the lower arrow. The original shape is reproduced at the end of each cycle (see Fig. 3.8 and compare with Fig. 3.3, computed for only two modes)

Fig. 3.8 Motion of a string
plucked at 1/7th of its length
over one complete cycle

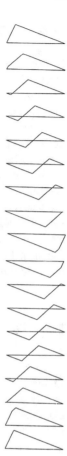

so far in the description.[6] In the words of Appendix A, the typical vibrating string
has a very high Q ("Quality Factor"). It decays very slowly over many cycles and
has a very sharp spectral width.

3.4   The Electric Guitar

One might think that the electric guitar would provide a good test case for examining
the actual motion of a plucked string experimentally for there is no bridge or
acoustic soundboard and the waveform is picked up electronically from the vibrating

---

[6]In addition, we have neglected the inharmonic character of vibrating strings, a subject that we will
come back to in the discussion of pianos. That property alters the periodic nature of the vibration,
but it is less important in harpsichords where the strings are relatively thin.

string directly. However, that turns out not to be the case. To be sure, the plucked strings in such a guitar do obey Eqs. (3.2) and (3.5) to good approximation. However, with most versions of the instrument, substantial changes are produced electronically from that which would be expected of the normal vibrating string waveform. First of all, the output voltage from a magnetic pickup coil responds to the velocity of the string and that process tends to emphasize higher harmonics. (As a consequence of Faraday's Law, the pickup signal is proportional to the time derivative of the string motion.) Because the pickup coils are fixed, the spectrum tends to be roughly independent of the actual plucking point. Further, there are usually two pickup coils placed at different points along the string and the output voltages from the two are often subtracted or added. Tone controls (not to mention loudspeaker response) produce additional coloration of the spectrum. Finally, the largest change in the waveform occurs with some instruments when the performing artist turns on the "fuzz" control. The latter produces distortion similar to that found in low-quality audio amplifiers. It is amazing how something so awful in one medium can be so artistically desirable in another.

3.5   The Steinberger Bass

One of the more interesting designs from a physicist's point of view is that used in the Steinberger bass. (See Fig. 3.9.) Here, the entire instrument is made of pyrolytic graphite and there is no false attempt to mold the shape into that of a classical acoustic guitar. (There is, of course, no soundboard or resonant cavity as in an acoustic guitar.) The strings vibrate laterally across two coils located above permanent magnets as shown in Fig. 3.10. Because of the large-diameter wrapped strings, the persistence times for notes on the instrument are comparable to those in the bass section of a concert grand piano. (Compare Table 3.1 with later Table 4.1.) The voltage from the two coils is shaped electronically to provide a warm sound quite different from what one might expect from the vibrating string alone. Perhaps

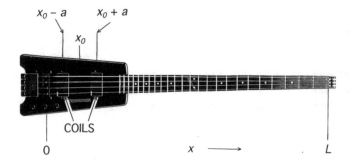

Fig. 3.9 A Steinberger bass (photograph courtesy of Ned Steinberger)

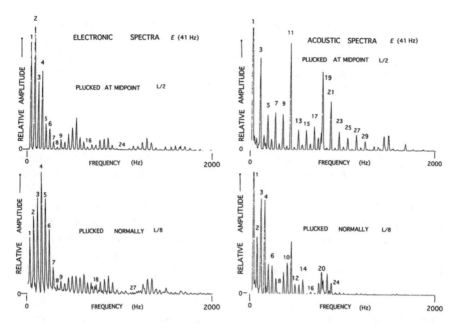

Fig. 3.10 Spectrum of the electronic output from a Steinberger bass (Left) compared with that picked up acoustically with a microphone placed above the string (Right). The tone controls were set in the mid-position (data taken by the author)

Table 3.1 Persistence times for the strings on a Steinberger bass

Note (Hz)	E (41)	A (55)	D (73)	G (98)
Diameter (in)	0.132	0.117	0.065	0.050
Decay time (s)[a]	≈ 25	35	38	33

[a]Time to decrease by about 60 dB (data taken by the author)

the most unique characteristic of the electric guitar is that the harmonic spectrum is largely determined by the location of the pickup coils, rather than the plucking point. As with most stringed instruments, the different harmonics decay at different rates and the spectra from individual notes vary with time after plucking.

The two pickup coils centered at $x_o \pm a$ in Fig. 3.9 are connected so that the output varies as the difference between the coil voltages. Among other effects, that connection scheme largely eliminates "hum" that might otherwise arise from stray 60-Hz magnetic fields. One can see how this connection affects the output spectrum by inserting the values for the coil locations into Eq. (3.2). The net output from the difference of the two coil voltages is of the form

$$V(t) \propto \sum_{n=1}^{\infty} A_n \left\{ \sin[n\pi(x_0 + a)/L] - \sin[n\pi(x_0 - a)/L] \right\} f_n(t) \qquad (3.7)$$

$$= 2 \sum_{n=1}^{\infty} A_n \cos\left(\frac{n\pi x_0}{L}\right) \sin\left(\frac{n\pi a}{L}\right) f_n(t)$$

where $f_n(t)$ is a function only of time for the nth harmonic and A_n is still given by Eq. (3.5).[7]

In addition to the nulls predicted by Eq. (3.5) which do depend on the actual plucking point, the spectrum described by Eq. (3.6) also goes to zero for the following harmonics [8]

$$n = m\frac{L}{a} \text{ and } n = (2m - 1)\frac{L}{2x_0} \text{ where } m = 1, 2, 3 \dots$$

For the Steinberger bass, $xo \approx 3.75$ in, $a \approx 2$ in, and $L \approx 34.1$ in. Hence, relations (3.7) predict nulls in the spectra at $n \approx 8, 17, 26, 34, \dots$ The electronic spectra on the left side of Fig. 3.10 show minima at about those same values which are roughly independent of the plucking point. (See the top and bottom spectra at the left in Fig. 3.11.) These minima also occur at roughly the same places minima are found in the acoustic spectrum when the string is plucked at $L/8$. (See the lower spectrum at the right side of Fig. 3.11.) Hence, one gets a tone quality from the instrument that approximates that found in an acoustic instrument that is plucked at the node for the eighth harmonic.

As is also evident from these spectra, even harmonics have been added electronically to the original waveform. Clearly, these result from some nonlinear element in the circuitry and probably arise from the pickup coil itself. (Because the magnetic field lines will spread above the coil region, a nonuniform value of the field will be encountered as the steel string vibrates back and forth. That, in turn, will introduce overtones of the periodic string motion, starting with the second harmonic.)[9] The

---

[7]The function $f_n(t)$ increases linearly with harmonic number, n, and can be obtained by taking the time derivative of Eq. (3.2):

$$f_n(t) = -2\pi n F_0 \sin(2\pi n F_0 t).$$

The further simplification in Eq. (3.7) occurs by use of the trigonometry identity

$$\sin(A \pm B) = \sin A \cos B \pm \cos A \sin B.$$

[8]That is, nulls occur that are independent of the plucking point for values of n such that

$$\frac{n\pi a}{L} = m\pi \text{ and } \frac{n\pi x_0}{L} = \frac{2m - 1}{2}\pi \text{ for } m = 1, 2, 3 \dots$$

[9]In general, the magnetic field (H), which is normal to the coil, will vary laterally over the string motion. The field can be expanded as a power series in the lateral coordinate (y), giving

$$H(y) = H_0 + H_1 y + H2 y^2 + \cdots$$

where $y = 0$ corresponds to the rest position of the string. The term involving y^2 will produce second harmonics of a sinusoidally oscillating string whose velocity varies as $\sin 2\pi F_0 t$. The latter follows from the trigonometry identity,

$$\cos 2A = 2\cos^2 A - 1. \tag{3.8}$$

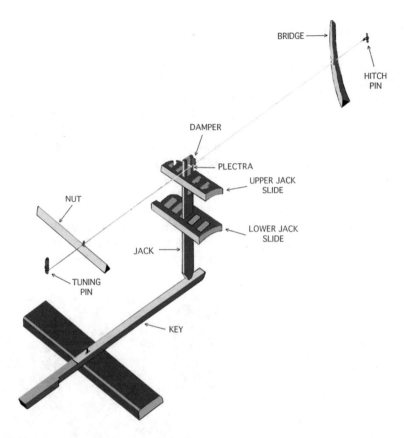

Fig. 3.11 Basic mechanism of a harpsichord. For clarity, only one stop is shown. In practice, a felt-lined flat strip of wood is placed above the row of jacks to stop them from flying up out of the instrument, and the bridge is glued to a triangular-shaped soundboard fastened to the case

fact that even harmonic production is involved is illustrated by comparing the spectra for the electronic output (left side of Fig. 3.10) with the acoustic spectra obtained by placing a microphone over the strings (right side of Fig. 3.10). The top two spectra were obtained when the bass string was plucked at the midpoint, a location for which we have shown that there should be no even harmonics present in the actual string motion. As expected, the acoustic spectrum is composed almost entirely of odd harmonics. (See the upper spectrum on the right side of Fig. 3.10.) In contrast, the electronic spectrum for this same plucking point has many strong even harmonic components. (See the top left spectrum in Fig. 3.10.)

In summary, the designer of the Steinberg bass has produced an instrument of modest size which provides a warm tone quality that has the sustaining power of a concert grand piano and produces a spectrum which is approximately independent of the plucking point.

3.6   The Harpsichord[10]

The action of any harpsichord is made up of a keyboard, registers (or stop
mechanisms), and at least one set of jacks. Each jack rests on the back of a key and
contains a retractable hinged mechanism holding the plectrum and a weak restoring
spring. In addition, one jack on each key has a cloth damper to deaden the string
when the key is not depressed. (See Fig. 3.11.) As the jack is pushed upward by
the rear end of the key, the damper comes off the string, and the string is raised
by the plectrum. After the string slips off the plectrum (i.e., is plucked), the hinged
mechanism permits the plectrum to retract so that the jack can then slide back down
past the string to its normal resting place. As verified by Giordano and Winans
(1999), the speed with which the key is depressed has negligible effect on the sound.

Harpsichords came in all sizes, ranging from the 4-ft pitch Italian ottavino shown
in Fig. 3.12 having only one set of jacks to enormous German instruments. Some of
the latter actually had three manuals, not to mention stops at 16-, 8-, 4-, and 2-ft

Fig. 3.12 Copy of an early
eighteenth-century Italian
ottavino (or treble virginal)
made by Paul Kennedy after
an original instrument by
Josef Mae del Coninus. Here,
the jacks (running diagonally
from lower left to upper right)
go straight through the
soundboard and the bridge is
at the right

---

[10]The historical information given here is largely based on Hubbard (1965), Ripin et al. (1980),
and helpful discussions with Richard Rephann of the Yale University Instrument Collection and
harpsichord maker, Paul E. Kennedy.

Fig. 3.13 The jacks on a French-style harpsichord made by Paul Kennedy; two 8-ft and one 4-ft stops are shown with one key depressed on the lower manual. Note the two raised jacks

pitch (using the terminology of organists)[11]—more strings than any sane person would want to tune.[12]

Most instruments have more than one set of jacks. (See Fig. 3.13.) Sometimes two jacks on the same key can excite the same string at different plucking points (with different stop settings), but more often the different jacks associated with a given key pluck different strings. Which jacks operate when the key is depressed is determined by the stop mechanism which generally involves moving the upper jack slides laterally so as to engage (or disengage) the plectrum from a particular string. The strings run from tuning pins at the front of the instrument over a fixed rail called the "nut," past the jack, over a bridge glued to the soundboard, and to hitch pins at the rear. The soundboard (usually made from thin spruce or cypress on Italian instruments) is mounted above a resonant cavity in the instrument which generally has an opening called a "Rose," which opens to the room (Fig. 3.14).

The plectrum of choice was the crow quill, and occasionally a carefully shaped piece of leather. However, quills made of "space-age materials" such as Delrin® and Celcon® have recently been substituted on many instruments with reasonable success. "Buff" stops (consisting of pieces of felt or leather moved in from the side) are sometimes incorporated to dampen the higher overtones on 8-ft strings and produce a pizzicato effect. "Lute" stops are sometimes incorporated in the form of

---

[11]Each stop is described in terms of the length of an open pipe required to produce the same pitch as the lowest C on the keyboard.

[12]Anyone who has ever tuned a harpsichord can imagine what a nightmare the 1710 instrument by J. A. Hass would produce with two sets of strings at 16-, 8-, 4-, and 2-ft pitch! (See Hubbard 1965, Plate XXVII.)

Fig. 3.14 The soundboard (complete with a Rose) on the same French-style harpsichord of Fig. 3.13 built by Paul Kennedy, showing both the 4- and 8-ft bridges

a second jack on the same string that plucks extremely close to the nut, producing a penetrating "nasal" quality.

Harpsichords with strings tuned to different pitch (e.g., 8- or 4-ft stops) have more than one bridge and, in some instances (especially, when 16-ft stops are present), more than one soundboard. The more elaborate instruments generally had two keyboards or "manuals," each with its own set of jacks and stops, and frequently with coupler "dogs" between the two manuals that are usually activated by pulling out (or pushing in) the upper keyboard. (The couplers, which were first introduced near the end of the seventeenth century, usually consist of vertical pieces of wood attached to the lower manual that can cause the keys of that manual to push against those of the upper manual when it is in the right position. They also tend to be broken off when one slides the upper keyboard back and forth to engage or disengage the coupler while a note on either manual is depressed.)

When the string is excited by the plectrum, a triangular pulse is launched down the string as shown in Figs. 3.8 and 3.9. When the pulse reaches the bridge (Fig. 3.12) and is reflected by it, the bridge receives an impulse pushing it upward. (The upward impulse given the bridge balances the downward impulse given the string at that point, as required to conserve linear momentum.) Because the bridge is glued to the soundboard, that board also receives an upward kick which in turn launches a sound (compression) wave upward into the air. The pulse given the board produces a running wave over its free surface much like that given to a kettledrum when it is struck with a mallet. The propagation of the pulse over the soundboard involves a complex two-dimensional problem that is formidable

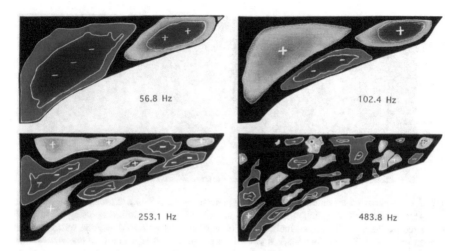

Fig. 3.15 Resonant modes in the soundboard of a Flemish-style harpsichord at four different frequencies. The nodal areas are outlined in white. Positive (up) and negative amplitudes (down) are indicated with "+" and "−" signs, and the areas of strongest amplitude are shown in darkest shading (drawn with some artistic license from data presented by Kottick et al. 1991)

to treat mathematically. However, as with a vibrating drumhead, the soundboard has normal modes of oscillation spread out over its surface area that resonate at different frequencies. Experimental studies of such modes were given by Kottick et al. (1991), who identified some 36 distinct modes below 600 Hz in the soundboard of a Flemish-style harpsichord. (See Fig. 3.15.) These modes have nodal lines that are determined by spatial boundary conditions: the board usually can have no appreciable vertical motion at locations where it is glued down—for example, at the edges of the board and along the supporting ribs underneath. However, exceptions occur when the modes also involve motion of the case.

The support regions have quasi-triangular shapes that vary from one instrument to another. The wood fiber of the soundboard is generally oriented parallel to the long dimension of the instrument. One reason is that it is easiest to get long pieces of wood in the direction of the grain. Because the stiffness of the board is much greater in the direction of the fiber than perpendicular to it (by a factor of about ten in the case of spruce), the velocity of surface waves in the long direction of the board, hence the separations between the maxima and minima in the mode patterns, are largest in that direction.[13] By comparing the separations between opposite polarity peaks for the modes shown in Fig. 3.15, it is apparent that the soundboard wave velocity is about three times greater in the long direction than in the perpendicular

---

[13]Modes of vibrating membranes are treated in Appendix B. The stiffness in the soundboard plays a role analogous to the tension in a string, where for the one-dimensional case the wave velocity is $c = \sqrt{T/\mu}$, T is the tension, and μ is the mass density per unit length. Stiffness varies as the 3rd power of the thickness.

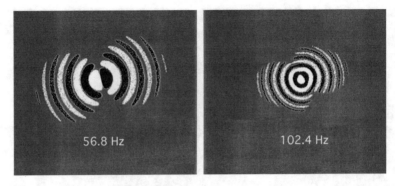

Fig. 3.16 Approximate radiation patterns in the plane of the soundboard expected for the 56.8 and 102.4 Hz modes of the harpsichord soundboard shown in Fig. 3.15. (Computed using the method discussed in Chap. 1.) Here, the actual distribution has been approximated by point sources of the correct polarity at the locations marked with + and − signs. Where more than one source was present within a principal resonance, a fractional source distribution normalized to unity was used. (For example, the large negative region in the 56.8 Hz resonance was made up of three sources each normalized to 1/3.) A more precise result would, of course, be obtained by integrating over a continuous source distribution

direction. That is, the resonant modes tend to be elongated in the direction of the grain. Of course, the thickness (hence, mass distribution and stiffness) often varies significantly over the surface of the soundboard.[14]

As shown in Fig. 3.16, the radiation pattern from the 56.8 Hz resonance in Fig. 3.15 corresponds roughly to that for a dipole and that for the resonance at 102.4 Hz is more like a combination of the patterns for dipole and quadrupole sources. (See the radiation patterns shown earlier in Figs. 1.17 and 1.19 of Chap. 1.) Of course, room reflections would produce changes in these patterns in practice and one would need an anechoic chamber to measure how closely these calculated results would come to the real case. As with the $n = 2$ mode of the vibrating string, the separation between negative and positive peaks for the resonance at 56.8 Hz must correspond roughly to a half-wavelength in the soundboard vibration. Although the coupling of the dipole in the soundboard to the air is substantially off resonance, one still gets a characteristic dipole radiation pattern. (Maximum radiation would occur if the wavelengths of sound in the air and in the soundboard were the same; here, the difference in the two wavelengths just changes the scale factor in the plot.) The resonant modes at 253.1 and 483.8 Hz produce weaker, nearly isotropic radiation patterns that are not shown here. As with many instruments, the different modes radiate in different directions.

---

[14]Not all sounds produced by a harpsichord originate with the soundboard. The mechanical motion of the jacks can create distracting noises. Once, when harpsichordist and composer Joyce Mekeel and I were recording a piece she had written called "Textures" for harpsichord and clarinet, she complained about strange noises in the recording. Finally, we put the microphone directly behind

3.7 Evolution of the Harpsichord[15]

The development of keyboard instruments from the harpsichord through the piano had a major effect on the evolution of western music. Reference to "an instrument like an organ which sounds by means of strings" dates to a private letter written in 1387 by John I of Aragon in Spain to Philip-the-Bold of Burgundy. Ten years later, a jurist from Padua wrote that one Hermann Poll had invented something he called a "Clavicembalum." Although the earliest known representation of a harpsichord appears to be in the sculpture for an altarpiece in Minden, Germany dated 1425, there is little doubt that northern Italy was the center of harpsichord production during the sixteenth century. Flemish instruments dating from somewhat later have a strong resemblance to their Italian counterparts (Fig. 3.17), suggesting that there was a migration northward of the technology from the early development of the instrument in Italy. Surprisingly, an Italian harpsichord made by "Gerolamo of Bologna" dated 1521 scarcely differs from other Italian instruments made as much as 150 years later. The Italians did not incorporate the innovations made later by their northern European competitors and continued to produce only single-keyboard instruments. Most Italian harpsichords had only two sets of 8-ft strings tuned in unison and were very long and narrow. Four-foot stops existed but were extremely rare.

The early Italian instrument makers went to some effort to keep the scaling constant. Because the pitch of a vibrating string varies inversely with the length

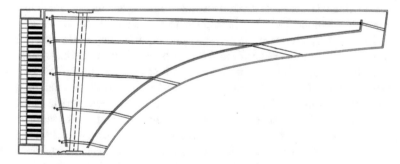

Fig. 3.17 Plan of an Italian Harpsichord made in 1677. Only the pairs of strings on the two 8-ft stops at different C's are shown. The jacks were placed in between the pairs of strings with the plectra pointing outward in opposite directions. The lowest *E* was often tuned to *C* (after Hubbard 1965)

---

her head and the "strange" sounds disappeared. The music desk had been shielding her ears from the jack noises.

[15]The background historical data quoted here have been taken primarily from Hubbard (1965) and Ripin et al. (1980) and helpful supplementary discussions with Richard Rephann and Paul E. Kennedy.

(i.e., as $1/L$), it was thought that an ideal instrument ought to have string lengths that doubled for each octave as one went down the keyboard. The trouble with doing that in practice is that the instrument gets to be pretty long as you approach the lowest bass notes. But, the Italians did succeed in that goal throughout about 5/6 of the instrument's range. The cases, which initially were just protective boxes, were made of unusually thin cypress and were quite light. One advantage of long, thin strings under low tension was that the cases did not need to be very strong. The soundboards were also made of cypress, which is not as stiff per unit weight as the spruce used in the later European instruments. The soundboards on some Italian instruments varied considerably in thickness, from about 1/6-in. under the bridge to perhaps 50% less at the edges. The soundboards move up and down quite easily when one pushes (gently!) on them with the fingers and behave somewhat like modern loudspeaker cones. Although these single-keyboard instruments were said by some to produce beautiful tone quality, they were primarily used to accompany singers.

Hubbard (1965, p. 9) noted that the tone quality of the Italian instruments began with a "plunk" rather than a "plink," meaning that it decayed rapidly (particularly in the treble) compared to the northern European instruments. However, he also implied that that can be of musical advantage for the accompaniment of singers because the Italian instruments excelled in "their dry sparkle." As shown by Sir George Stokes in the mid-nineteenth century, the viscous damping rate for a vibrating string varies inversely as its diameter. Since the strings in Italian harpsichords were significantly thinner, as well as longer (especially in the bass), the shorter decay times of the Italian instruments no doubt have their origin in higher loss from air resistance.[16] Stokes also showed that the higher frequency harmonics on a given string will decay faster than the lower ones by an amount which increases by about 70% when the frequency is doubled. (See Crandall 1926 and Wood 1955.)

As also noted by Hubbard, no provision of any kind was made in the Italian instruments for changing either of the 8-ft stops while playing. Instead, the Italians put their efforts into building harpsichords with absurdly large numbers of keys. Amazingly, some were built with as many as 31 keys per octave (Hubbard, p. 31). These strange instruments were made at the request of music theorists of the time who wanted perfect harmonic intonation in many different keys. More modest versions with split sharp keys (so as to provide a difference between sharps and flats) were more prevalent. But, keyboards of even that complexity were nearly impossible to play with much facility. Fortunately, adoption of the "Well-Tempered Scale" did away with most efforts in that direction.[17]

---

[16]Hubbard (1965, p. 327) suggests that thicker gut stringing was probably used on some Italian instruments. But, he was "bound to report that gut strings on Italian instruments sound very badly." He also noted that Bach had written for the Lauten Werck, or lute harpsichord, an instrument using gut strings.

[17]Willard Martin of Bethlehem, Pennsylvania is still making Italian-style harpsichords with 19 notes per octave as of the present writing. An extra key is also added in between B and C natural and between E and F natural. (See Hubbard 1965, Plate V, and Brookes 2002, p. 11.)

3.8 The Spiral of Fifths

It is useful to illustrate the problems produced by harmonic tuning in more detail. The basic trouble is simply one of arithmetic. Starting at least with the ancient Greeks, intervals on the musical scale were defined in terms of the harmonics of a vibrating string. As we showed before, these form a series in which the nth harmonic has a frequency that is simply n-times that of the fundamental, where $n = 1, 2, 3, 4, 5 \ldots$ By listening to "beats," two strings can be adjusted in relative pitch so that the fundamental frequency of one closely matches a harmonic of another.[18]

For $n = 2$, we get the simple octave relationship in which the fundamental frequency is doubled. As implied by the word octave itself, the Greeks postulated that there should be eight notes distributed over that span for each scale or mode. The possible intervals between notes on the scale were derived by taking the third harmonic of each note in succession and then lowering it by an octave (or more, as required). The result gives a series of "Fifths" that approximates the twelve different notes on the chromatic scale. (The Greek modes used just eight of those notes, as in the case of the current major and minor scales.) For example, starting on C yields G (up a fifth), starting on that G gives D, and so on. When you get up to B (step 6), the result is F♯.

This process gives rise to the well-known "Circle of Fifths" with which piano teachers torture their students. As one goes on in the sequence C, G, D, A, and so on (clockwise in Fig. 3.18), one traces out all the major key signatures with increasing numbers of sharps, each step going up a fifth on the scale. Similarly, starting at the top of Fig. 3.18 (now, A minor) and going around the circle in the counterclockwise direction involves going down in steps of fourths (i.e., up a fifth and down an octave) and yields the signatures for all the minor keys. By the time you get halfway around the circle (F♯ or G♭), most people opt for a key signature with the least numbers of sharps or flats. But, there are exceptions: Brahms liked to slip in a movement in seven flats now and then in his chamber music for the piano (possibly, just to tease the string players, who generally hate flat keys.) Mussorgsky, who had a loathing for sharp keys (especially, E-major) from childhood, wrote out the first entrance of the

---

[18]The "beating" process described here is a linear one that arises simply by adding two waves at different frequencies. Consider two equal amplitude sine waves whose frequencies are proportional to A and B. Adding the two trigonometry identities,

$$\sin(x \pm y) = \sin x \cos y \pm \cos x \sin y,$$

and letting $x = (A + B)/2$ and $y = (A - B)/2$, one gets the result,

$$\sin A + \sin B = 2\cos\left[\frac{A - B}{2}\right]\sin\left[\frac{A + B}{2}\right].$$

Thus, when $A \approx B$, the resultant wave is at the average of the two frequencies and is amplitude-modulated at half the difference frequency—an effect called "beating." One tunes the strings by adjusting the beat frequency, $(A - B)/2$, to zero.

MAJOR KEYS

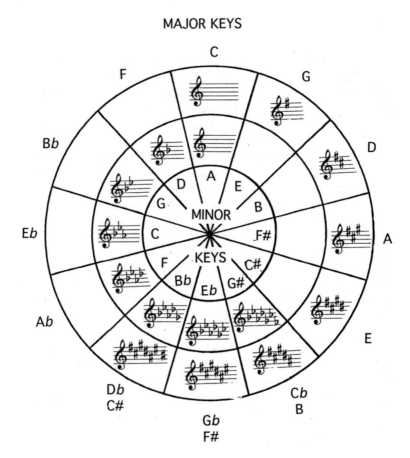

Fig. 3.18 The Circle of Fifths

"Church Motif" from the *Great Gate of Kiev* in nine flats![19] Of course, the "Circle of Fifths" also contains the basis of nineteenth-century tonality.

The main difficulty with the method is simply that the process does not close on itself. (See Table 3.2.) If you go around the "circle" in the clockwise direction a second time, the frequencies are over 1% (about 20 "cents") higher than those obtained on the first pass, and things get still worse on higher passes. (The discrepancy arising after one trip around the circle is sometimes called "the comma of Pythagoras.") The result is that instead of a circle of fifths, one gets a never-ending spiral.

---

[19]Mussorgsky was emulating the hexachordal mode of Russian church music given in the Obikhod, which may have forced him into that large number of flats. He actually did slip into E-major once in his lifetime in the "Prelude" to Khovanshchina, but that was probably forced upon him by the minor third present in the church bell on Red Square signaling matins. (Mussorgsky's church bells usually tolled in C♯.)

Table 3.2 Relative frequencies in the "Circle" of Fifths

Step	Note	Pass		
		(a)	(b)	(c)
1	C	1	1.013644	1.027473
8	C# (Db)	1.067871	1.08244	
3	D	1.125	1.140349	
10	D# (Eb)	1.201355	1.217745	
5	E	1.265625	1.282892	
12	F	1.351524	1.369964	
7	F# (Gb)	1.423828	1.443254	
2	G	1.5	1.520465	1.541209, etc.
9	G# (Ab)	1.601807	1.623661,	etc.
4	A	1.6875	1.710523	
11	A# (Bb)	1.802032	1.826618	
6	B (Cb)	1.898438	1.924338	

Many string players still use harmonic intervals (especially in tuning their instruments) and some harpsichordists use compromise tunings in which some keys are more harmonically in tune than others. In view of the differences in tuning resultant from the basic inharmonic nature of strings discussed in Chap. 4, the compromise provided by the Well-Tempered Scale seems like a minor concession.

The problem was resolved by the time of Bach by the adoption of the "Well-Tempered Scale."[20] Here, one simply defines the ratio of frequencies between successive notes on the chromatic scale mathematically to be given by $2^{1/2} = 1.05946310\ldots$; that is, raising that number to the 12th power gives precisely a factor of two. (See Appendix E.)

3.9 Northern European Developments

Because of their early start in the field, one would have thought that the Italian makers could maintain a monopoly. Yet, Guild records in Antwerp show quite a number of harpsichord makers already present in that city by the mid-1500s. One stipulation of that Guild called for the "maker's marks" to be affixed to all harpsichords made in that city—hence, the use of the "Rose" as a kind of gilded ("guilded"?) logo containing the maker's name or initials. These were sometimes made from a slice of pipe (or cast from a mold) coated with gold and often were accompanied by elaborate floral designs. (See Fig. 3.19.)

According to Hubbard, the addition of the rose had no discernible effect on the sound quality. Although some might think that the Rose provided an opening to

---

[20]Curiously, the Well-Tempered Scale appears to have been a Chinese invention. Early approximations have been traced to Ho Che'eng-t'ien in the fifth century. But, the first use of a scale in which the intervals were based on the 12th root of two has been credited to Chu Tsai-yü in the sixteenth century (Yung 1980, p. 261).

Fig. 3.19 The Rose from a
two-manual harpsichord by
the eighteenth-century French
maker Louis Bellot, a fervent
admirer of the Flemish school
(drawn by the author from an
instrument in the Crosby
Collection of the
Metropolitan Museum of Art
in New York)

Fig. 3.20 Computed spectra
from Eq. (2.5) for the motion
of a string plucked at the
different points given for the
Ruckers instrument in
Table 3.3. The phase is
included in the sign of the
amplitude here. In practice,
coupling to the soundboard
results in very different
spectral amplitudes of the
radiated sound, especially
with strings near the edge of
the soundboard

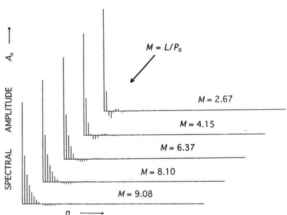

a kind of "Helmholtz resonator" below the soundboard, the air resonances in that space are more like those from a closed organ pipe with an opening at the action end of the instrument. (The area of the Rose is negligible by comparison.) Italian harpsichords were the only instruments that were totally enclosed except for the Rose opening. Kottick et al. (1991) suggest that the main role of such air resonances is to enhance the springiness of the soundboard (Fig. 3.20).[21]

---

[21]When the two-manual 1770 Taskin harpsichord first came into the Yale University Collection in 1957, there was a rectangular hole in the underside of the instrument large enough to permit

Fig. 3.21 Single-manual harpsichord by Andreas Ruckers, Antwerp, 1640. Courtesy of the Yale University Collection of Musical Instruments (photograph by Joseph Szaszfai/Carl Kaufman)

Probably, the most famous name among the Flemish makers was that of the Ruckers family. An organ tuner named Hans Ruckers was admitted to the Guild in 1579. The two-manual harpsichord appears to have been invented by some Flemish builder at about that time. The idea was no doubt inspired by the design of pipe organs. Two-manual harpsichords of a rudimentary sort appear to have been made in the Ruckers workshop as early as the 1590s. Whereas the Italians often covered the inside of their harpsichord lids with beautiful paintings and added floral designs to the sound boards, the Flemish instruments were usually covered on the inside with a kind of block-printed wall paper, often including appropriate mottos in Latin.[22] The outside of the case was usually painted in an imitation marble surface. The two red tassels at the side of the 1640 harpsichord by Andreas Ruckers shown in Fig. 3.21 control the 8- and 4-ft stops. The inscription on the lid, MVSICA

---

inserting one's head. One heard quite a glorious sea of sound in there when the instrument was played.

[22]This technique was extended to pianos by an interior decorator living in Princeton during the 1940s. She covered her entire Steinway inside and out with the same blue wallpaper used in the rest of the living room, not to mention the closet and entrance doors. Once inside the room, one might not find the way out, let alone locate the piano! Hubbard commented (1965, p. 22), "The Ruckers harpsichords are charming, but their naïve crudity is to the sophisticated Italian harpsichord as a cuckoo clock to Brunelleschi's Duomo." (One Ruckers motto was AVDI VIDE ET TACE, which might be freely translated as "Listen, look, and be quiet.")

Table 3.3 Comparison of fractional plucking points of Italian and Flemish instruments (for the first jack)

	Italian (Bononiensis, 1521)	Flemish (Andreas Ruckers, 1648)
d'''	3.25	–
c'''	3.45	2.67
c''	3.86	4.15
c'	5.17	6.37
c	8.33	8.1
C/E	11.32	–
C	–	9.08

Values of $M = L/P_0$ are given for different octaves
Source: Hubbard (1965, p. 8)
Note: In Hubbard's notation c' = middle c; C is two octaves below middle c. C/E means bottom note (E) tuned to C

LETITIAE COMES MEDICINA DOLORVM ("Music is the companion of joy and the medicine of grief"), occurs on many Ruckers instruments of that period. These instruments were favorite models for the Flemish painters, especially Johannes Vermeer.[23] In addition to the harpsichords, a virginal built in 1640 by Johannes Ruckers shows that same motto and block-printed paper and appears in several of Vermeer's paintings.

There is an interesting connection between Vermeer and the Ruckers shop. It has been concluded that a certain harpsichordist and composer named Constantijn Huygens (1596–1687) both knew the Ruckers craftsmen and introduced Vermeer to the camera obscura. (See Steadman 2001.) That particular Huygens was the father of Christiaan Huygens (1629–1695), who is well known among physicists as the discoverer of "Huygens' Principle" in wave motion. (See Chap. 1.) The elder Huygens had not only purchased a camera obscura in London but had also ordered a virginal in 1648 from the Ruckers shop that was made by Jean Couchet, a nephew of Johannes Ruckers. (See Broos 1995.) It seems probable that the elder Huygens also introduced Vermeer to the Ruckers instruments. His other son, Constantijn Jr. (brother of Christiaan) was apparently more interested in art than physics and may well have been the one who actually provided the camera obscura lens to Vermeer. Possibly, Vermeer merely borrowed the Ruckers instruments, which were said to have cost about half the price of one of his own paintings. These were positioned under the windows in his studio at his favorite corner opposite from the camera obscura where models for nearly all of his well-known indoor paintings were posed.

The similarity in design between the Italian and Flemish instruments is illustrated by the comparison of plucking points on 8-ft stops shown in Table 3.3. A study by the author of the transient decay of the sound from the 8-ft stop on an instrument

---

[23]For example, see Vermeer: "The Music Lesson" (1662–1664), "The Concert" (1665–1666), "Lady Standing at the Virginal" (1672–1673), and "Lady Seated at the Virginal" (1675). Jan Steen's painting entitled "The Music Master" (1660) almost certainly shows a Ruckers instrument, but Steen signed his own name on the harpsichord. Also see Gerrit Dou, "Woman at the Clavichord" (1665).

Fig. 3.22 Transient decay of the low C on the 8-ft stop of the Ruckers 1640 instrument shown in Fig. 3.21. The C(N) are magnitudes of the relative harmonic amplitudes normalized to the maximum component at $T = 0$ (the fifth harmonic in the upper-left figure. Compare with Table 3.3 and Fig. 3.20 (the data were taken in collaboration with Jean Bennett)

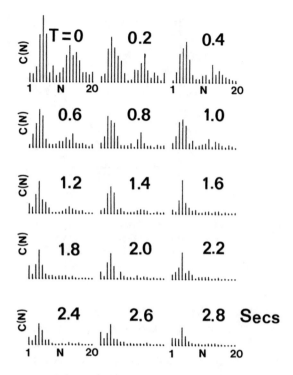

in the Yale Collection made by Andreas Rucker in 1640 is shown in Fig. 3.22. A restoration of this instrument by Frank Hubbard was completed in 1972 before the present data were taken, using crow quills and 0.020″ steel wire on the low C.[24] The instrument was tuned to $a' = 409$ Hz, with quarter comma meantone temperament. Note the near absence of the ninth harmonic, which is to be expected on the basis of the fractional plucking point listed in Table 3.3. Because that particular string is located above a position on the soundboard nearest to the edge of the case, the fundamental and first few harmonics are greatly attenuated and the maximum amplitude occurs at the fifth. Some harmonics appeared to oscillate in intensity with time as the waveform died out, probably due to slow beating effects between harmonics of the 8- and 4-ft strings—the latter being undamped because the C-key was kept depressed during the measurements. Ignoring the coupling effect between these two sets of strings, one can estimate the Q of the string as a function of frequency from the data in Fig. 3.23. As shown in Appendix A, the Q ("Quality Factor") is given as a function of frequency (F) by

$$Q = 2\pi F/\gamma \text{ where } \gamma = 2/\tau. \tag{3.9}$$

---

[24] A later restoration by Frank Rutkowski and Robert Robinette (done in March 1987) using 0.023″ yellow and red brass (90% copper and 10% zinc) on the low C was made after the present data were taken.

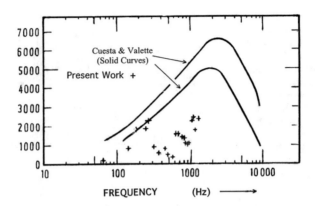

Fig. 3.23 Variation of Q with frequency for some harpsichord strings. The solid curves are limits obtained by Cuesta and Valette (1988) for freely vibrating 0.33-mm diameter steel strings supplied by the Zuckermann Company in a harpsichord kit. The points marked "+" were computed by the author from the data in Fig. 3.19 based on the decay of sound from the lowest note on the 8-ft stop of the 1648 instrument by Andreas Ruckers in the Yale University Collection

Here, γ is the energy decay rate and $\tau (= 2/\gamma)$ is the time the amplitude takes to decrease by $1/e(= 0.3678794\ldots)$ of its initial value. The different frequencies are given by $F = nF_0$ where n is the harmonic of the fundamental frequency, $F_0 \approx 61.73\,$Hz. (The harpsichord was tuned to $a' = 409\,$Hz to match the original conditions for which the instrument was designed.) Applying Eq. (3.6) to the results in Fig. 3.22, one gets the scatter of points labeled "+" in Fig. 3.23. The scattering is no doubt due partly to the beating between the harmonics of the 8- and 4-ft strings. The two solid lines in Fig. 3.23 represent the upper and lower limits on Q measurements made on three 0.014-in diameter steel harpsichord strings ("fer de Zuckermann") of different length by Cuesta and Valette (1988). Their measurements were made on freely vibrating strings unimpeded by attachment to a harpsichord. One would, of course, expect to find that the Q of the string "loaded" by the bridge and soundboard would be lower because energy is then also coupled from the string to the instrument. The data in Fig. 3.24 support that interpretation. The values obtained at 200–400 Hz reach the lower bound of the measurements by Cuesta and Valette, whereas from 200 to 1200 Hz about half of the decaying energy from the string appears to be coupled from the string to the Ruckers instrument (as opposed to air resistance and internal viscoelasticity loss in the string alone).

Cuesta and Valette also concluded that although damping from air resistance is primarily responsible for the decay rate of the isolated string at low frequencies, internal loss in the string through viscoelasticity and domain dislocations in the wire were the main factors at high frequencies (above $\approx 1\,$kHz). These losses vary a great deal from one string to another and depend strongly on the string's metallurgical state. They found that the Q of the unloaded string typically peaks in the range from about 1000 to 4000 Hz—the region where the ear is most sensitive. Maximum Q's of about 6500 were found at frequencies of about 2–4 kHz and fell off by a factor of about six by 10 kHz. By comparison, the maximum Q's from freely vibrating nylon strings (having more viscoelasticity loss) were in the order of 1400.

Fig. 3.24 Two-manual harpsichord by Pascal Taskin, Paris, 1770. Courtesy of the Yale University Collection of Musical Instruments (photograph by Joseph Szaszfai/Carl Kaufman)

Harpsichord building in Paris did not occur as early as it did in Italy or Antwerp and no French instruments have survived that were made before the middle of the seventeenth century. According to Hubbard, the French efforts were divided between two groups: those who produced innovative models of their own and others who spent their time rebuilding Ruckers instruments. Of the former group, the most eminent were members of the Blanchet family, starting with Nicholas Blanchet (1660–1731) and ending with Pascal Joseph Taskin (1723–1793), who married the widow of François Etienne Blanchet II. Taskin is regarded by some as the Stradivarius of harpsichord making. Indeed, a two-manual instrument of his design with one 8-ft stop on the upper manual and an 8- and 4-ft stop on the lower manual, together with a buff stop has come to be regarded as the epitome of the "French harpsichord." Unfortunately, most of the original instruments by Taskin were burned after the French Revolution. According to Hubbard (p. 116), "There can be no doubt that the violence of the Revolution, particularly since it came at the moment of transition to the piano, was responsible for the destruction of a large number of harpsichords." One of the old employees of the Paris Conservatory recalled that those "harpsichords served to heat the classrooms, a purpose for which some were taken from time to time from the garret."

In any case, only six of the original two-manual instruments by the great French master are still known to exist: Two (dated 1764 and 1769) are in the Russell

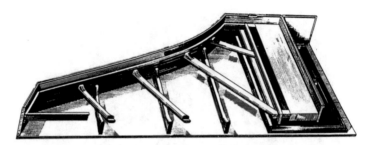

Fig. 3.25 Structural supports in the case of a Taskin instrument (after Hubbard 1965)

Collection at Edinburg University in Scotland; one is in Paris (1769) owned privately by Mlle Guerelle; one dated 1770 is in the Yale University Instrument Collection in New Haven (Fig. 3.23); one dated 1780 is in the Gelsen Collection in Milan; and a remaining one built in 1782 is owned by the Marquisa Olga de Clotilde of Portugal. (See Boalch 1974.) Hubbard felt that even the framing structure used by Taskin had special esthetic appeal. (See Fig. 3.25.) The meticulous measurements reported in Hubbard's 1965 book have enabled a number of people to make replicas of the Taskin instruments.[25]

On some instruments, Taskin added jacks with leather plectra behind the ones equipped with crow quills to produce a solo stop. In addition, he sometimes added a "Peau de Buffle" (literally, "buffalo skin") register. The relative striking points on the two sets of 8-ft strings were different, with the result that one manual would sometimes have a more "nasal" quality than the other. Certainly, the spectra from the two-manual Yale instrument are rich in harmonic content (Fig. 3.26).

The spectra from the lowest C on each of the two separate 8-ft stops are shown in Fig. 3.27, where 16 harmonics below 1000 Hz were encountered. In each case, the spectrum peaked at the second harmonic. However, the lower manual 8-ft stop must have been plucked near the node for the seventh harmonic and that for the upper manual in between the node for the fifth and sixth harmonics—thus producing significantly different tonal color from the 8-ft stops on the two manuals. According to Ripin et al., p. 230, the subtlety of tonal color and sweetness of sound in Taskin's harpsichords may have been introduced primarily to attract attention to the instrument and away from the "less-substantial [French] music" of the period.

The time-dependent decay of the harmonics on the 4-ft stop of the lower manual (Fig. 3.27) showed the same sort of slow beating effects with harmonics of the 8-ft strings encountered before with the Ruckers instrument. Note especially the sharp

---

[25]Hubbard himself established a company (in a barn in northeastern Massachusetts) that sold precut parts as kits. Kits are also available from the Zuckerman company and various premade parts are sold by the B & G Instrument Workshop in Ashland, OR. However, "do-it-yourselfers" should be warned that it takes about 300 expert-person hours to build a two-manual Taskin and that you have to complete five or six before you start turning out good ones. Back in the 1970s, I asked a Hubbard associate what he could do for a kit that was poorly completed by one of my children. (At that time, the two-manual "Taskin" kit sold for about $2400.) The reply was, "We take them up to the second floor of our barn and toss them out the hay-loft window."

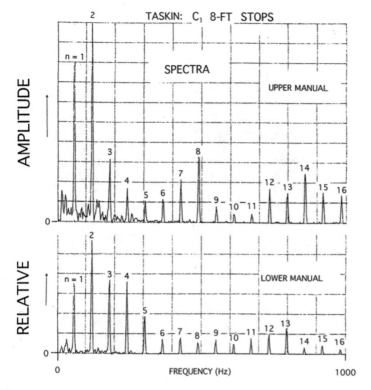

Fig. 3.26 Spectra from the lowest C on the 8-ft stops for the upper and lower manuals of the 1770 Taskin Yale instrument (data taken with Jean Bennett)

minimum in the second harmonic amplitude at about 1.2 s after plucking the string and the subsequent maximum at about 3 s. The bass strings were evidently tuned by minimizing the beat of the fundamental of the 4-ft stop with the second harmonic of the 8-ft stop. As discussed in Chap. 4 in regard to piano tuning, the beats are different between different pairs of overtones that normally would be expected to be in tune. The difference is due to nonlinearity in the vibrating strings and the data for the second harmonic in Fig. 3.27 constitute a direct demonstration of this phenomenon.

One of the most beautiful sounding replicas of a French-style harpsichord I have heard personally was a copy of a two-manual instrument by Henri Hemsch started by Frank Hubbard and completed by Richard Rephann after Hubbard's death. That instrument was outfitted with carefully annealed wire and plucked with crow quills.[26] However, as with the aroma of fine cigars, not everyone appreciates the tone quality of a fine harpsichord.[27]

---

[26] A stereo recording of the Hubbard–Rephann instrument made by the author of several Bach pieces played by harpsichordist Lola Odiaga is available on CBS Discos. (See Odiaga 1974.)

[27] Oboist Robert Bloom once told me that he thought "they all sounded like old bed springs." On the other hand, Sir Thomas Beecham felt that the sound was more "like two skeletons copulating

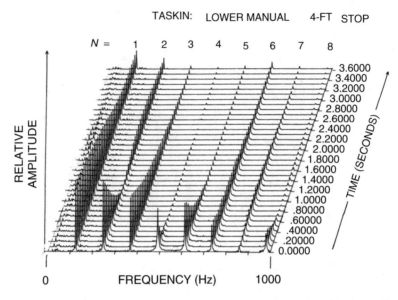

TASKIN: LOWER MANUAL   4-FT STOP

Fig. 3.27 Spectral-surface showing the time-dependent decay of harmonics of the lowest C on the 4-ft stop of the lower manual on the Taskin instrument in the Yale collection. Note the prompt start of each harmonic at $t = 0$ and the subsequent minimum and maximum in the second harmonic amplitude with time after plucking the string (data taken with Jean Bennett)

There were several major deficiencies of the harpsichord as a musical instrument: The sound was weak compared to that of many other instruments, the sound decayed in a very short time, the dynamic range was very limited, and the sound level could only be changed significantly by adding or removing stops. Finally, the intonation tended to be unstable and the instrument required frequent tuning—often in the middle of performances. (Rephann once warned me that even bringing a cup of hot coffee into the same room with a harpsichord would throw it out of tune because of the change in humidity!)[28]

According to Ripin et al., the last harpsichord in the classic tradition was made in 1809. By that time, development of the piano had reached such a point that general interest in the harpsichord had waned. By then, Beethoven had composed 24 of his 32 sonatas for the piano, had broken lots of strings, and had spilled several bottles of ink on his succession of inadequate pianos.

---

on a corrugated tin roof" (Atkins and Newman 1978, p. 34.) But, Beecham was probably referring to one of the later English models which Ripin et al., p. 234 say sounded like loud brass bands compared to the woodwind-like quality of the French instruments.

[28] Yale chemist Martin Saunders (private communication) proposed a possible solution to the tuning problem: A computer-controlled electric-power source feeds current through each string separately at about 1/2 Watt maximum per string. Increasing the current heats the string, lowering its pitch (and vice versa), providing required tuning changes within a 1- or 2-s thermal response

The harpsichord remained dormant until the late nineteenth century when both Arnold Dolmetsch and Hans Richter began presenting concerts featuring the instrument. Shortly afterward, two French piano companies, Pleyel and Erard, independently brought out elaborately decorated harpsichords modeled after a classic Taskin instrument and showed them at the Paris Exposition of 1889. A decade later, a harpsichord dubiously associated with the composer J. S. Bach was built by William Hirl of Berlin (and eventually by J. C. Neupert of Bamberg). Meanwhile, the Erard instruments had diffused at least as far south as Vienna. Wanda Landowska, in her role as Professor of Harpsichord at the Berlin Hochschule für Musik, persuaded the Pleyel company to produce its "Landowska" model starting in 1912. Instruments of this vintage were essentially pianos outfitted with plectra instead of hammers and featured two keyboards with lots of pedal-operated stops, 16-ft pitch, metal frames, and thick strings under high tension. Many used only a plain soundboard (as in a piano) with no enclosed volume of air. The tone was bright and penetrating, but without much bass and could not be heard across the hall.

In 1905, Arnold Dolmetsch moved to the USA and established a short-lived harpsichord department with the Chickering piano company of Boston. The Dolmetsch-Chickering instruments were lighter in construction and modeled after the earlier Taskin design, producing more mellow tone quality. Following financial troubles at Chickering, Dolmetsch departed for Paris in 1910 and harpsichord building in the USA went into hibernation until 1931 when John Challis returned to America from Europe. Challis was an inventive fellow and quick to adopt current methods of technology. In his later years, he designed an instrument made entirely of metal, including the soundboard! Those instruments had remarkably stable intonation, but as Ripin et al. (1980) put it in mild understatement, "the tonal quality was not to everyone's taste." After World War II, Frank Hubbard and William Dowd established their joint company in Boston and started a return to the seventeenth- and eighteenth-century traditions of harpsichord making that was much lauded by Ralph Kirkpatrick and other harpsichordists of note.

Although the decay times from the best harpsichords were much longer than from those in the early Italian instruments, they were still at least a factor of two shorter than those of the larger diameter strings on a typical piano. Indeed, of the various ways of simulating the sound of a harpsichord on a piano (including thumb tacks in the hammers), one of the closest is to record a piece an octave lower on the piano keyboard and then play the recording back at twice the original speed. In this way, one also picks up a factor-of-two in technique on fast passages that can be rather startling to the uninformed.

---

time. A computer-controlled FFT continuously samples the output from a microphone above the strings and is used to monitor the pitch of each note continuously. The computer then controls the current through each string to keep the entire instrument in tune. Saunders suggested that one could also have the program provide continuous harmonic tuning as the music changed keys. But, it is not clear how the method could distinguish between two harmonics on different strings that were of the same frequency. (The method might just result in a puddle of molten strings on the soundboard.)

3.10   The Clavichord

Although the clavichord is neither plucked nor struck with a hammer, it represents
a kind of intermediate stage between the harpsichord and the piano. Like the
harpsichord, it was also mostly in use between the fifteenth and eighteenth centuries.
Although ingenious in its simplicity, it is probably the least efficient way known to
man to obtain sound from a vibrating string. As illustrated for one note in Fig. 3.28,
a Λ-shaped piece of metal called a "tangent" is attached to the end of each key and
engages the string at right angles and lifts it up in the air. The active length of the
vibrating string is determined by the tangent at one end and by the bridge at the
other. As with the harpsichord, the bridge is fastened to a soundboard in order to
couple the string vibrations to sound waves in the room. The section between the
tangent and the hitch pins is heavily damped by cloth. That section of the string
is prevented from vibrating when the tangent is engaged and the damping material
also deadens vibrations in the entire string when the key is released; hence, one does
not need an additional damper on each key, as with the harpsichord. The mechanism
by which the string is set in motion is a little puzzling at first glance. Because the
tangent hits the string at a node ($x = 0$ in Fig. 3.29) for all harmonics of the active
length of string, one might think that none of the harmonics of the string would be
excited at all. However, that is not completely true, unless the key is depressed very,
very slowly. Some excitation of the string actually does occur, but by a much more
indirect process than in the harpsichord or the piano.

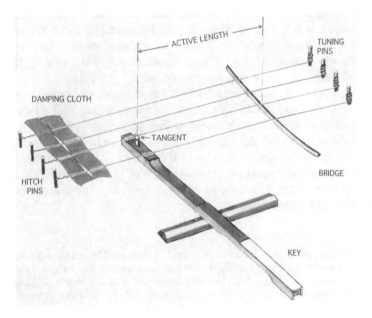

Fig. 3.28 Diagram of the action of a clavichord

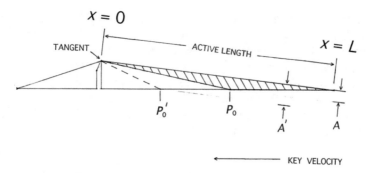

Fig. 3.29 Diagram illustrating the excitation of a clavichord string

3.11 Method of Excitation

If the tangent hits the string with any appreciable speed, the inertia of the string results in the string being bent slightly at points between the tangent ($x = 0$) and the bridge ($x = L$). In the real case, the bending will occur continuously over part of that region. Nevertheless, one can approximate this process by assuming that the bending occurs at just one effective point ($x = P_0$ in Fig. 3.29). When the key is depressed with near zero velocity, P_0 is coincident with the bridge location ($x = L$), and no sound is produced. But, as the initial key velocity increases, that effective point moves toward the tangent (at $x = 0$). The result is that the string is distorted into a quasi-triangular shape similar to that of a harpsichord string at $t = 0$, except that the triangular displacement is upside down. For example, when the bending point is at P_0 in Fig. 3.29, the string is initially distorted into the shape of the shaded triangle shown with initial amplitude, A. As the initial key velocity increases further, the effective bending point moves to another location $x = P_0'$ nearer to the tangent where the new triangle has a larger negative amplitude A', and so on. The harmonic spectrum of the string in this approximation is just the same as that calculated for the plucked string in Eq. (3.5) for $M = L/P_0$, except that the triangle is upside down. The pulse then travels down the string as illustrated in Fig. 3.8, except that the displacement is negative. One important consequence of the negative amplitude of the initial pulse launched down the string is that after the "hard" reflection by the bridge, the returning pulse is positive. When this returning positive pulse arrives back at $x = 0$, it tends to make the string jump up off the tangent. From the present excitation model, one can see that both the amplitude and the number of higher harmonics will increase as the initial key velocity increases. Since one can give the key a large variation in initial velocity, the dynamic range of the clavichord is immense. Yale clavichordist Richard Rephann[29] believes that the dynamic range (not the loudness) of a clavichord is actually greater than that of a concert-grand piano. Although the thought is counterintuitive, I believe that he is actually right. Even a well-regulated piano action has a definite threshold level below which the

---

[29] Private communication.

hammer does not have enough momentum to hit the string. The same is not true of the clavichord. However, the sound produced throughout much of its range is below the threshold of hearing and when struck very hard the pitch goes very sharp. (The clavichord is just the thing for city apartment dwellers who worry about disturbing their neighbors or for composers who want to try out their ideas in private.) One other advantage of the instrument is that by varying the pressure on the key after the tangent is engaged, one can produce vibrato on the note.[30] Figure 3.30 shows a clavichord similar to one that Mozart owned.

Fig. 3.30 Clavichord made by Johann Christoph Georg Schiedmayer, circa 1796 (courtesy of the Edwin Ripin Collection at the Museum of Fine Arts in Boston, Massachusetts)

───▶

Fig. 3.31 (continued) (also called a Rubell); (13)–(14) Marine Trumpets; (15) and (16) seventeenth-century Pochettes; (17) and (18) twelfth- and thirteenth-century Italian lutes; (19) and (20) tenth- and eleventh-century Gigues (a family of Rebecs). Row 3. (21) and (22), fourteenth- and fifteenth-century Rebecs; (23) and (24), eleventh- and tenth-century Rebecs; (25) pimitive viol; (26) fifteenth-century Lira da Brassio; (27) fifteenth–eighteenth-century Viola d'Amore; (28) sixteenth-century viole; (28) and (29) Vielles a roue from the Middle Ages ("Hurdy-Gurdies"); (30) Organistrum. Row 4. (31) German Cithare; (32) Guzla; (33) Hawaiian Guitar; (34) American Guitar; (35) Indonesian Taranj; (36) sixteenth-century Viol; (37) fifteenth–eighteenth-century Quinton; (38) Bastard Viol (Middle Ages); (39) Modern Plate Mandolin; (40) Modern Guitar; (41) Neopolitan Mandolin; (42) Ancient Guitar. Row 5. (43) Spanish Mandolin (Banduria); (44) and (45) Banjo and Alto Banjo. Row 6. (46) Russian Balalaika; (47) Jazz Guitar; (48) Trapezoidal Violin (1830); (49) Moon Violin (1850); (50) sixteenth-century Violin; (51) Midget Violin; (52) Belleville Violin (1828); (53) Chanot Violin (1840)

──────────────────

[30]Amazingly, Ralph Kirkpatrick once did a recording of Bach's Well-Tempered Clavier on the clavichord without using any vibrato at all.

Fig. 3.31 Collection of stringed instruments that can be plucked. (Courtesy of E. Richard, Master Luthier of Rouen, France.) Identification, row-wise from top left: Row 1. (1) Wrinnor (Portable Oriental harp); (2) Greek Lyre; (3) Roman Lyre; (4) eighth- or ninth-century Trojan Crouth; (5) ninth-century Lyre Guitar (plucked with a plectrum); (6)–(8), ninth-, twelfth-, and fifteenth-century Psalters. Row 2. (9) tenth- and eleventh-century bowed Crouth; (10) thirteenth-century Rotta; (11) fifteenth-century Minstrel Harp; (12) eleventh- and twelfth-century Arabic Rabab

3.12   Other Plucked Instruments

A variety of other plucked instruments is shown in Fig. 3.31. Space limitations prevent a detailed discussion of these instruments, but it will be appreciated that they operate in a plucked mode much the same manner as those discussed earlier in this chapter.

Problems

3.1 Suppose there were only one coil located at $x_0 = 3.75$ inches in the electric bass shown in Fig. 3.10. Where would the nulls in the electronic spectra occur? (Note $L = 34.1$ in.)

3.2 Suppose the coils in the Steinberger bass (Fig. 3.10) were connected so that the voltages added. At what harmonics would minima in the spectrum occur?

3.3 From the mode pattern at 56.8 Hz in Fig. 3.16, estimate the surface wave velocity in the long dimension of the harpsichord soundboard. Take the long dimension of the soundboard to be about 4.69 ft (corresponding to a Ruckers single-manual instrument.)

3.4 On a certain gut-strung harpsichord, the string for middle C is 63.6 cm long and has a diameter of 0.029 in. Assuming that the instrument is brought up to $A = 440$ Hz, what is the tension on the string? (The mass per unit length of the gut string is 0.0060 g/cm.)

3.5 Suppose someone wanted to restring the gut harpsichord in the previous problem using steel wire of the same length chosen so as to produce the same tension. What diameter should the wire be for middle C? (Take the density of steel to be 7.83 g/cm^3 and the density of gut to be 1.4 g/cm^3.)

3.6 Research Problem: See if you can design a virginal in which all the strings are plucked in the middle, hence producing clarinet-like tone quality. (A virginal has the keyboard parallel to the strings.)

Chapter 4
The Struck String

4.1 The Piano[1]

> Some say that pianists are human and quote the case of Harry Truman.—Ogden Nash

The piano is classified as a percussion instrument because it uses hammers to hit the strings. Still, many pianists think of it as a "singing instrument"—perhaps because they can't refrain from singing themselves while playing. The enormous dynamic range in present compact disc recordings has made that habit something of a liability. Audio engineers have been known to shout, "Will someone please get that drunk out of the studio?"—not realizing that the sound actually came from the soloist's mouth. Some pianists wiggle the keys and others make various grimaces and bodily gestures to fool the audience into thinking it really is a singing instrument. But in the cynical words of music critic Bernard Holland, "At the end of a day, it is still a drum with different pitches." Nevertheless, the piano has been the backbone of musical development over the past two centuries and in many instances can sound quite warm and beautiful.

The original version of this chapter was revised: Equation on page 108 was corrected. The correction to this chapter is available at https://doi.org/10.1007/978-3-319-92796-1_8

[1]For the history of the piano, see Dolge (1911), Ripin (1980), and Gill (1981). The author has also benefitted from helpful discussions with Edmund Michael Frederick and Patricia Frederick of the Ashburnham Massachusetts Historical Piano Collection, master piano technician Christopher Robinson and the late Edward Deutsch, former curator of pianos at Yale University.

© Springer Nature Switzerland AG 2018
W. R. Bennett, Jr., *The Science of Musical Sound*,
https://doi.org/10.1007/978-3-319-92796-1_4

4.2   From Cristofori to Mozart

Just as the late nineteenth-century harpsichords were essentially pianos outfitted with plectra, the first piano made by Bartolomeo Cristofori (official keeper of musical instruments for the Medici family in Florence) in 1698 was basically a harpsichord using hammers. Indeed, the sound of the Cristofori instrument is strongly reminiscent of that of a harpsichord because of the thin strings and use of a hard hammer. To be fair, Cristofori did address (and solve) many of the major problems that were faced over and over again by other piano makers in later years.

One major problem in the design of a piano is satisfying the requirement that the hammer escape from the action (the mechanism that launches the hammer after depressing the key) immediately before hitting the string; otherwise, it will lift the string. Another problem is in amplifying the motion so that a much larger velocity is given to the hammer than that given to the key by the finger. Additionally, a damper must be raised off the string when the key is depressed. Finally, some provision has to be made to catch the hammer after it bounces off the string so that it does not rebound and hit the string a second time. To facilitate a rapid rebounding process, thicker wire was used than in the harpsichord and under higher tension. That in turn, also provided a louder sound and a longer decay time than had been obtained on the harpsichord.

These requirements were largely met by the action Cristofori developed by the 1720s—a period from which three of his original instruments have survived. The damper was moved to the end of the key and mounted on what resembles an older-style harpsichord jack, together with the old upper and lower jack guides. A new "piano jack" was mounted at the old plectra location. In Fig. 4.1, downward force on the key at the left raises the damper off the string at the right and causes the jack to push upward on the motion—amplifying lever which is hinged at the right. After relaying the key impulse, the jack disengages from that lever by slipping off to the left ("escapement"). The left end of the lever pushes upward on the hammer shank which rotates about a hinge at the left, hurling the light (hollow) hammer at the right upward against the string at a velocity some eight times larger than that initially given the key downward. The hammer is caught by the back check after bouncing off the string, thereby preventing it from rebounding to hit the string a second time and the damper remains raised until the player's finger is removed from the key. Two of Cristofori's surviving instruments had inverted (upside down) pin blocks. Although they must have been awkward to tune, the hammer blows in that case would tend to tighten the tuning pins rather than to loosen them, as in most modern pianos.[2]

---

[2]Chris Robinson (private communication) noted that one reason the early pins tended to move upward was that they were not threaded. Hence, the early tuning keys had a hammer head on them to permit driving the pins back down—thus the origin of the term "tuning hammer" which persists to the present day.

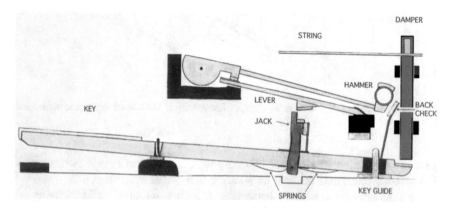

Fig. 4.1 The piano action designed by Cristofori circa 1720

4.3 String Vibration and the Una Corda Mode[3]

The tradition of Italian harpsichord building (with pairs of strings on every note tuned in unison) served Cristofori well. In two of his surviving instruments, the keyboard could be slid sideways so that the hammers would hit only one string on each note, providing the una corda effect (soft pedal) much used by Beethoven and other later composers. This development not only provides a difference in sound level and tone quality, but also results in tones with longer persistence. Because each note has two strings tuned in unison, energy is coupled between the two strings through the bridge. The coupling process results in two normal modes of vibration for each harmonic with slightly different frequencies. In one mode the two strings vibrate up and down together. That mode is excited initially when the hammer hits both strings at once (left-hand pattern in Fig. 4.2). There, because the pulses in each string are in phase, energy is coupled strongly to the soundboard when the two pulses arrive at the bridge. However, in the other normal mode of the system, which is excited preferentially when the hammer hits only one of the two strings (the una corda pedal position), the strings vibrate out-of-phase (right-hand drawing in Fig. 4.2). With the hammer shifted to the right and both strings undamped, the pulse generated by striking the right-hand string alone is inverted upon reflection at the bridge. From Newton's "Law of Action and Reaction," the bridge itself goes up slightly as the right-hand string goes down, thus imparting an upward pulse to the left-hand string that runs back towards the front of the piano. Both strings quickly settle down to oscillate in a mode in which out-of-phase (by 180°) running waves on the two strings hit the bridge simultaneously, and hence impart less energy per

---

[3]See Weinreich (1977) for references to previous work on this problem and a mathematical treatment of the mode-coupling process. A coupled electric circuit model is derived in Appendix A.

Fig. 4.2 The circulating pulses on the strings when excited in the normal symmetric mode (left) and in the anti-symmetric mode (right)

pass to the soundboard. Evidence of these pulses may be seen by using stroboscopic illumination tuned near the note frequency at one end of the strings.

As time goes on, the higher harmonics damp out the most quickly, leaving the string vibrating in one (or both) of the two fundamental modes of the first harmonic. Again, since the even symmetry mode transfers energy more rapidly to the bridge, the energy will eventually decay through the longer-lived odd-symmetry mode regardless of the initial position of the hammer.[4] Similar things occur in modern pianos, except that three strings are used on most notes instead of two.

There is an interesting possibility that Cristofori *could* have tried, although there is no evidence that he actually did so. One might obtain a *crescendo* after a given note has been struck in the out-of-phase mode by incorporating a second set of dampers for each note. The second set would normally be off the strings, but could be used to dampen the left-hand strings suddenly (say, with another pedal or knee lever) after a note was struck with the una corda pedal down (hammers shifted to the right). The process would work as follows: when the two strings are vibrating in the 180° out-of-phase (una corda) mode, very little loss of energy occurs at the bridge and the tone persists much longer than with the in-phase mode. But if you dampen one string suddenly after the note was struck in the una corda position, there would no longer be a 180° out-of-phase component of the string motion at the bridge and the tone should decay more rapidly because of the greater coupling efficiency of the single pulse to the sound board. For that reason, the intensity should go up just because the rate of energy transfer has increased. Weinreich (1977) reported an increase in the late decay of the string using this technique of nearly 20 dB.[5] In an experiment by the author using middle C on a modern Steinway concert grand (in which the left string was damped completely with a felt wedge and the una corda pedal was adjusted to hit only the right-hand string), the increase in loudness was

---

[4]The above description, of course, assumes that the strings are only vibrating in the vertical plane. As Weinreich (1977) noted, if the top of the hammer is not perfectly horizontal, the strings may also be excited to vibrate in the horizontal plane, in which case there is much smaller coupling (e.g., ≈ 10 dB less) to the bridge. Hence, a longer-lived component with very much smaller energy may also be excited through that mechanism with poorly voiced hammers.

[5]Chris Robinson (private communication) noted that some piano tuners deliberately keep the unison strings slightly out of tune to preserve the decay time of the instrument. [The necessary detuning at middle C would only be about 0.5 cents. WRB]

quite perceptible, but not much more than 5 dB.[6] In a similar experiment using the two eight-foot stops on a French style harpsichord by Paul Kennedy, the effect could not be detected at all. However, that may have been due to differences in the string position for the two stops on the eight-foot bridge. [See the discussion of the coupled mechanical circuit model given by Weinreich (1977) and the coupled electric circuit model discussed in Appendix A of this book.]

Gottfried Silberman built pianos in Germany in the mid-1700s with light actions that were nearly identical to Cristofori's design and of which Johann Sebastian Bach approved. (Silberman used hollow square hammers, rather than hollow cylinders as did Cristofori.) But piano building in Vienna was something of an anomaly. In contrast to what had happened in Italy, France, England, and Germany, there was not much of a harpsichord-making industry in Vienna to fall back upon. Pianos seemed to have evolved there during the last decades of the eighteenth century almost through spontaneous creation in a manner isolated from developments in other countries. Although the pianos of Mozart's time did look a lot like harpsichords (complete with knee-levers to work the dampers), the Viennese actions were quite different from those developed in the northern European cities and remained that way through the time of Brahms. The characteristic action in the Viennese school during Mozart's life was the "Prellmechanik" (or "Rebounding mechanism") illustrated in Fig. 4.3 that is said to have been invented by one Johann Andreas Stein (1728–1792).

As shown in Fig. 4.4, the Stein mechanism had a minimum number of levers and the hammer shank itself provided the main mechanical amplification of the hammer velocity. As shown in the figure, a beak-like structure at the end of the hammer shank was released by a spring-restored-single-escapement mechanism. Mozart seemed to like the Stein mechanism and complained that the hammers "blocked and stuttered" on other instruments. The hammers on the Stein action were very small and light, but only covered by a thin layer of leather.[7] Anton Walter (1752–1826), the Viennese maker who built Mozart's own piano developed a modification of the Stein action which included larger hammers and a back-check rail to prevent the hammer from rebounding and hitting the string a second time after the key was struck. Unlike the Cristofori instruments, many "Mozart pianos" had triple stringing and incorporated a "moderator" which was operated by a knee lever that would insert a sheet of material between the hammers and the strings to soften the sound. Although certainly possible to include in principle, the "una corda" shift of the hammers was not incorporated on the Stein instruments I have seen. Pianist Lili Kraus owned a Mozart piano, but complained about its dynamic range: "If you play

---

[6]Taub (2002, p. 34) points out that one can also obtain a slight crescendo by depressing the normal right-hand pedal after the notes are struck, but that, of course, is a very different effect than the one discussed in the text. Taub (p. 47) also suggests obtaining a "crisp fp" by depressing the keys rapidly to create the forte, immediately releasing them and then depressing them again immediately so that the dampers rise quickly allowing the strings to vibrate further at decreased amplitude. That is really the inverse of the crescendo effect discussed in the text (which perhaps should be designated pf.)

[7]The ones I've heard personally sounded as if the strings were actually struck by "steins" (i.e., rocks).

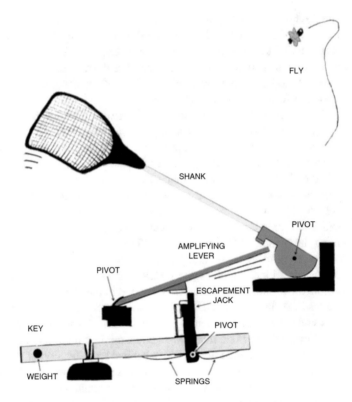

Fig. 4.3   The single escapement fly swatter (inspired by the Cristofori piano action)

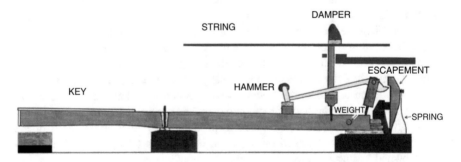

Fig. 4.4   The "Prellmechanik" action designed by Stein

it too loudly, the hammers will break; if you play it too softly, it simply won't speak."
(Elder 1982, p. 190.) That may explain Beethoven's observation that Mozart "had a
fine but choppy (zerhacktes) way of playing, non ligato." (Thayer and Forbes 1973,
p. 88.) The notion that Mozart couldn't play legato passages is certainly counterintuitive, judging from all the beautiful ones written in his music. But Beethoven's
nephew Karl wrote in one of the "conversation books" that "Mozart's fingers were

Fig. 4.5 Mozart, his clavichord, his piano, one of his compositions, and his mother. The piano was made by Anton Walter circa 1780 with a Viennese ("Prellmechanik") action and a five-octave range starting on low F. It was once equipped with a large sustaining pedal and was transported to Mozart's various concerts in Vienna (the figure is a pastiche drawn from several period paintings)

so bent from playing that he couldn't even cut his meat [at the dinner table]." (Solomon 1995, p. 301.) Perhaps Mozart was just suffering from writer's cramp? (He wrote about one major piece by hand every 2 weeks on the average throughout his lifetime.)

Mozart's own piano originally had a knee lever to raise all the dampers at once, although it is said that he later had a large foot pedal added for that purpose. (Possibly Walter Gieseking's insistence on avoiding the pedal in playing Mozart was because he didn't realize that such a thing existed on a Mozart piano.) (Elder 1982, p. 10) Other Viennese pianos, especially those by Conrad Graf, were equipped with various damper-lifting pedals, some of which would only operate on the dampers in the bass, or in the treble, separately (Fig. 4.5).[8]

---

[8]The pedal mania reached a climax in the instruments built by Conrad Graf in Vienna during the early 1800s, some having as many as six or seven pedals. Usually, two would split up the dampers between bass and treble. According to Taub (2002, p. 42), Beethoven only mentioned that device once (in the score for the Sonata, Opus 53) and said not to use it. Some pedals were designated "bassoon stops" and pushed a tissue-paper-like structure onto the strings. Another, the "Janizary stop," would thump the sound board with a leather glove and in some cases ring a bell for use in "Turkish marches." There is currently a multi-pedal Graf instrument in Ferdinand Schubert's

Some pianos of the period made with triple stringing by the Clementi firm did incorporate a una corda pedal which shifted the hammers. (A small lever at the side of the keyboard controlled the distance the hammers went laterally so that one string could be excited alone.) It is doubtful that Mozart, Beethoven, or any of the other composers of that time understood how the sustaining power provided by the una corda pedal actually worked.

4.4   The Myth of the "Authentic" Beethoven Piano

As described by Robert Taub,[9] Beethoven owned or borrowed at least 14 different pianos during his lifetime, including ones having just about every new development. These included Viennese models made by Stein and Streicher (to whose firms he felt great allegiance) and non-Viennese pianos made by Erard (Paris, 1803), Broadwood (London, 1818), and Vogel (Budapest). The earliest ones, with a five-octave range starting on F, had hand stops or knee levers as a carry-over from the harpsichord tradition and were of the two-string-per-note design in which the term una corda had literal meaning. Later models incorporated two pedals (the left, to shift the keyboard and the right, to raise the dampers), triple unison stringing, and extended keyboard range. One made by Graf in 1825 (now in the Beethoven House in Bonn) actually had four strings per note, not to mention a special resonator intended to compensate for Beethoven's loss in hearing. (But by then, his hearing was almost completely gone, and he never warmed to the instrument.)

It is clear that Beethoven was dissatisfied with every instrument he ever tried and was constantly in search of pianos with greater dynamic range and keyboard extent. Comments such as "Clavicembalo miserabile" and "It remains an inadequate instrument" are sprinkled in his manuscripts and correspondence. (Taub 2002, p. 89). Although he liked the fast response of the early Viennese pianos, he evidently advised piano maker Andreas Streicher "to abandon the soft, yielding repercussive tone of the other Vienna instruments and give his instruments greater resonance and more elasticity, so that the virtuoso who plays with strength and significance may have the instrument in better command for sustained and expressive tone."[10] As Taub traced in detail throughout the 32 piano sonatas, Beethoven was quick to incorporate (and ask for) new improvements. For example, the 1803 piano given him by the Erard firm had a keyboard range of five and a half octaves that was

---

apartment in Vienna, but the composer himself was too poor to afford any piano. Franz spent his last days in his brother's apartment. (See Neumayr 1994, p. 407).

[9]Taub 2002. (See, especially, Chapter 7, "The Myth of the Authentic Pianoforte.")

[10]Karl Reichardt in his Vertraute Briefe of February 7, 1809 (see Thayer and Forbes 1973, p. 461.) It is not clear how this request fits in with the octave glissandos Beethoven wrote in the Waldstein Sonata and the C-Major Piano Concerto. I know one pianist who spent many weeks pricking her fifth fingers just to develop calluses sufficient to permit playing those glissandos on a modern Steinway. Artur Schnabel and Rudolph Serkin would discretely lick their little fingers to ease the slide. Rosen (2002) suggests adding lubricant to the vertical pins at the front of the keys. Teflon sprayed on the hammer-shank knuckles also helps.

extended in the top to c4. Beethoven immediately incorporated those new notes in the Waldstein and Appasionata Sonatas and the Fourth Piano Concerto. (However, he complained that the Erard action was too heavy.) The Broadwood piano given him in 1818 by the London firm incorporated a six-octave range going down to low CC (rather than F), which Beethoven had already incorporated in his (1816) Sonata Opus 101 two years earlier. (But Beethoven commented to his friend Johann Stumpff that the Broadwood instrument had not fulfilled his expectations.) When Beethoven composed the "Hammerklavier" Sonata Op. 106 in 1818, no piano at his disposal was able to handle both the extreme bass and a six-and-one-half-octave range. The demands he continually placed on pianos through his writing forced the instrument makers to try to keep up!

In view of these observations, the comments made by some "purists" are rather puzzling. A musicologist once told me, "you really can't understand the meaning of the Waldstein Sonata until you've heard the opening bass chords played on a fortepiano." It seems absurd to assume that a man who boasted that he was "writing music for future generations" wouldn't also have been thinking of future generations of pianos on which to play it. Although Beethoven was impaired by deafness from about 1802 on, his inner ear was unaffected by the limitations of his piano. As Taub noted, the equivalent thing to the use of a "period instrument" in the art world would be to insist that a painting by Rembrandt or Vermeer be viewed only by candlelight. In a New York Times interview, Taub commented, "The day I play a concert on a fortepiano will be the day when the whole audience arrives by horse-drawn carriages and the hall is lit with candles..."[11] Rather than seeking to reproduce "the authentic Beethoven instrument" (which we must agree really didn't exist), it would seem more profitable to concentrate on understanding the intentions Beethoven implied in his written notation. As Beethoven commented to Czerny after hearing him play his Quintet for Piano and Winds in E-flat, "...you must pardon...a composer who would have preferred to hear his work exactly as he wrote it, no matter how beautifully you played in general." (Thayer and Forbes 1973, p. 641) To that end, Robert Taub and Charles Rosen have made substantial contributions in their recent books published in 2002 on the interpretation of the Beethoven piano sonatas. Rosen (2002) gives many hints on solving the technical problems faced in pursuing that goal.

Piano building in England started late in the eighteenth century, but grew on the tradition of English harpsichord building—i.e., the instruments were large, loud and heavy. According to Ripin, the piano was also much promoted in England by J. C. Bach, who had gone to London in 1762 to become the Queen's teacher and "to direct London's most expensive non-theatrical musical events." It was also influenced by the fact that Clementi had set up his firm in London and that a number of key patents on piano design were filed in England (e.g., those for heavier hammers and tubular iron frames and ones by Sebastian Erard on the agraffe and double-

---

[11] The New York Times, May 14, 2000, p. AR 37. But according to The New York Times, June 15, 2003, p. NJ 9, Mr. Taub may have changed his mind.

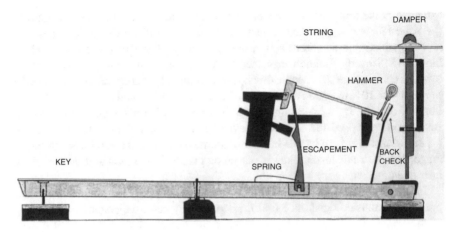

Fig. 4.6 The Broadwood single-escapement action circa 1799

escapement action), not to mention the merger of the Shudi harpsichord firm with that of Broadwood. (For example, James Shudi Broadwood patented one of the early bracing schemes for pianos using iron.)

The early Broadwood action (Fig. 4.6) had a fragile non-Viennese single-escapement mechanism, complete with back check and a rather anemic-looking hammer. Beethoven's 1818 Broadwood probably included some improvements and is shown in Fig. 4.7. That instrument is of the usual parallel-strung variety prevalent before the mid-nineteenth century, but did have both a una corda pedal and a second pedal to raise all of the dampers at once.

4.5   The Erard Double-Escapement Action

The French piano maker Sebastian Erard (1752–1831) contributed the next major advances: namely, the agraffe, a metal stud (generally brass) through which the strings for a given note are run before attachment to the tuning pins. The agraffe determined the boundary for the vibrating string at the keyboard end of the instrument and added stability to the tuning by keeping the string from riding up on the tuning pin.[12]

---

[12]The agraffe also serves another beneficial purpose that is not generally appreciated: the strain on the string is usually largest at that point, where there is a bend in the wire. Consequently, when a string breaks under the high tension in a modern piano, it is usually at the player's end of the instrument and the whip action of the loose wire goes toward the back of the piano rather than hitting the player in the face. (One piano maker left a short piece of a thick bass string imbedded in a beam in the ceiling of his shop as a grim reminder of the danger inherent when a string breaks.)

Fig. 4.7 Beethoven's 1816 Broadwood piano, incorporating a six-octave range starting on low *C* (courtesy of the Hungarian National Museum in Budapest)

The double escapement design is illustrated in Fig. 4.8, where the action is shown after the key has been depressed once, but not released. Initially, the jack was under the knuckle and the mechanical advantage hurled the hammer upward at the string with great velocity. In the figure, the hammer has already bounced off the string and has been captured by the back check. With a slight lift of the key at this point, the springs push the jack back under the knuckle, ready for the next note. Of course, the damper remains raised during this process. Because the hammer does not go all the way down to the normal hammer rest, the action can respond rapidly to a subsequent note. The mechanism at the right of the figure permits the dampers to be operated by a pedal. Because he was living (and dying) in Paris at the time, Chopin could easily have had an Erard piano. But it was said by Liszt that that great composer of fast passages for the piano preferred the older-style Pleyel instruments equipped with a Viennese-style single-escapement action because of "their silvery and somewhat veiled sonority and . . . easy touch." (Hedley and Brown 1980, p. 300). Nevertheless, the Erard was his second choice. The Erard double-escapement action shown in Fig. 4.8 is the direct precursor of all modern grand piano actions.

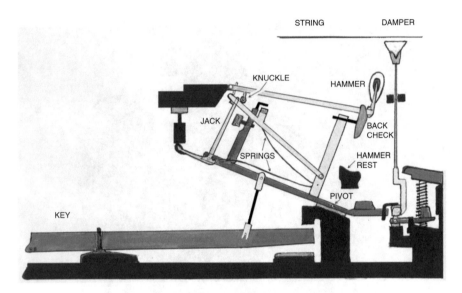

Fig. 4.8 The Erard double-escapement action of 1822

The remaining developments in the nineteenth century were largely devoted to making the piano louder and more durable. Still-thicker strings were employed, producing much louder sounds of longer duration. The modern piano has three strings per note over most of its range, and the strings are wrapped with copper wire in the extreme bass to avoid the need for excessive length. Typically, the total force from tension on the strings amounts to over 18 tons—45,373 pounds on a Steinway D concert grand. Much of the nineteenth century was devoted to strengthening the piano so that it wouldn't fly apart during performance. Developments in that direction were greatly assisted by that well-known smasher of pianos, Franz Liszt. (See Fig. 4.9) Liszt usually took two pianos along with him on tour, one of which was said to be knocked out of commission by intermission. V. V. Stasov, who attended the first concert given by Liszt in Petersburg (in April, 1842) had a different interpretation in his Memoires[13]: Liszt, heavily bedecked in medals arrived at the hall where 3000 people were seated and instead of using the steps, leaped to the stage, sat down at one of the pianos, tore off his white kid gloves (tossing them to the floor), threw back his enormous mane and plunged into the William Tell Overture.

After that, Liszt moved swiftly to the second piano facing in the other direction and, alternating between the two instruments, played such things as his Fantasy on Don Giovanni and his transcriptions of Schubert's Ständchen and the Erlkönig. From Stasov's description, the role of the second piano was to provide equal exposure to both sides of the audience and not just a source of spare parts.

---

[13] See, Jonas (1968, p. 121). Liszt, himself, boasted that he "practiced not less than ten hours a day" (Sitwell 1967, p.14); evidently, he really could play a piano for a long time without demolishing it.

Fig. 4.9 Liszt at the piano (after an 1845 caricature by Alexandre)

In between pieces, Liszt would sit on the edge of the stage chatting with people in the first few rows. As with Horowitz, Liszt preferred to give his concerts in the afternoon (as one nineteenth-century wag said, "in order to avoid the cost of candles").

4.6 The Struck String

4.6.1 Harmonic Content and Motion

As discussed in Appendix B, the initial boundary condition for the motion of a piano string is on the velocity distribution given to the string by the hammer at the striking point. To be sure, the normal modes of the string are involved and must satisfy the requirement that the amplitude be zero at $x = 0$ and L (the end points of the string). To be completely rigorous, one would have to include the dynamics of the hammer motion while in contact with the string. However, as a first approximation we will assume that the hammer flies away from the string in a time small compared to its vibrational period. In that case, the string suddenly acquires the velocity distribution of the hammer at the point of contact before the string has had a chance to move. (The approximation should work best for the lowest strings on the instrument, where the period is longest and for hammers that are fairly rigid.) As shown in Appendix B,

Fig. 4.10 Spectrum of the
string vibration from Eq. (4.1)
for $M = 7$ (no 7th harmonic).
The assumed shape of the
velocity distribution $V(x)$
initially given the string is
shown in the inset

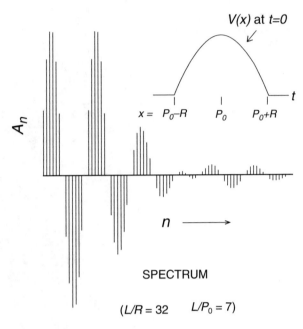

one obtains a fairly complicated expression for the spectral amplitude distribution
(A_n) even in that simple case:

$$A_n = \frac{2H^2}{n^4\pi^4 F_0} \left\{ -\cos n\pi \left(\frac{H+M}{MH} \right) + \cos n\pi \left(\frac{H-M}{MH} \right) \right.$$
$$\left. -\frac{n\pi}{H} \sin n\pi \left(\frac{H-M}{MH} \right) - \frac{n\pi}{H} \cos n\pi \left(\frac{H+M}{MH} \right) \right\} \quad (4.1)$$

where n is the harmonic number, $H = L/R$ (the string length divided by the
hammer radius) and $M = L/P_0$—the ratio of the string length to the striking point,
in analogy to the case of the plucked string. As in the case of the plucked string,
Eq. (4.1) gives only odd harmonics when $M = 2$, or the string is struck in the
middle. Also, $A_n = 0$ for $n = M$ (or the string is struck at a node for the Mth harmonic). The initial spectrum produced when the string is struck at the 7th node for
one choice of hammer radius is shown in Fig. 4.10. In this approximation, voicing
the hammer (changing the spectrum) corresponds to adjusting the value of H.

As in the case of the plucked string, the motion of the string after impact by the
hammer is obtained by substituting the values of A_n in the general solution for the
vibrating string, which in this case is:

$$y(x, t) = \sum_{n=1}^{\infty} A_n \sin(n\pi x/L) \sin(2\pi n F_0 t) \text{ where } F_0 = c/2L . \quad (4.2)$$

Immediately after the hammer hits the string, a narrow pulse (of width 2τ at the
left of Fig. 4.11) pops up at the striking point ($x = P_0$, occurring at $t = 0$ in

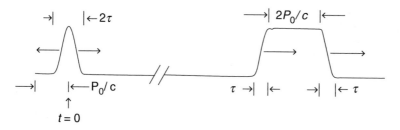

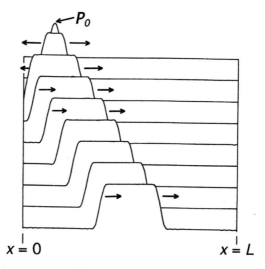

Fig. 4.11 Pulses launched on the string by the striking process. Left: The initial pulse at the striking point. Right: The broader pulse running down the string to the right after reflection of the initial left-running wave

Fig. 4.12 Evolution of the pulse described by Eq. (4.2) and Fig. 4.13 computed using the harmonic coefficients and initial velocity shape illustrated in Fig. 4.10

the figure). This narrow pulse consists of two equal-amplitude, oppositely directed running waves. As time increases, the initial pulse broadens until the wave running to the left bounces off the support at $x = 0$, where it undergoes a "hard" reflection and changes sign. Now negative, it travels back in the $+x$ direction, canceling out its previous positive portion. Meanwhile, the running wave initially launched to the right has continued on its way. The result of adding these two running waves together is an isolated broader, positive pulse (at the right in Fig. 4.12) that runs the full length of the string. The rise and fall times (τ) of this wider pulse are each equal to half of the initial narrow-pulse duration.

However, the breadth of the wide pulse is determined by the time delay taken for that half of the initial pulse that bounces off the support at $x = 0$ to get back to the striking point at $x = P_0$. Hence, as indicated in Fig. 4.12, the broad pulse time duration is $2P_0/c$ in this approximation, where c is the velocity for transverse wave propagation. After reaching the point $x = L$, two "hard" reflections occur (in succession for each running wave) which send an inverted pulse back toward $x = 0$. The initial formation and motion of the pulse is illustrated in Figs. 4.11 and 4.12. With modern felt hammers, which remain in contact longer with the string, the pulse shape ends up being more rounded.

Fig. 4.13 Motion of a string
struck at 1/7th of its length
over one complete cycle.
Time runs from top to bottom

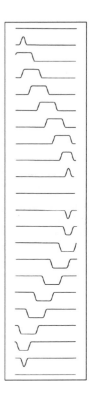

4.6.2 Piano Hammers

Many improvements in the modern piano occurred in the United States.[14] One
problem that plagued early pianos was the fragile nature of the hammers. The early
Christofori pianos used small hollow wooden cylinders covered with soft leather.
(Top of Fig. 4.14) As greater sound was required, a form evolved in which Λ-shaped
pieces of wood tapering toward the top were covered first by a layer of sheepskin
and then by a layer of hard leather. With the advent of the iron frame and heavier
wire strings, the leather hammer proved insufficient. In 1833, Alpheus Babcock
of Boston introduced hammers covered with felt. The hand-made hammers of that
period started out with a layer of hard leather over a tapered wooden form; it was
then covered with felt and held together by an outer layer of buckskin. (One can
readily imagine such a device exploding in the hands of a pianist like Liszt.)

---

[14]Thomas Jefferson was probably the first American to purchase a piano. According to Ripin
(1980, p. 702), during the Spring of 1771 Jefferson asked Thomas Adams in England to purchase
a piano instead of the clavichord he had originally ordered for his fiancée. Alas, he sold it to a
captured and paroled Hessian general in 1779 for £100 (Salgo 2000, p. 10, footnote.) The first
piano made in colonial American was by John Behrent (alias Johann Behrend) in Philadelphia in
1775. (Dolge 1911, p. 49, and Brookes 2002, p. 4).

Fig. 4.14 Piano hammers through the ages. From top to bottom: a late Cristofori hammer; a hammer of the type used on Mozart's piano; a hammer used on a Broadwood grand piano such as that owned by Beethoven; a single-coat felt hammer developed by Dolge and used by Steinway on concert grands in New York starting in 1873; and a contemporary hammer made by the Renner Company in Stuttgart used on several European pianos (e.g., German Steinway, Bösendorfer, Falcone, Fazioli, Bechstein and Blüthner)

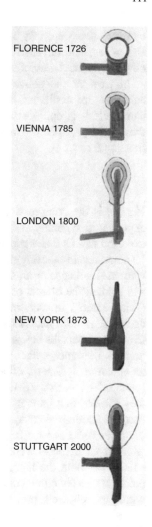

The ever-increasing thickness of the strings and the demand for louder tone necessitated development of something more durable than the hand-made variety of hammers, which were used in Europe until about 1867. Two Americans from New York, Rudolf Kreter (in 1850) and Alfred Dolge (in 1871), patented machines that permitted bonding felt onto wooden forms under high pressure. One version of the Dolge machine could actually handle layers of felt that were 1.75-in. thick and was used to make hammers for Steinway concert grands as far back as 1873. Dolge's early "single-coat" hammers proved to be a little too heavy and tended to flatten out at the top. To combat the flattening, Steinway & Sons began to saturate the felt about halfway up with a chemical solution that hardened the lower half. Improvements of Dolge's technique have been used to make hammers by Steinway in New York and

by the Renner company in Stuttgart for use on contemporary European pianos.[15][16]
Use of felt with unusually long and tough fibers by the Renner company evidently
obviates the necessity of soaking the hammers in chemicals.

In view of the above discussion, it is hardly surprising to learn that piano
hammers are nonlinear. That is, the force $F(z)$ exerted on the string when the
hammer felt is compressed does not obey Hooke's Law. Many people studying
the elasticity of piano hammers experimentally have fit their data arbitrarily to the
functional form

$$F_h(z) K z^p .$$ (4.3)

Here, z is the compression due to the applied force and measured values of the
exponent p typically fall in the range 2.5–4, as opposed to the value ($p = 1$)
expected from a linear medium obeying Hooke's law.[17]

Some recent measurements have been reported by Giordano and Winans (2000),
who attached accelerometers to a set of grand piano hammers provided by Knabe
and studied the effects of their impact on a solid object. They found that for soft
("pianissimo") impacts at hammer velocities of about 0.6 m/s, the values of the
exponent p were nearly the same (≈ 3) over the entire range of hammers studied
(from three octaves below middle C to two octaves above middle C). But, for large
impacts (hammer velocities of 1.0 m/s) corresponding to loud tones, they found
an apparent hysteresis effect in the curves of force versus hammer compression
(plots of F_h versus z). Using additional accelerometers on the hammer shanks,
they showed that most of the apparent hysteresis was caused by oscillation in the
(maple) hammer shanks; i.e., the shanks bowed in the middle as the hammer hit
the solid force sensor. In a continuation study by Giordano and Millis (2001) in
which measurements were made with hammers hitting a string (with accelerometers
attached to both the hammer and the opposite side of the string from the impact
point), it was concluded that there actually was a small amount of hysteresis created
within the felt itself, probably due to internal motion of the fibers.

Those measurements (made on middle C on a Steinway grand piano) showed
that the hammer typically remained in contact with the string for ≈ 5 ms. During
that time the acceleration of the string (from successive traveling pulses reflected
from the end points of the string) was out-of-phase with the acceleration of the
hammer, resulting in rapid compression of the felt.

---

[15]See Dolge (1911) pp. 97–106; also, the Renner web site at www.rennerusa.com.

[16]Piano craftsman Christopher Robinson always does a spectrum analysis of the tone quality
produced by a new hammer before gluing it onto the shank. He found that the spectrum often
differed significantly when the hammer was rotated 180° about its long dimension. They are clearly
not quite symmetric.

[17]See Giordano and Winans (2000) and the numerous references given in that paper. They
determined the compression by integrating the measured time-dependent acceleration twice: the
first integral gives the velocity and the second gives the compression as a function of time. The
maximum compression occurs when the velocity goes to zero.

4.6.3   Hammer Voicing

The hammers are among the few components of a piano that may be easily altered and that also have a major effect on tone quality. New hammers often produce an undesirable transient sound, somewhat like "chiff" on a low-pressure pipe organ. In addition, prolonged playing of an instrument (especially if the hammers are not often shifted horizontally with the "soft pedal") results in localized hardening in the hammer tips, which in turn makes the notes louder and harsher. Ridges can develop in the top of the hammer that must sanded off, after which softening is usually required. The process of adjusting the hammer is called "voicing."

   "Voicing" can be achieved by sanding the felt (if the tone is too weak or lacking in harmonics and the hammer is too soft) or by pricking the felt with a fork-like object containing three $\approx$ 1-cm long sharp needles in place of tines (if the tone seems too loud and the hammer is too hard). The second treatment is illustrated in Fig. 4.15. All the do-it-yourself books warn the reader to avoid working on the top of the hammer whenever possible. The sanding is to be done in upward strokes from the two sides of the hammer and the needle-pricking is to be done by pushing the angled fork down into the felt at points substantially away from the top—for example, between 9:00 and 10:30 AM and 1:30 and 3:00 PM (where "noon" is the top of the hammer).

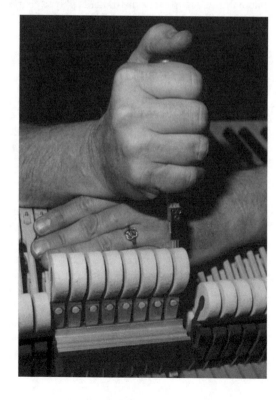

Fig. 4.15 Softening a hammer the conventional way in the skillful hands of Christopher Robinson. Note that the three needles are being pushed down at the upper side of the hammer (at about 10 o'clock as viewed from the left side) and not at the top

I was therefore shocked one time when an "expert" technician came to work on my own instrument. Without hesitation, he jammed the needles straight down from the top through the heart of the hammer, as if he were killing a vampire. Seeing me cringe in horror, he said, "I know all the books say not to do that. But unless you go right through the top, it doesn't do any good!" One trouble with that method is that it is irreversible. Of course, when done by just the right amount, it can provide lasting beneficial effect. Still another approach, which is at least reversible, is to soften the hammer felt with detergent soap, and then to harden it again using acetone. The belief here is that the chemicals cause the fibers (normally wound up tightly inside the felt) to unwind in the former case and rewind in the latter. However, it is my impression that this voicing approach does not last as long as the conventional needle-pricking method described above and shown in Fig. 4.15.

Another technique, which is especially useful for bringing out the tone in the top few hammers in the extreme treble, is to put a drop or two of shellac on the crown. Here again the books say to proceed with caution. Again, I was startled to see the expert virtually bathe the top three hammers in that elixir—but with good results. (Those notes only contain one or two overtones, which are so high that tonal color is a meaningless concept anyway—except perhaps to small dogs and bats.) Finally, voicing the hammers can also work wonders for the overall sound when the instrument is in a room with strong acoustic resonances. (Of course, voicing the piano doesn't help other instruments played in the same room.)

4.6.4   The One-Piece Iron Frame and Over-Stringing

A major development in terms of tuning stability was the one-piece cast iron frame. Although full cast-iron frames for square pianos date to their invention by Alpheus Babcock of Philadelphia in 1824, it took quite a while before the innovation was adapted to grand piano construction. Jonas Chickering, who had a firm in Boston, patented a full one-piece cast iron frame for grand pianos in 1843. Other makers (Steinway in New York, Erard in Paris) were reluctant to adopt the iron frame immediately and Chickering remained the only important producer of grand pianos to use this approach for over a decade. (A later suggestion by Thomas A. Edison that pianos be made of concrete was ignored.) Interestingly, Liszt acquired a Chickering concert grand. Actually, Liszt was given several pianos by different companies, including a second Chickering concert grand with elaborate engravings which is presently in the Liszt museum-home in Budapest.

The start of the New York Steinway firm dates to 1850, when a family of German piano makers emigrated to New York in pursuit of better business conditions. (Fostle 1993). In 1855, one of the four sons, Henry Steinway, obtained a patent for an over-strung grand piano. (See, Fig. 4.16). But neither the cast iron frame nor the over-strung bass strings were actually adopted by Steinway until the 1870s. (See Fig. 4.16.)

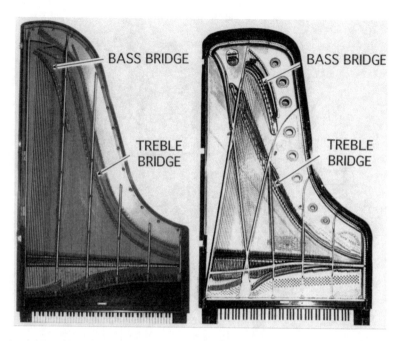

Fig. 4.16 Comparison of pianos with straight-bass (left) versus overstrung-bass (right) strings. The 1892 Concert Grand Piano by Steinway & Sons on the right is similar to that used by Paderewski (the dimensions of the over-strung piano are closely similar to those of a contemporary Model D Steinway)

One might suppose that the over-stringing technique illustrated in Fig. 4.16 would permit using longer bass strings within the same overall piano length. Surprisingly, that conclusion seems to be wrong. In practice, the length of the straight-strung bass strings in many pianos ranging from those made by Steinway to ones by Erard and Streicher in the nineteenth century were actually longer than those adopted in the "over-strung" models. One must then ask, "Why was the change in stringing made at all?" The main reason was based on the tone quality obtained. As can be seen from Fig. 4.16 (left side), the bass bridge on the straight-strung piano is closer to the edge of the sound board on both the long side and the end of the piano and does not overlap with the long treble bridge at all. In contrast, the over-strung bass bridge is closer to the center of the board laterally and does overlap the treble bridge. As a result, a very different set of modes tends to be excited in the sound board by the bass bridges in the two different cases. (See the later discussion of sound board modes.) With straight stringing, fewer low-frequency modes of the sound board are efficiently excited and more of the sound arises from the higher harmonics of the bass strings. With over-stringing (the right side of Fig. 4.17), more central modes of the sound board, which include those having the lowest resonant frequencies, are excited by the bass strings and one gets a warmer, more "lush" sound. In that case, the tone quality is also more uniform over the entire range of the piano because many of the same modes are excited efficiently by both the

Fig. 4.17 An 1877–78 Erard piano with parallel stringing and massive metal bracing (photographed by the author at the Fredericks' Collection in Ashburnham, Massachusetts)

bass bridge and by the long bridge (which extends up into the treble end of the instrument).

Not everyone agrees that the sound resultant from over-stringing is more desirable. There apparently were some nineteenth-century pianists who liked the sound from straight stringing just because it produced more contrast over the range of the instrument. E. M. Frederick argues that some of the inner notes on bass chords in the music by Brahms (whose 1868 Streicher was a straight-strung piano) become muddied and lost when played on a modern Steinway because of the over-stringing. As might be expected from the different mode patterns excited, straight stringing also produces more of a "stereo effect" across the sound board. To satisfy the different tastes, Erard made both parallel-strung and over-strung instruments during the late nineteenth century. Although the dark mahogany Erard grand with parallel stringing shown in Fig. 4.17 had a very impressive dynamic range, the tone at high volume seemed a little raucous to the present author. Part of that might be due to resonances in the long metal frame members. Of course, the tone quality of any piano depends very much on how the hammers have been voiced. [18]

---

[18]The sound of an 1881 Erard with straight stringing is preserved on a digital CD recording by Madeleine Forte (2002).

4.6.5 Brahms' Piano

While standing, admiring the elegant 1871 Streicher piano in the Fredericks'
collection (Fig. 4.18), I wondered to what extent the sound of that instrument would
really have influenced Brahms' writing. It may well be that the slightly strident
quality in the bass resulting from parallel stringing does help to bring out the
opposing, broken arpeggios in something like the first movement of the Bb-major
piano concerto (especially in the "heroic" section where the piece goes into *F*-
minor). On the other hand, I thought the piano had a somewhat mournful tone
quality more suitable for the Horn Trio or perhaps the foreboding section at the
end of the Edward Ballade where Brahms was literally imitating the sound of blood
dripping on the keys.[19] Of course, these are not especially objective thoughts. If

Fig. 4.18 An 1871 Streicher piano similar to the one owned by Brahms (photographed by the
author from the Fredericks' Collection)

---

[19]The Edward Ballade is one of the few "programmatic" pieces by Brahms and is based on one
of the Scottish tales in Herder's "Stimmen der Völker." After anxiously calling for her son and
being reassured that he was merely playing in the barn, Edward's mother discovers that he has
slaughtered his father.

you doubt that the development of the piano had little effect on his compositional technique, try playing something by Brahms on a harpsichord.

One technical development shown on the 1871 Streicher piano became more and more important toward the end of the nineteenth century: The legs are on wheels! As concert grand pianos became longer and heavier (some current ones are as much as 10 ft long and weigh nearly 2000 pounds), moving them without wheels became increasingly difficult. But there is a lot of inertia involved when those instruments start rolling. Many stages slope outward toward the audience, and it is important to lock the wheels after a piano is moved on stage. An example of what can happen when you don't do that occurred during a performance by George Bolet. While he was playing, the piano started to roll toward the audience. Mr. Bolet tried desperately to stop it, but that's hard to do while playing Liszt. He followed along with the piano for a while, but the instrument gathered speed, rolled off the edge of the stage and was destroyed during the crash.[20] (Perhaps one should have a brake pedal in addition to the others.)

One consequence of the changes in piano design was an increase in stiffness of the action. When Clara Schumann returned to the concert stage after many years to support her institutionalized husband, the critics noted that her tempos had slowed down considerably. Although she maintained that that was to provide a more sensitive interpretation of the music she brought back to the concert hall, others suspected it was mainly because the action on the newer pianos (see Fig. 4.19) was too hard for her to manage at higher speed.

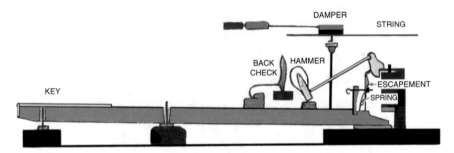

Fig. 4.19 The single-escapement action developed by Johann Streicher (circa 1845) and similar to that in the pianos owned by both the Schumanns and by Brahms

---

[20] A similar thing happened during a concert by pianist Christopher O'Riley. While looking down at the wreckage, a cellist overheard him exclaim, "I hate it when it does that!" Some composers actually wrote a "crashing piano" into the musical score. For example, the sound of a crashing piano in E major is contained at the end of the Beatles recording of Sgt. Pepper's Lonely Hearts Club Band. (EMI Stereo SMAS X-2-2653, NEM Enterprises Ltd., 1967, Side. 2.)

4.6.6   The Sostenuto Pedal

By 1870 Steinway had invented the "Sostenuto Pedal"—now the middle pedal on many pianos. The idea behind this device was that one could sustain one note (or notes) after striking the key by pushing down the middle pedal. The only problem with their first design (which was incorporated in a Steinway once owned by the author) was that the dampers in question were to be held up by a silken cord that engaged pins sticking out of the front of the damper mountings.[21] The cord was supported at several points across the keyboard and would sag in between. Consequently, relying on this version of the Sostenuto Pedal during a performance would be like playing "Russian Roulette" with several chambers loaded. Later versions by Steinway replaced the silken cord with a more reliable mechanism based on felt components. Although one might think that music by Debussy and Musorgsky (which frequently had long sustained pedal tones) would have benefited from this innovation, neither of those composers had instruments equipped with the middle pedal. As can be seen from the famous portrait by Willie von Beckerath of the bearded Brahms playing his Streicher piano while smoking a cigar, Brahms did not have one either.[22] From the complex, above-the-strings mechanism used on the Streicher piano dampers, it is hard even to imagine how a sostenuto pedal action could have been added.

The Streicher action seemed to have lagged behind the northern European developments of the mid-1800s. In addition to the lack of a sostenuto pedal,[23] Brahms had only a single-escapement instrument made by the Streicher firm with an action similar to that shown in Fig. 4.18. Here, as with the earlier Viennese pianos, the mechanical advantage was all contained in the hammer shank and only a spring-loaded single-escapement was provided. By then, the pianos were more heavily strung and had 88 keys (with white naturals and black sharps).

To stimulate American interest in the piano, William Steinway launched a program to bring famous European musicians to tour the United States starting during the 1872–73 season with Anton Rubenstein. The great Russian pianist gave

---

[21] The piano in question was a large rosewood, late-1860s-vintage Steinway grand, similar-looking to that shown at the left in Fig. 4.16. It had 85 keys, parallel stringing, and did not have a one-piece cast iron frame.

[22] This fact was obscured by a color photograph published by The New York Times (Thursday, November 22, 2001, The Arts, p. E1) showing an 1871 Streicher piano similar to the one Brahms owned that seemed to have a third pedal. Investigation showed that the "third pedal" in the photograph was actually the shadow of one of the two real pedals on the instrument. Someone had cut the piano out of its background in the photograph and thought that the shadow was a third pedal. The same piano is shown in the photograph taken by the author in Fig. 4.18.

[23] As with the earlier French harpsichord builders, the Viennese piano builders were inter-related. For example, Nanette Streicher, who with her husband Andreas Streicher tried to incorporate Beethoven's suggestions in pianos built by their company, was the sister of André Stein. Her son, Johann Baptist Streicher (1796–1871) provided Brahms with his 1868 piano, similar to the 1871 instrument shown in Fig. 4.17. But by then, Johann's son Emil had taken over the Streicher firm (which was finally dissolved in 1896).

over 218 concerts in about as many days in places like Deadwood, Montana and felt so over-worked when he left the country that he swore never to return. The program culminated with tours sponsored for Paderewski in 1891 and 1892. During the first tour, Paderewski injured the fourth finger of his right hand on a Steinway with an extremely stiff action—an injury from which he never recovered. However, for his second tour in 1892, an instrument with lighter action was provided, not to mention a private railroad car for his entourage (wife, secretary, valet, tuner, manager, chef, two porters, and, of course, the piano).

4.6.7   Scaling and Bridge Shape

If we assume that one objective is to build a piano in which the tension on the strings is the same from one note to the next (assuming constant wire density), a simple exponential curve arises for the shape of the bridge. Keeping the ratio T/μ constant means that the length of the string should be doubled every time the frequency is halved. That prescription was used by the early Italian harpsichord makers, who succeeded in following it over some 5/6th of the keyboard range. Of course, one gets into trouble at the bass end, just as doubling the number of grains of wheat on successive squares of a chessboard requires more wheat than exists in the world long before you fill the whole board. In the case of a modern concert grand piano starting with string lengths of about $2''$ at the highest C, by the time you got to the lowest A, the strings would be over 25-ft long. Compromises therefore need to be made to keep the piano within the dimensions of a typical room. It is simple to derive an equation that satisfies the doubling requirement. Let the x coordinate be parallel to the keyboard and let S be the distance in that direction over which the string length is halved. It doesn't matter what units are used to measure x and S as long as they are the same in each case. Since the keys are constant in width, they provide a convenient unit for measuring the distance in the x direction. From the basic halving requirement, one immediately sees that

$$L = L_0 2^{-x/S} = L_0 \exp(-0.693147x/S) \tag{4.4}$$

where L_0 is the string length at $x = 0$ and we have used the fact that $\mathrm{Log}_e 2 \approx 0.6931472\ldots$ That is, at $x = 0$, $L = L_0$. When $x = S$, the original length is halved, and so on. Ideally, one would want $S = 12$ notes (one octave on the keyboard) or about 6.62 in. on most pianos. ($S \approx 6$ in. on harpsichords.)[24]

---

[24]J. B. Hayes (1982) concluded that the optimum shape for a piano bridge was a catenary curve But just why he thought that remains a mystery. The common form of the catenary is given by the equation

$$y \propto [\cosh(kx) - 1] = \frac{1}{2}[\exp(+kx) + \exp(-kx) - 2]$$

where k is an adjustable constant. [See Synge and Griffith (1949), pages 99–104.]

It is of interest to see how closely the string lengths on a real piano satisfy Eq. (4.4).[25] Figure 4.21 (left) shows a plot of all of the active string lengths on a Steinway D concert grand piano as a function of notes on the keyboard. The long treble bridge starts out in the bass at $F_2 = 87.3$ Hz with $L_0 \approx 71.87$ in. On this type of semi-log plot, an exponential relationship will appear as a straight line. The solid line in Fig. 4.21 has been drawn for Eq. (4.4) in the case where $S = 12$ (i.e., string length halving every octave). Although the measured string lengths on the treble bridge do follow this decaying exponential fairly closely, the slope on the graph changes slightly after the first octave. Above $F_3 = 174.6$ Hz, the exponential slope changes to one where the halving point is at $S \approx 13$ notes on the average, rather than twelve. The technique spreads out the required departure from ideal scaling over most of the instrument. Of course, below F_2 (in the bass bridge region), the lengths have to saturate quickly with decreasing frequency to keep all the strings inside the piano.

The results at the left in Fig. 4.20 do not necessarily give the actual bridge shape. On the Steinway D, the strings are not all perpendicular to the keyboard and the agraffes curve exponentially toward the keyboard as one goes down the scale in order to optimize the hammer striking points. The actual bridge shape on a Steinway D is shown on the right side of Fig. 4.20, where a new x-axis was defined by extending a line from the treble end of the capo d'astro bar over the

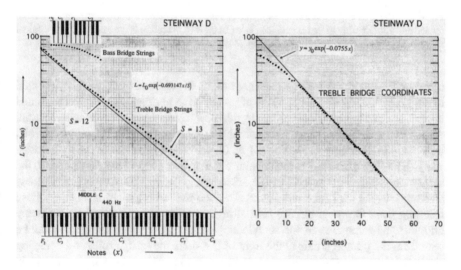

Fig. 4.20 Left: Active string lengths as a function of notes on the keyboard of a Steinway D concert grand. Right: Long treble bridge coordinates on a Steinway D

---

[25] The Steinway Company feels such data constitute "proprietary information." Of course, they are available to anyone with a Steinway piano and a tape measure. Measuring the lengths of string terminating under the capo d'astro bar is the trickiest part, but can be done easily by slipping narrow strips of paper along the strings to the end points under the bar.

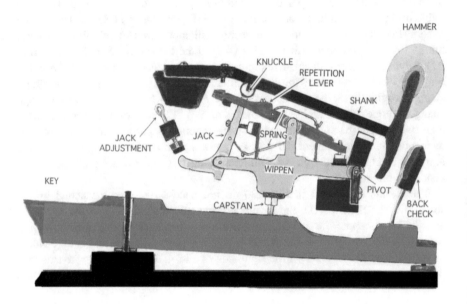

Fig. 4.21 The action of a modern (Mason and Hamlin) grand piano. Although the action is more refined and considerably more rugged, the basic principle of operation is the same as in the original Erard double-escapement mechanism. Compare with Fig. 4.12

range of the piano covered by the long bridge. (This line is not precisely parallel to the keyboard.) The y coordinates of points on the bridge were then determined by constructing perpendicular lines to this new x axis from each bridge location. Although the strings start out perpendicular to the new x axis in the extreme treble, they begin to fan out to the left below F_3 (174.6 Hz). By the end of the long bridge at F_2 (87.3 Hz) where the perpendicular construction defined the point $x = 0$, the strings are going at about $8°$ to the left of the y axis. This fanning-out process tends to balance the horizontal forces from the treble-bridge strings against those resulting from the (over-strung) bass strings and also permits coupling the treble bridge to the same area of the soundboard excited by the bass bridge. As can be seen from the plotted points at the right of Fig. 4.20, the bridge follows a decaying exponential curve very closely except in the region where the strings fan out in the bass. Small fluctuations in the measured points mirror discontinuities in the cast iron frame.

4.6.8   The Modern Piano Action

A diagram of a contemporary grand piano action is shown in Fig. 4.21, which works as follows:

1. When the key is pushed down (in the region not shown at the left of the diagram), the capstan screw pushes upward on the wippen, which is pivoted at the right side where the action is screwed into the piano. (The "action" is often regarded as the combination of the wippen, jack, and repetition lever and is attached by a single screw at the right.) The damper is raised by a mechanism (not shown) at the right-end of the key and remains up above the string while the key is depressed at the left.
2. As the wippen rotates about the pivot at the right, it pushes the jack up against the felt-covered knuckle underneath the hammer shank, causing the hammer to fly upward off its normal felt-covered resting place in an arc centered on a pivot at the left end of the shank. (The jack is constrained to move in a slot cut into the left end of the repetition lever.)
3. The lower left-hand tip of the L-shaped jack hits the felt at the bottom of the jack adjustment screw causing the top of the jack to slip off to the left side of the knuckle.
4. The hammer bounces off the string and is caught by the back check (which has been raised by the right end of the key), thereby keeping the hammer from rebounding against the string. If the key had been released at this point, the hammer would merely have fallen back to its normal resting place on the felt attached to the stationary part of the piano to the right of the repetition lever.
5. If the key is held down after the impact of the hammer with the string (at which point the hammer has been grabbed by the back check), the upper part of the jack remains slightly to the left of the knuckle. At this point, a slight release of the key allows the springs attached to the wippen to pull the jack back under the knuckle. The hammer (still supported by the back check) is then ready to strike the string once more if the key is depressed again slightly. In this way a second note may be sounded rapidly without the key and hammer falling completely back to their normal resting positions.

As should be obvious from Fig. 4.21, the hammer velocity is determined by the initial impulse given the key and, in contrast to the clavichord, no amount of wiggling of the key afterward in the manner of Claudio Arrau or Glenn Gould has any effect on the tone quality produced. Similarly, grand sweeping gestures of the right arm over the performer's head in the manner of Peter Serkin have no effect on the sound. (They may, of course, provide useful relaxation of the arm muscles.) The tone quality is determined entirely by the design of the instrument and the impulse provided to the hammer. It doesn't matter whether the initial impulse given the key is given by a human finger, a falling lead weight, or an umbrella stick: the same tone quality is obtained for the same initial hammer momentum. (The relative size of the hammer impulse between one note and the next—not to mention the use of dampers and lateral shift of the hammers—is, of course, controlled by the player and is what determines the musical quality of the successive notes.)

The piano action has evolved considerably since the time of Cristofori to take advantage of human physiology. The initial "thumbless" fingering promoted prior to *C*. P. E. Bach (basically a carry over from harpsichord technique) has been transformed to the point where pianists now sometimes wish they had even more thumbs. The length of the keys has increased to almost 6 in. (enough to permit the flat-fingered technique used by Vladimir Horowitz, Radu Lupu, and Eubie Blake) and the octave span has been increased by about 1/2 in. from that of the harpsichord.[26] These changes, of course, make it easier to use the thumb and in the process permit more rapid trills (especially, with 1–3 fingering) and impressive sforzandos. (According to one eyewitness, Paderewski loved to create a forzando by coming down hard with his left thumb in the bass rather than to begin an arpeggio "more logically" with his fifth finger.) A dynamic range of over 80 dB can be obtained with the modern action on a Steinway concert grand, going from the thundering octaves of Horowitz[27] to tones described by pianist Ray Hansen "as soft as the sound made by a snow flake hitting the ground." However, as demonstrated by the composer/pianist Robert Schumann, attempts to make the human physiology evolve have been much less successful. Schumann's efforts to strengthen the fourth finger of his right hand with a gadget of his own design left him crippled, and ended his piano-playing career altogether. (Figure 4.22 shows one of several different designs

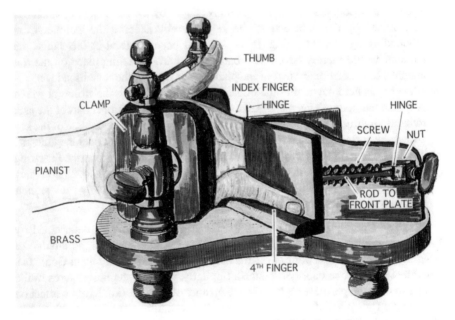

Fig. 4.22 A nineteenth century device designed to "strengthen" the fourth finger

---

[26]It is said that Steinway made a special piano with a narrower octave span for the legendary pianist Joseph Hoffman because his hands were so small.

[27]The absolute sound levels at maximum from a concert grand can exceed the limits recommended by OSHA for prolonged exposure by a significant factor and may well cause deafness.

to achieve the same crippling effect.) There is no one hand position or technique that works best for everyone. Turning the thumb under the third finger is a basic technique for playing legato scales stressed by most teachers and piano methods. Rosen (2002, p.2) notes that because his thumbs were very short, he mastered a technique in which he "displace[d] the arm quickly to the right when shifting from the third finger to the thumb" to obtain the same legato effect. (Hanon, who even stressed the importance of turning the thumb under the fifth finger as an "exercise of the highest importance"[28] would probably spin in his grave at this thought.)

Surprisingly, Paderewski's 1892 Steinway piano (apart from the bulbous legs, curlicue music desk, and a slight alteration of the metal frame) was very close to the present form of the Steinway "Model D" concert grand. With minor adjustment, it is possible to put an entire modern keyboard and action into an 1899-vintage Steinway concert grand, the difference in keyboard width being less than 1/2 an inch. Although the capo d'astro bar,[29] a metal rail that determines the front end points of the vibrating strings on the upper two octaves of the instrument (starting at D above $A = 440$ Hz) and replaces the Erard agraffe in that section of the piano, had been incorporated during the late nineteenth-century, there was one slight difference in the stringing in that portion of the piano in the 1890 models. The pin assemblies on the bridge in the capo d'astro region were rotated slightly so that each of the three strings on a given note was of different length. (The difference amounted to about 3/8-in. on an 11-in. string.) Although each group of three strings was tuned to the same pitch, the difference in length meant that the fractional striking points, hence harmonic production, on each of the three strings, would be somewhat different. For unknown reasons, that approach (which ought to have given a more interesting sound to the extreme treble strings) was abandoned early in the twentieth-century. The other major innovation, the "accelerated action," was introduced in the mid-1930s. The latter consisted mainly of moving the balancing weights on the key toward the fulcrum so as to reduce the moment of inertia.[30] (Some rounding of the fulcrum area was also added.)

The introduction of plastic key covers and Teflon bushings in the period after World War II was more controversial. Although preferable to slaughtering elephants, the early plastic key covers were non-absorbent and had peaks and valleys

---

[28] Hanon (1900), exercise No. 35. Several professional pianists have told me that the first thing they have had to teach their students is to forget what Hanon had taught them.

[29] The capo d'astro bar is of great practical value when the bridge is properly adjusted, but it also contributes to what one piano technician described to me as "Steinway Disease." The problem occurs when the sound board sags, pulling down the bridge with it. Under this condition, the geometry is such that when the returning pulse from the bridge (which is peaked downward) reaches the capo d'astro bar, it can pull the string down off the bar, producing a sound much like overload distortion in a "Hi-Fi" set. (The phenomenon is similar to the returning pulse in a clavichord pulling the string up off the tangent.) Cure usually involves raising the bridge, hence restringing the instrument.

[30] The moment of inertia about an axis a distance r away for a point mass M goes as Mr^2, whereas the balancing torque goes as Mr. Hence, putting a larger mass closer to the balance point can speed up the response of the action.

which lined up; hence, a small amount of oil or perspiration tended to make the fingers skid off the keys. It is said that Artur Rubinstein used to rub rosin on his fingers tips to counteract the slipperiness. With ivory, the peaks and valleys were more random and the material had small pores which absorbed oil from the fingers. Later, more porous versions of plastic with irregular surfaces solved the problem.[31] The adoption in 1962 of the Space-Age lubricant Teflon® in place of the old-fashioned felt bushings made the actions noisier, although the bearings would last longer and vary less with humidity changes. The main problem with Teflon® bushings arose because they did not change dimensions with humidity and hence fell out when the wooden parts became enlarged. The original felt bushings had the opposite problem: they would swell up with humidity and increase the frictional loss in the action.[32] By 1982 a compromise solution was adopted in which felt bushings were impregnated with Teflon® particles.[33] (Lenehan 1982)

By 1900 more than half the World's piano production occurred in the United States and many of the European makers were making pianos in the "American tradition." During that period, nearly every household had a piano and the main exposure most Americans had to classical music was in four-hand reductions of the "great works." (It is said that even Richard Wagner had his first exposure to Beethoven's Ninth in a four-hand version.) In Europe, the recognized leaders were Bechstein, Blüthner, Bösendorfer and the Hamburg branch of Steinway. Blüthner introduced a new feature called "Aliquot Stringing," in which a second set of (single) strings was added and tuned one octave above the main set—somewhat akin to the 4' stop on a harpsichord.[34] These additional strings would vibrate in sympathetic resonance with the main set, but were not struck directly by the hammer. (They also added excitement to the piano tuner's life.) A different version of Aliquot Stringing was introduced by the relatively new piano company formed in 1978 by Paolo Fazioli in the town of Sacile (60 km northeast of Venice). Here, provision

---

[31] See, Malcolm W. Brown, "With Ivory in Short Supply, Pianists Tickle the Polymers," The New York Times, May 25, 1993; p. C1.

[32] See, "A Humid Recital Stirs Bangkok" in the Washington Post, July 23, 1967. This hilarious review describes a concert in which the pianist became so exasperated with sticking keys and other problems with a concert grand that he grabbed a fire axe and started to smash the piano to pieces. Although masterfully written by Kenneth Langbell for the English Language Bangkok Post under the title "Wild Night at The Erawan" on May 27, 1967 and made available to the Washington Post by Martin Bernheimer of The Los Angeles Times. The story is, alas, just humorous fiction according to the Web service, Urban Legends.

[33] This frictional loss can be extremely important. For example, it can increase the weight necessary to push down middle C from about 52 g (standard) to more than 70 g. For comparison, Horowitz had his Steinway adjusted so that a weight of only 35 g was needed. (That decrease in required force increases the ease of playing fast passages enormously; however, it also makes controlling the loudness of successive notes harder.)

[34] Blüthner probably couldn't think of a good German word for it. Aliquot is Latin for "some" or "several." In Italian, ali'quota comes from the French and means "quota" or "share." According to the O.E.D., the English word "aliquot" means "a part contained in the whole an integral number of times." In German, "aliquot part" becomes ohne Rest aufgehende Teil.

is made to have the string sections between the bridge and hitch pins precisely tunable to harmonics of the fundamental pitch. Those sections are usually out-of-tune in older pianos and often covered with felt damping material. It is not clear how much that innovation contributes to the wonderful tone quality of the Fazioli instruments. (The company boasts that the soundboards are made from the same close-grained red spruce used by the great old Italian violin makers.) As with most pianos, the soundboards are also tapered at the edges in the manner used earlier on harpsichords to improve their acoustic response. In the Steinway D, the close-grained sitka spruce is about 9 mm thick in the center and tapers to 6 mm at the rim in a double-crowned fashion. The Fazioli concert grand appears to have set the world's record on length at slightly more than 10 ft. The Bösendorfer "Imperial" is a close second at about nine and a half feet, but has additional notes in the bass (lacking in the Fazioli) that get down to low CCC (16.3 Hz) instead of $A = 27.5$ Hz as on most contemporary pianos.[35] In addition to allowing one to play all the pedal notes in the organ music of Bach, the extra notes on the Imperial (which have inaudible fundamental frequencies in the range of earthquake tremors) meant that the normal low notes were moved farther away from the edge of the case and into a region where the soundboard could respond more fully to the fundamental components of the vibrating strings. In addition, when the dampers are raised with the "loud" pedal, one obtains reinforcement by sympathetic resonance between the higher harmonics of these low strings and the normal notes on the bass end of the keyboard. The relative fractional energy content at the fundamental pitch in the low notes on the "Imperial" does indeed seem larger than in a Steinway D; however, the harmonic content in the latter appears to be more abundant (Fig. 4.23) and the low notes seem to carry better in large concert halls, possibly due to different directional radiation patterns.

The Japanese firm of Yamaha began producing copies of Steinway pianos early in the twentieth century and by the mid-1970s had surpassed the largest American companies in the number of instruments produced annually. Sadly, the great American family interest in pianos for the home that had characterized the early part of the twentieth century and which was rejuvenated by the Japanese during the 1970s has now largely subsided in favor of electronic instruments.

One version of an electronic piano was invented by Harold Rhodes (1912–2001) and used a piano keyboard to make hammers hit small metal tines whose motion was sensed by electromagnetic pickups. The device was marketed by the Fender Company. It was said that its "pure tones were fed into electronic amplifiers and special effects generators that could produce a stereo vibrato effect in between loudspeakers or bite like a wah-wah guitar" and that "'Acid Jazz' producers and

---

[35] One version of the Bösendorfer Imperial (the Bösendorfer 290 SE System) came complete with a computer that registered the impulse given the hammers optically at a rate of about 800 samples per second. Together with solenoids connected to each note, the system provided a kind of "High Tech" player piano with expression on which mistakes could be edited with the computer. In a demonstration of the system, pianist Charles Rosen said that the playback provided an exact mirror of his touch and dynamic expression. (Audio, January 1986, pp. 20,21.)

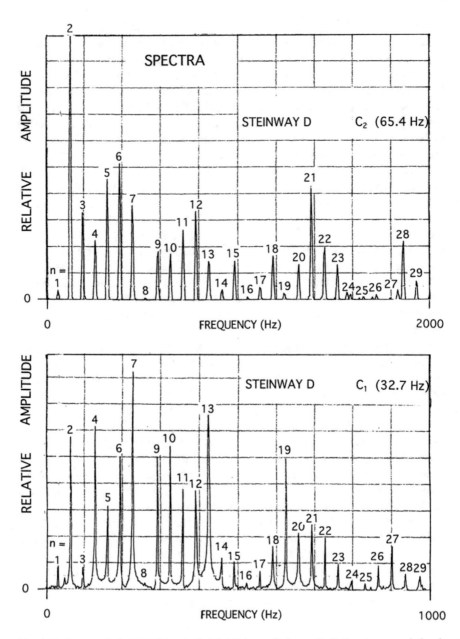

Fig. 4.23 Spectra of C_1 (32.7 Hz) and C_2 (65.4 Hz) on a Steinway D. Both notes were obviously struck at about 1/8 of the string length. Note the missing harmonics at $n = 8$, 16, and 24 (data taken by the author)

Funk revivalists embraced the Rhodes sound." [36] The instrument was produced in small, suit-case sized pianos and lead to the demise of the 1970s piano market. As if to hedge their bet, CBS took over both the Fenders Company (in 1965) and the Steinway Company (in 1972). Nevertheless, it is thought (and hoped) that there will always be a market for the concert grand. In the post-CBS period, Steinway & Sons has changed hands twice: once to Selmer Industries, the band instrument company which had been taken over by investment bankers working for Drexel Burnham Lambert (home of the infamous junk bond king, Michael R. Milken). Most recently, Selmer Industries changed its name to Steinway Musical Instruments, using a stock symbol LVB (Beethoven's initials).[37]

More recently, electronic pianos have emphasized the use of digitally sampled tones from real pianos controlled by MIDI (for "Musical Instrument Device Interface") keyboards that sense the impact ("touch") given by the fingers. In at least one case, an unusually high-frequency and high-resolution sampling system was used to record every note on a new Steinway digitally.[38]

4.6.9 Transient Build-Up and Decay

As we have seen, a major advantage of the piano over the harpsichord is in the persistence of tones after the key has been struck. Although the decay of a note may involve complex time-dependent behavior due to the exchange of energy between different coupled modes of vibration, one can simply measure the time for a note to decrease by a given amount. The criterion generally used to measure reverberation times in concert halls (a decrease of 60 dB, or a factor of one million in intensity) is a useful bench mark. In practice that is about the maximum range over which one normally hears the persistence of tones under concert conditions. Often, a more practical approach is to measure the time for a sound to decrease by 30 dB and then double it to get the 60 dB equivalent. Noise levels in a reasonably quiet room are seldom down by as much as 60 dB from the level at which music is normally played. Persistence times for different notes on a 9-ft Steinway Model D (concert grand) are shown in Table 4.1. The persistence times are longest for the bass notes and fall off in a quasi-logarithmic fashion as one goes up the scale. The decay times, of course, are determined by the energy loss mechanism of the strings and sound board and as previously discussed, can involve double (or even multiple) exponential decay rates.

Much shorter time intervals are involved for the sound to build up after striking a key. But in contrast to the harpsichord waveform, where the sound level starts

---

[36] Jon Paroles, The New York Times, January 4, 2001, p. 7.

[37] James Barron, "Why Today's Piano is Much Like Yesterday's and Last Century's," The New York Times, July 19, 2003, p.A11.

[38] One piano technician commented to me after hearing it, "It's too bad they didn't voice the piano first!"

Table 4.1 Time to decrease by 60 dB for different notes on a Steinway D

Note	C_1	C_2	C_3	C_4	C_5	C_6	C_7	C_8
Frequency (Hz)	32.7	65.4	130.8	261.6	523.2	1046	2093	4186
Time (s)	≈ 36	48	37	28	26	10.5	3.7	1.5

Data taken by the author

off at near maximum volume when the string is plucked (see Fig. 3.28), there is a significant time delay before the piano note reaches its maximum intensity. These time intervals typically range from almost one second on the lowest note, through a quarter of a second in the middle of the keyboard, to a small fraction of a second for the highest notes. These time delays obviously involve additional mechanisms to those found in the decay of the sound; otherwise, the build-up time would be the same as the decay time. In contrast to the case of the harpsichord, the piano string is initially undeflected. Because of mechanical inertia, there is some time delay required for the string to start vibrating after being hit by the hammer. But one would expect that delay merely to take a few vibration periods. The main effect is probably due to the time taken for energy setup in the vibrating string modes to couple to the soundboard. A few examples of the amplitude build-up in a Steinway D are shown in the spectral surfaces in Fig. 4.25. (Some of the harmonics for C_2 appear to oscillate with time, probably due to slow beating effects between harmonics of the different strings tuned to the same note. Although C_1 involves a single, wrapped string, all other C's from C_2 through C_8 use triple stringing on the Steinway D.) The delayed peak in the piano sound results in a short, but unavoidable crescendo after the key has been struck. Some pianists try to enhance this natural crescendo effect by depressing the "loud" pedal after the tone has reached its maximum level. (Taub 2002, p. 34) It is likely that these time delays on the lower notes determine an optimum tempo for some pieces.

4.6.10   The Piano Soundboard and Bridge

As with the harpsichord, the soundboard in the piano is the primary medium for coupling energy from the vibrating strings to sound waves in the air (Fig. 4.24). But with the piano, the soundboard itself is just a single, thicker piece of wood (usually spruce) made by gluing a large number of narrow strips together with the grain at about 45° to the long dimension of the piano. The soundboard is tapered at the edges, and is not the top part of a resonant air cavity as in the harpsichord. On the underside of the board, a series of ribs tapering off at the ends run perpendicularly to the grain with a spacing of about 5.5 in. (See, Fig. 4.25) As well as determining the bowing and providing mechanical support, the ribs increase the stiffness of the board in the direction perpendicular to the grain and, hence, tend to make the velocity of surface waves in the soundboard more isotropic than in the harpsichord. (See the later figures of mode patterns.) The vibration of the strings is coupled to the bridge

Fig. 4.24 The underside of the soundboard form a Steinway B grand piano showing the ribs placed at right angles to the grain. The position of the short bar at the upper left of the figure tunes the main soundboard resonance. The white strings were put in temporarily to check the bowing of the board (photo taken at Robinson's Acousticraft factory by the author)

Fig. 4.25 Underside of the compound bridge from a Steinway B grand piano. Note the undercutting in the area where dowels attach the bridge to the soundboard ribs (the four white circles). The bass bridge is at the bottom and a small section of the long treble bridge is at the top (photograph by the author)

using the simple pin arrangement shown in Fig. 4.25. The bridge is fastened to the soundboard by dowelling through the soundboard ribs.

Although the one-piece cast-iron frame stabilizes the piano enormously, the swelling and shrinking of the wood in the soundboard and bridge still affects the pitch because the wood pushes upward on the strings, thus changing their tension. Humidity changes are much more important than temperature variation. Hence, the pitch tends to rise in the summer when the humidity is highest and go down in the winter. If the room is unheated in the winter, the piano tuning can be remarkably stable. I know of one Steinway D that is left in a New England house kept without heat all winter in which the intonation changes by less than 1 or 2 cents on successive summers.

Fig. 4.26 Method used by
most makers to couple the
strings to the bridge

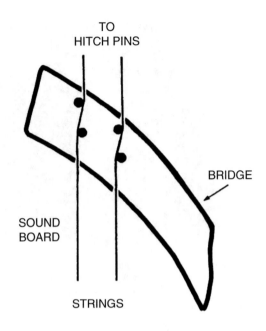

TO
HITCH PINS

BRIDGE

SOUND
BOARD

STRINGS

The bridge itself is made from hard rock maple and is glued together as one piece
that includes both the long treble bridge and the bass bridge. As shown in Fig. 4.26,
the bottom of the bridge is undercut in the region where the dowels fasten it to the
ribs of the soundboard so as not to inhibit vibration of the board in that area.

4.6.11   Sound Board Modes of Vibration

The modes of vibration of the soundboard are rather different from those found in
the harpsichord and have been studied by a number of people. (See Kindel 1989;
Giordano 1998a,b; and references in those papers.) The measurements are made
by applying a periodic driving force to the sound board at different frequencies
and positions and by analyzing the spatial distribution of sound at different regions
above the soundboard with a microphone and phase-sensitive detection. Relative
phase was determined in Kindel's work with a dual-trace oscilloscope in which the
driving voltage was used as a reference.

The experimental results presented by Kindel (1989) for the soundboard from a
9-ft Baldwin concert grand are extremely comprehensive and most relevant to our
purposes (Fig. 4.27). He examined two ribbed sound boards made from quarter-cut
sitca spruce, tapering from 11/32" at the middle to 9/32" at the edges, mounted in
the piano body without the metal frame. Although the presence of the metal frame
and stringing of the instrument will no doubt change the resonances somewhat,
it is expected that Kindel's measurements provide a good first approximation to

Fig. 4.27 Vibrational modes in the sound board of a 9-ft Baldwin concert grand (drawn from data taken by Kindel 1989)

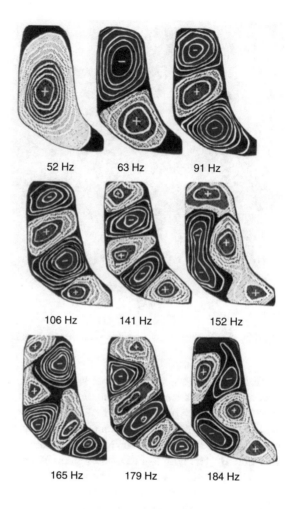

the problem. Ignoring the resonances he encountered from the piano body, several modal patterns of the sound board have been reconstructed in Fig. 4.29 from his data. These results are illustrative of the dozens of resonances he studied and are useful for estimating the radiation patterns from the soundboard at different frequencies. For example, the monopole resonance at 52 Hz where there is just one peak will radiate in all directions with equal intensity above the board (assuming the piano lid is removed, or wide open). The dipole resonance at 63 Hz where there is both a positive and negative maximum amplitude will radiate preferentially along a line drawn through the positive and negative extrema, and the higher resonances will have radiation patterns corresponding to the various higher-order multipole source distributions shown in Chap. 1. In estimating the directions of propagation. It is useful to remember that sound waves traveling through the air consist of alternate compressions (positive amplitudes) and rarefactions (negative amplitudes).

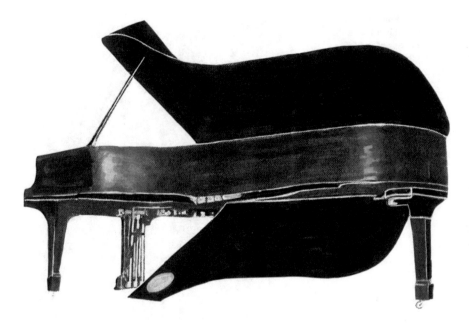

Fig. 4.28 The Revenaugh modification to the concert grand (drawing by the author)

Hence, for example, with a dipole sound source such as that at 63 Hz in Fig. 4.28, maximum propagation should occur along the line through the plus and minus signs. (See the discussion of dipole and multipole radiation patterns in Chap. 1, especially Figs. 1.16–1.20.) Many of the resonances shown in Fig. 4.28 should have maximum radiation in the direction of the long dimension of the piano, although those at 52, 152, and 184 Hz, will also have strong components normal to the long dimension and in the direction toward the audience when the piano lid is opened at about 45°.

Kindel found that these basic patterns were representative of whole families of resonances, some of which could not be resolved experimentally. He also determined the distribution patterns for these resonances and studied them all numerically using finite element analysis. Identification of the modes was based on spatial patterns rather than resonant frequencies. Measured and calculated values for the frequencies typically varied by about 10%.

One novel suggestion to alter the normal radiation pattern from a piano was made in 1997 by pianist Daniell Revenaugh.[39] Noting that roughly half the sound is radiated downward toward the floor by the sound board under normal conditions, Revenaugh added a second, lower lid to his piano. (See Fig. 4.29)

Although Revenaugh and pianists Andre Watts and Peter Serkin thought the sound in the vicinity of the keyboard was more interesting, the implied assertion

---

[39]See A. Tomasini, "Not Even Practice Gets a 2-Lid Piano into Carnegie," New York Times, December 8, 1997, p. 1.

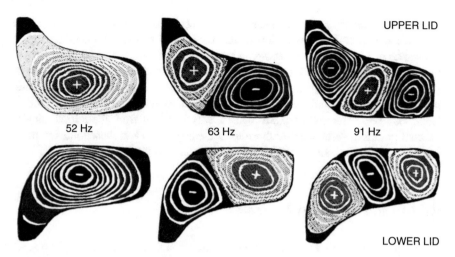

Fig. 4.29 Radiation source distributions for the Revenaugh piano modification determined using the data in Fig. 4.28

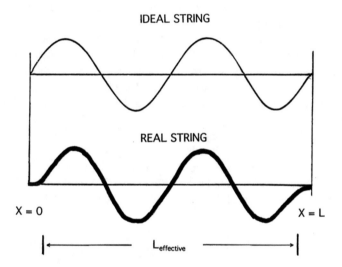

Fig. 4.30 Comparison of bending in an ideal string versus that in a real string with significant stiffness

that the second lid provides more sound to the audience neglects the effects of phase. The radiation going downward from the soundboard is 180° out of phase with the sound radiated upward, and this out-of-phase relation is preserved by the sound reflected by the two lids at 45°. Consequently, the radiation sources as viewed from the audience should look like those shown in Fig. 4.30 for the first few modes, rather than the ones in Fig. 4.28.

The isotropically radiating resonance at 52 Hz has now turned into a dipole that will have maximum radiation in the *vertical* direction; the original dipole resonance at 63 Hz has now become a quadrupole radiator with two directions of emission maxima—one in the vertical direction and the other along the long axis of the piano; and so on. In general, because of the out-of-phase relationship, the sound radiated toward the audience will be reduced, although what the pianist himself hears may be louder and certainly will be very different from the normal single-lid configuration. Perhaps the authorities at Carnegie Hall had these things in mind when they refused to allow the double-lid instrument in the hall.

4.6.12 Inharmonicity in Strings

There are two main ways in which nonlinearities can detract from purely harmonic production in the overtones of vibrating strings. One is by violating the low amplitude approximation upon which the linear wave equation itself rests. In a loose qualitative sense, over-zealous plucking or banging hard on the string with a hammer momentarily increases the distance the string must travel during a cycle and hence decreases the initial resonant frequency. This can result in an initial "clanging" sound during the decay of the tone. Although that effect can be a problem with early nineteenth century pianos, the modern Steinway and other high-tension instruments are fairly forgiving in this regard. (The effect amounts to about a 1 cent flattening during the first 1/4 s of the transient in the midrange of a Steinway D and is mainly a problem in accurate tuning of the instrument.) The more major source of inharmonicity in contemporary pianos arises from the effects of stiffness just because the instruments use thick strings under high tension.

As illustrated qualitatively in Fig. 4.31, it was tacitly assumed that the ideal vibrating string was able to sustain a discontinuity in slope at the support points. But the real string cannot bend sharply because of stiffness. Hence, at the end points (and

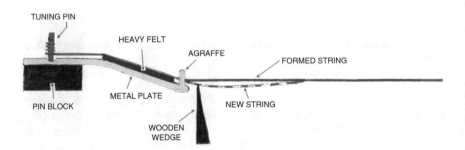

Fig. 4.31 Technique to "form" a new string. With the string under tension (tuned, say, one-half step below normal pitch) one gently taps upward on the wooden wedge from beneath the string. The process is similar to another one called "chipping," in which one picks at the strings with a wooden dowel to relieve the strain

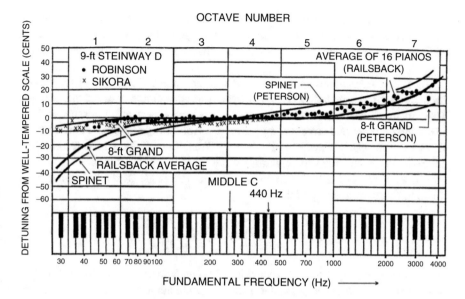

Fig. 4.32 Departure from equal temperament found in various pianos that were tuned aurally. The individual data points (solid circles) are measurements made on a 9-ft Steinway Model D concert grand piano tuned aurally by Christopher Robinson. The points marked "*x*" were on a different Steinway D tuned by Greg Sikora. The data for the Robinson tuning were taken by Christopher Haberbosch in collaboration with the author, using both a Sanderson Acutron® tuner and a Reyburn Cybertuner® on each note. The two measurements for each note are superimposed with solid dots in the plot. In most cases the two sets of values agreed within ±0.1 cent. The pitch on $A = 440$ Hz drifted upward by 0.2 cents during the measurements and the data were not corrected for that systematic error. Meaningful data were not obtained with these instruments below E_1 and above B_7. The data for the Sikora tuning were taken by the author using a Korg Model DTM-12 tuner. Those measurements were only made for notes below $A = 440$ Hz. The middle solid curve represents an average of 16 different pianos measured by Railsback (1938). The stretch curves recommended by Peterson (1999) for a spinet and for an 8-ft grand are also shown

also at maxima along the string), the bending is different in a real string than in the ideal case. That, in turn, means that the real string has its maxima compressed over a shorter distance and that its active vibrating length is shorter than the length between supports. Hence, the real string will vibrate at a slightly higher frequency than in the ideal case. (Generally, greater curvature in the wire results in a higher vibration frequency.) The frequency increase becomes still larger for the higher overtones.

The problem is especially severe with a newly strung piano. As indicated by the dashed line in Fig. 4.32, the new string will resist bending sharply at the agraffe and tend to introduce an extra minimum in the contour. Eventually, repeated use of the action will cause the string to flatten out, but that process can take a long time. A way of speeding up that process is to tap (gently) upward on a wooden wedge placed underneath the string and just beyond the agraffe after the new string is under tension. This technique also increases the stability of the piano after it is tuned. The experts like to "baby" the piano up to pitch starting a semitone low, then 50 cents low, then to pitch, and then 10 cents high until the piano begins to hold a tuning. This process allows the stresses to be distributed gradually over the whole piano. In addition to tapping upward near the agraffe, they also tap the strings downward at the bridges.[40]

There are several consequences of the stiffness effect. First, the overtone series for the real string is no longer harmonic in the presence of non-zero string thickness. That, in turn, means that the waveform is no longer perfectly periodic. Strictly speaking, one cannot analyze the spectrum precisely using a discrete Fourier series. However, one can always use the Fourier Transform method to determine the approximate spectrum for the vibrating string as discussed in Chap. 1.

Allowing for stiffness, the frequency for the nth mode of the string has been found to be approximately of the form[41]

$$ F_n \approx n \left(\frac{c}{2L} \right) \left[1 + n^2 \left(\frac{\pi^3 a^4 Y}{8L^2 T} \right) \right] \tag{4.5} $$

where a is the string radius, Y is known as "Young's modulus" for the material, L is the length, T is the tension, and n is the harmonic number. As $a \to \infty$, Eq. (4.5) reduces to the normal harmonic series for the ideal vibrating string. One can see that the fractional effect of the inharmonicity goes up with n^2 and is largest for thick, short strings under low tension. However, the strings in the bass section that are wrapped with copper wire are more flexible than solid strings of the same diameter; hence, the inharmonicity is less with wrapped strings than with solid ones of the same length and thickness.

A derivation of Eq. (4.5) would be beyond the scope of the present book as it follows from an approximation to a fourth-order differential equation. Nevertheless, a few comments about the properties of Young's modulus and steel strings are in order. Young's modulus (described here by Y, but sometimes designated E in reference books) is defined in general for homogenous isotropic elastic solids by the stress–strain relation:[42]

$$ \frac{\text{Force}}{\text{Area}} = Y \times \frac{\Delta L}{L} \quad \text{or} \quad \text{Stress} = (\text{Young's modulus}) \times (\text{Strain}) . $$

For a cylindrical wire, the stress, or force per unit area, is just $T/\pi a^2$ where T is the tension and a is the radius of the wire. The strain, $\Delta L/L$ for a cylindrical wire, would be the fractional elongation that would occur due to tension applied to it, say by hanging a weight on the wire. Since the ratio $\Delta L/L$ is dimensionless, it is obvious that the dimensions of Y are force per unit area and that Eq. (4.5) is dimensionally correct. The pertinent question is, "How much does the square bracket of

---

[40]Christopher Robinson (private communication.)

[41]Young (1957), p. 3–102, Table 3b-2.

[42]See Feynman et al. (1963) Vol. 2, Chapter 38, for an explanatory discussion of elasticity in solids.

Table 4.2 Some properties of steel piano wire (in cgs units)

Density	≈ 7.83 g/cm^3.
Breaking point	$\approx 10^8$ dynes/cm^2
Young's modulus	$\approx 2 \times 10^{12}$ dynes/cm^2

Note: 1 dyne $= 2.247 \times 10^{-6}$ lbs force
Source: American Institute of Physics Handbook (McGraw-Hill, 1957)

Eq. (4.5) deviate from unity in practice?" For convenience some properties of steel piano wire are summarized in Table 4.2. But one should note that all of the measured values vary somewhat with the alloy composition.[43]

4.6.13 The Steinway Model D Concert Grand as an Example

The longest unwrapped string on a Steinway D is for F_2 at 87.3 Hz, having a length of 71.9 in. $= 182.6$ cm and a diameter of about 0.047 in., or a radius of $a = 0.0597$ cm, and a density per unit length of $\mu = 0.00886$ g/cm. The tension on the string is $T = (2LF)^2 \approx 9 \times 10^7$ dynes ≈ 202 lbs. The cross-sectional area $\pi a^2 = 0.0113$ cm^2 and the force per unit area $\approx 8 \times 10^9$ dynes/cm^2, about 20% below the elastic limit.[44] In this case,

$$\frac{\pi^3 a^4 Y}{8L^2 T} \approx 3.3 \times 10^{-5} \text{ for Low} F = 87.3 \text{ Hz}$$

and the inharmonicity is quite negligible.

However, at the top of the keyboard it is another matter. There, C_8 at $F = 4186$ Hz has a string length of $L = 1.962$ in. $= 4.983$ cm and a radius of $a = 0.0165$ in. $= 0.0419$ cm, or a cross-sectional area of 0.00552 cm^2. The string density per unit length is $\mu = 0.00434$ g/cm.[45] If the same tension were applied here as to the low F, the force per unit area would be 1.63×10^{10} dynes/cm^2 and the string would break. But, because the string is much shorter, the actual tension is quite a bit lower for the highest C. As a first approximation, one can estimate T from the ideal string equation, obtaining

$$T = (2LF)^2 \mu = 7.5 \times 10^6 \text{ dynes.}$$

---

[43] See Trent and Stone (1957), 2–70.

[44] According to Hayes (1982), piano strings are normally tightened so much that bringing the string up another whole step, or a frequency increase of 1.122—hence, a tension increase of 1.26 would break the string.

[45] Young (1957, p. 3–101) provided a table of mass per unit length for steel and gut strings of different diameter.

However, this actually overestimates the tension because the inharmonicity itself plays a role in the tuning process. Rewriting Eq. (4.5) for $n = 1$, we get

$$F_1 \approx \left(\frac{\sqrt{T}}{2L\mu}\right)\left[1 + \left(\frac{\pi^3 a^4 Y}{8L^2 T}\right)\right] \qquad (4.6)$$

which in turn may be rewritten as a quadratic equation in T whose solution is[46]

$$T = \frac{1}{2}(A^2 - 2B) + \frac{A^2}{2}\sqrt{1 - 4B/A^4} \approx A^2 - B(1 + 1/A^2) \approx A^2 - B \qquad (4.7)$$

$$\text{where } A^2 = (2LF_1)^2 \mu \text{ and } B = \frac{\pi^3 a^4 Y}{8L^2}.$$

Note that A^2 is just the value of the tension calculated from the ideal string equation and that the correction to this value from inharmonicity is contained entirely in the term B. Inserting appropriate numerical values, we get $A = 2749$ and $B = 7.499 \times 10^5$ in cgs units. Substituting the values of A and B into the expression for the tension results in

$$T = 6.8 \times 10^6 \text{ dynes},$$

a somewhat smaller value than our initial rough estimate. Using this more accurate value of T, the inharmonicity coefficient for the top note on the piano becomes

$$\frac{\pi^3 a^4 Y}{8L^2 T} \approx 0.141 \text{ for the highest } C = 4186 \text{ Hz},$$

a value that is substantially larger than that determined for the low F at 87.3 Hz. The presence of anharmonic overtones in the top treble note is not of great esthetic importance because they fall at frequencies where the ear's sensitivity to pitch is decreasing. In fact, anything that adds to the loudness of the top few notes is an asset.

4.6.14   The Effect of Inharmonicity on Tuning

The effects of inharmonicity are reduced by decreasing the string diameter as one goes up the scale. In practice one wants to have a smooth transition in the wire diameter changes, or "scaling" of the instrument. The major consequence of the inharmonicity in piano strings is on the tuning of the instrument. For example, the

---

[46]The quadratic equation is $T^2 - (A^2 - 2B)T + B^2 = 0$.

first overtone of any given note will no longer be precisely an octave higher than the fundamental. Tuning a piano then consists of trying to find the least roughness in the beat patterns between overtones of the different notes, and the approach used is often characteristic of both the individual piano and of the tuner. In the bass, the fundamental tone on small pianos is often missing or very weak and the notes fall in a frequency range where the ear itself is losing sensitivity. Consequently on very small pianos, tuners sometimes adjust the fourth harmonic of one note in the bass to coincide with the second harmonic of the note one octave higher. As observed by Railsback (1938), piano tuners who start by setting the temperament over the middle octave centered at $A = 440\,\text{Hz}$ generally tend to tune the notes in the bass so as to fall progressively below those on the well-tempered scale and those in the treble to be progressively sharper. That this effect results naturally from the nonlinear overtones of the strings was shown by Schuck and Young (1943). That the resultant stretched octaves are actually preferred on small pianos by both music students and by piano technicians alike was demonstrated in carefully controlled listening tests by Martin and Ward (1961).

The effect is illustrated in Fig. 4.32, which summarizes data provided by several authors. The solid data points in that figure represent measured values for a Steinway D concert grand piano that had been selected by Steinway in New York City for use as a concert rental instrument for many years. Those measurements were made by piano technician Christopher Haberbosch in collaboration with the present author on the single middle strings of each note (with the outer strings damped) immediately after it had been tuned aurally from a pitch fork (for $A4 = 440\,\text{Hz}$) by Christopher Robinson (Master Piano Technician at the Acousticraft Company in Connecticut). Typically, after hitting a note the pitch starts out flat for the first 1/4 s (presumably due to momentary stretching of the string) and then becomes sharper. Consequently, average values were recorded in each case over the interval from 1/4 to 2 s after striking the note. Two electronic instruments were used for each of these measurements, a Sanderson Acutron®and a Reyburn Cybertuner®, that gave agreement within about ±0.1 cent in most cases. (A few notes exhibited a slight instability that gave rise to a discrepancy in the order of 0.5 cents.) A drift of about +0.2 cents in the note at $A = 440\,\text{Hz}$ occurred over the period of these measurements. The data were not corrected for that systematic error. The points marked "x" represent similar measurements made by the author using a Korg Model DTM-12 tuner on a second Steinway D immediately after it had been tuned by Greg Sikora (Chief Tuner for the Philadelphia Orchestra).

Stretch-tuning curves reported by Peterson (1999) for a small spinet and for an 8-ft grand are also shown in Fig. 4.32. In addition, the solid curve in the middle of the other two represents an average of 16 different pianos measured by Railsback (1938). Noting that one hundred cents is defined to be a half-step on the chromatic scale, it is seen from the figure that notes at the upper and lower ends of the keyboard may be tuned off by a quarter of a step from the frequencies of the well-tempered scale. The stretching effect is usually about the same in the treble for both small and large pianos because the strings in that region are of about the same length and thickness.[47] However, as shown in the figure, very large differences occur between pianos in the bass. Because the strings in the bass section of a concert grand are generally thinner and longer by a factor of about two than those in a console or upright piano, the tuning departure for a concert grand in the bass is generally very much less than shown for the Railsback average in Fig. 4.32. Thus, the strange sounding intonation in the bass of a crapaud (French for "toad" or "baby grand piano"), console, or upright is primarily due to the inharmonic nature of the shorter, thicker strings.[48] (See Eq. 4.5.) But the problem is further exaggerated by the human ear: first, because the ear's sensitivity to the actual fundamental frequency falls off rapidly at low frequencies (by more than 40 dB at low A); second, because it tends to synthesize a fundamental tone from the difference frequencies between modes that are an octave apart, which intervals all vary substantially below middle C on a small piano. The advantage of a concert grand is not just in its superior tone quality in the bass, but also in its superior intonation. Stretched tuning on a small piano may make it sound better when it is played by itself, but it also makes playing the piano with another instrument unsatisfactory. Although stretch-tuning curves are available for different size pianos from several companies that make electronic tuning devices, I have yet to hear a piano tuned electronically that sounded as good as the same one tuned aurally by an expert. Pianos differ in small ways from one another and using an average curve based on a number of instruments of the same type is seldom the best solution. The musically trained ear is sensitive to deviations of less than one cent in relative tuning and it still seems to require an expert human being to bring out the best sound in individual instruments.

Problems

4.1 From the scale drawing in Fig. 4.3, estimate the gain in velocity resulting from mechanical advantage in the Stein "IJprellmechanik" action. What would it be for the Broadwood action in Fig. 4.5 and for the Streicher action in Fig. 4.18.

4.2 Suppose you wanted to imitate the Beatles and have a crashing piano sound a chord in E-major. How could you do it?

4.3 What would the normal modes be in a triple strung piano?

4.4 It has been found experimentally that piano wires of the same material and length will break at about the same pitch, regardless of their diameter. Why does that occur?

---

[47] A table of string diameters for nearly every model piano currently made is given in Travis (1982), Part II.

[48] That may explain why Babe Ruth threw his upright piano into Willis Pond near Boston in 1918. (See The New York Times, 9/30/2002, p. A17 and 11/5/2002, p. WK 5.)

4.5 The lowest unwrapped (steel) string on a Steinway concert grand piano is about 71.9 in. long (F_2 at 87.3 Hz). What is the wave velocity on the string?

4.6 What is the tension on the string in the previous problem (noting that the string diameter is about 0.0472 in. with a density per unit length of about 0.0886 g/cm). The main difficulty with this problem is in the units. Putting the string length in cms gives the tension in dynes, where 1 dyne $= 2.247 \times 10^{-6}$ lbs of force.

4.7 A tuning hammer has a shaft length of 10 in. The top of the tuning pin has a width of 0.225 in. What force do you have to apply to the end of the tuning hammer to break the string on F_2 (normally 87.3 Hz) of a Steinway D?

4.8 From the mode pattern of the resonance at 63 Hz in Fig. 4.28, estimate the wave velocity in the soundboard. Assume the soundboard was 80 in. long.

4.9 What's the ratio of the third "harmonic" to the fundamental for F_2 (87.3 Hz) on a Steinway D? For C_8 (at 4186 Hz)?

4.10 Research Problem: Borrow a stroboscopic light source to study the vibrational modes of the lowest unwrapped strings on a piano, with and without the soft pedal depressed.

4.11 Research Problem: Is it possible to obtain something analogous to the una corda effect in a double-strung Italian harpsichord? Perhaps construct a model with two strings attached to the bridge on a sound board and try plucking and dampening them with your fingers.

Chapter 5
Violins and Bowed Strings

5.1 Early History of the Violin

The use of vibrating strings in musical instruments was certainly known to the ancient Greeks. However, bowed string instruments probably originated in Asia. One author suggests that the invention of the bow on that continent may be traced to the fact that Asiatic peoples were among the first to use horses as domestic animals—horse hair being the most suitable material for stringing a bow. The southern Slavs actually used horsehair for the bow and the string in a primitive instrument called the gusle. The earliest instruments had crude sound boxes consisting of a drum or hollow wooden tube. The two-stringed Chinese erhu, which used two silken strings with the bow hair passing between them, is still in use today and may even be heard in the New York subways.[1] The instrument produces a slightly whining sound said to resemble the human voice and uses a drumhead made from python skin. The single-stringed Arabian rebab was probably the first instrument to incorporate a trapezoidal resonant cavity and was a direct ancestor of the European violin. It was brought to Spain by the Moors in a form known as the rebec in Western Europe.[2]

The early Arabian instruments were played with the performer bent over in a stooping position with the instrument resting on the ground, and instruments with more than one string had to be turned to the right or left when bowed. That resulted in a rocking motion called geigan in Old High German, a term which evolved into the modern German word geige to stand for "string" in general. In order to make the

---

[1] See Yilu Zhao, "Asian Music Accompanied by the A Train," New York Times, July 6, 2004, p. B1. The sound may be heard on a CD recorded by erhu virtuoso YU Hon-mei (Wind Records SMCD-1010, 1998).

[2] According to Heron-Allen (1885), the rebab was in use in Spain for centuries before 1200 AD and still exists in the Basque districts of the Pyrenees under the names "rabel" and "arrabel." He adds that it was seldom mentioned because of its "rudeness."

© Springer Nature Switzerland AG 2018
W. R. Bennett, Jr., *The Science of Musical Sound*,
https://doi.org/10.1007/978-3-319-92796-1_5

bowing easier, the Germans made the instruments narrower and deepened the inward curves. The term geigen came into use throughout Germany for the predecessor of the violin, whereas fiedel or "fiddle" took on a derogatory meaning. (A "fiddler" was someone who played the geige badly.) There were three basic sizes of geige used in Germany, the smallest of which became a predecessor of the violin and was called geigen-rubeben—an outgrowth of the three-stringed rebec, which was used in France until the beginning of the eighteenth century.

Another development incorporated two openings on top of the instrument that later evolved through various /, C, and S-shapes into what we now call f-holes. Farga (1950) suggests that the f-shape of the sound holes may have started as a tribute to francis I, King of France from 1515 to 1547, an enthusiastic patron of the musical arts. Farga suggests that a man living in Lyon by the name of Gasparo Duiffoprugghar (alias, Kaspar or Gaspard Tieffenbrucker or Dieffenbrucker) had more claim than others to the invention. Little is known about the man with certainty except that he lived in Lyon around 1553 where he made instruments that at least resembled violins.[3] However, according to Faber (2005, p. 116), the violins accredited to Gaspard Tieffenbrucker "were nineteenth century impostors ...antiqued to look old."

Ironically, although the violin is depicted in a few paintings by Italian artists in the mid-sixteenth century, bowed instruments were imported from France and Germany to Italy where they served as a source of inspiration for the early Italian makers. The Italian viols of that time had six strings and fretted fingerboards covering one and a half octaves. These instruments came in various sizes: the viola da bracia (predecessor of the modern tenor viola), the viola da gamba (or "knee viola" and predecessor of the modern cello, albeit with five to seven strings), and the violone (a six-stringed double bass). Still another, the viola bastarda, that had metal sympathetic strings underneath the more usual gut strings of that period turned into the viola d'amore. In Leipzig, it is thought by some that Johann Sebastian Bach invented a five-string viola played while resting it on the arm (rather than under the chin) known as the viola pomposa (also known as the "bassoon viola") which was constructed by the German maker Johann Christian Hoffmann (1683–1750) in Leipzig circa 1720.[4] According to Gaillard (1939), the extreme difficulty in playing Bach's Sixth Solo Cello Suite arises because it was originally written for the five-stringed viola pomposa.[5] It is tempting to suppose that that viola might have been the one that Carl Philip Emanuel Bach said was his father's favorite instrument.[6]

---

[3]See Heron-Allen (1885), Farga (1950), Silverman (1957), Sacconi (2000), and Faber (2005) for more extended discussions of the history and development of the violin.

[4]An 8-ft "viola pomposa" stop was found on some pipe organs of Bach's time.

[5]Apel (1972, p. 907) asserts the attribution of this invention to J.S. Bach is unfounded and based on unreliable sources dating from 32 to 42 years after Bach's death in 1750.

[6]Letter from C.P.E. Bach to his father's biographer Johan Nicolaus Forkel in 1774 (quoted by Allan Kozinn in the New York Times, 7/16/2003, page AR 24).

5.2   The Cremonese Makers

The phenomenon of Cremona may be appreciated by looking at the map in Fig. 5.1. The workshops of nine of the most famous violin makers of all time were located within about one city block of each other! It is said that 20,000 violins were made here and shipped to all parts of the world during the golden age of violin making in Cremona. Alas, a little over a century and a half later, the shops and even the Church of Santa Domenico on which they bordered were all gone. Paolo Stradivari, son of the most famous of the Italian violin makers, had tried repeatedly to get the town council to set up a museum to contain his father's tools and some of his instruments, but his efforts were in vain. At his final defeat, a councilman told him that "not only would Cremona refuse to entertain the thought of a Stradivari memorial, but that he himself would use his influence to see that the magnificent San Domenico Church was razed to the ground." (Silverman 1957, p. 27.) According to Farga, the threat to destroy the church was not actually carried out until 1869.[7] Stradivari had purchased a tomb in the church of San Domenico in which he was buried. But when the church was torn down, his bones were scattered to the winds (the Hill brothers, 1902). Farga (1950, p. 50) notes that "the street which had housed Cremona's famous violin makers was rebuilt and made into a large, dull block of tenement flats." In the words of the Hill brothers (1902, p. 209), "It is pathetic to think that Cremona contains . . . nothing which witnesses to the glory of that splendid age of violin-making for which her name will ever be famous." More recently, a modest museum containing such artifacts as could be found was established in the town (although a MacDonald's stand now marks the location of Stradivari's shop) and an attempt has been made to restart a violin-making community. (See Faber 2005 and Steinhardt 2006.)

Fig. 5.1 The violin-maker shops of Cremona (source: Silverman 1957)

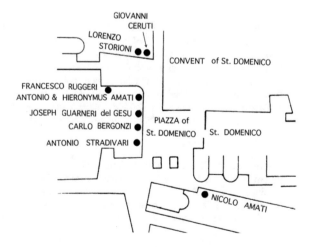

---

[7]Silverman (1957, p. 244) has "before" and "after" photographs of San Domenico.

Andrea Amati (1505–1577) was the first of these major craftsmen and designed the first instruments that closely resemble the modern violin. He started as a maker of rebecs and viols and switched to violin-making later in life. He constructed violins of two sizes, not to mention large violas and cellos of extraordinary beauty. Many of these were richly decorated and set a precedent for later makers to imitate. He, as did most of the Cremona violin makers, used Dolomite spruce of the type that grows in the Italian and German Alps for the top plate (or belly) of the instrument. This variety was found in rocky, mountainous soil which prevented fast growth of the trees and hence resulted in very uniform and closely spaced ring structure. That, in turn, gave rise to pieces of wood that had remarkably close and uniform grain. He was one of the first to use Balkan maple (acer-pseudo platanus) for his back plates of a type that was especially beautiful due to the flaming marks running roughly perpendicular to the direction of tree growth. His instruments had an exceptionally high arch on the top plate and produced a tone that was very clear, but soft; it was especially good for chamber music, but did not carry as well in the concert hall as some of the earlier instruments by Maggini of the Brescia School (Fig. 5.2).

When I was Master of Silliman College at Yale University in the 1980s, I persuaded a collector and Yale alumnus named Lawrence Witten to loan his quartet of Amati instruments for a concert by the Muir String Quartet celebrating the 50th Anniversary of the College.

These instruments had been in obscurity for some time and had only recently been uncovered by Mr. Witten.[8] The sound was very mellow, but according to the musicians their real advantage was in the ease with which they could be played. As violist Steven Ansell (now principal violist of the Boston Symphony Orchestra) put it, "They played like melting butter."

I quote here from the program notes written by the late Lawrence Witten for this concert (given in the newly renovated acoustics of the Silliman Dining Hall on October 20, 1983):

Andrea Amati made two world-famous sets of instruments: one for the Italianate court of King Charles IX of France, son of the Florentine princess Caterina de' Medici, painted and guilded with the mottoes and devices of the young king who died in 1574; the other for an unidentified nobleman was similarly decorated. In an age of magnificence, they were among the most splendid instruments of their time, and unquestionably the most important of their type, if not indeed the only ones. Except for bricks and mortar, they are the solitary artifacts of the Italian Renaissance which have been in more or less constant use since their creation four centuries ago. Their qualities are unsurpassed.

Witten went on to say

The Silliman performance utilized three of about seventeen instruments by Andrea Amati that were still extant. The cello was the best preserved of its size of those created after 1574 for the French court. The exact extent of this set, looted from the royal collections in the Revolution of 1789, is not known. It was the ancestor of the Sun King's 'Vingt-quatre violins du roi', directed by Lully a century later. Both the first violin and the viola of the quartet are

---

[8]Lawrence Witten is more widely known as the person who sold the Vinland Map to Yale University.

Fig. 5.2 Violin by Andrea Amati made in the mid-sixteenth century (courtesy of Bein and Fushi)

from the other decorated set, identical in construction and made about the same time. The second violin was dated 1628 by Andrea's grandson, the renowned Nicolò Amati…

I made a high-fidelity four-channel stereo recording of this concert (which included the first Razumovsky Quartet by Beethoven) for later analysis. Each of the four condenser microphones used was located over a different instrument.

5.3 Spectrum from an Andrea Amati

The waveform and harmonic spectrum from a violin by Andrea Amati are shown in Fig. 5.3 compared with that from a very ordinary contemporary instrument when both were played on open G by violinist Syoko Aki, of the Yale School of Music. The data were analyzed using the methods described in Appendix C. One can see from a casual glance at the numerous wiggles in the waveform that the Amati was very much richer in harmonic content or "overtones."

Andrea Amati had two sons, Antonio and Hieronymous (a.k.a, Girolamo), who took over their father's workshop after his death and continued for a while to make violins in their father's style. Of the two, Hieronymous (1556?–1630?) was the more inventive and changed the design of the top plate to produce a concave chamfer toward the edges, thus making the tone sweeter. Hieronymous died of the plague, as did his first wife and two daughters. However, one of the nine children he had had by his second wife was a son named Nicolò (1596–1684) who lived to be 88 years old and became the greatest craftsman in the Amati family. He then went on to teach many others the art, including his own son Hieronymous II (1649–1743), Andrea Guarneri (1626–1698), Francesco Ruggieri (1645?–1700?) and, possibly, Antonio

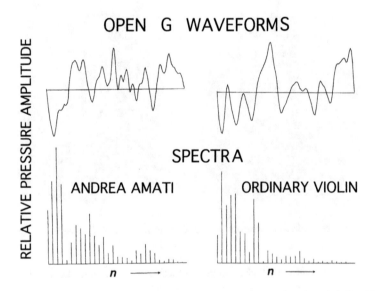

Fig. 5.3 Comparison of the waveforms and spectra of a violin by Andrea Amati (left) with those from an ordinary present-day violin (right). Both were played on open G without vibrato by the same violinist and analyzed using the same Sennheiser condensor microphone. The waveforms are shown as a function of time over one cycle at the top of the figure, normalized to the same peak amplitude. Relative amplitudes of the spectra are shown as a function of harmonic number at the bottom for each instrument

Stradivari (1644–1737).[9] Ruggieri, in turn had two sons, Giacino and Vincenzo who also made splendid violins. Vincenzo Ruggieri made the viola in 1690 that was owned by Beethoven. Ludwig Spohr (inventor of the chin rest, among other things) supposedly told his outstanding students to buy a Ruggieri instrument, if they were not in a position to afford an Amati, Stradivari or Guarneri.[10]

An outline of the family bloodlines and teaching lines among the Cremona violin makers is shown below. In at least a creative sense, these makers were all descendants of Andrea Amati. He not only taught several of them personally, but his influence permeated the entire group. Of his direct descendants, Nicolò Amati was not only a great craftsman but also the greatest teacher of the art of violin making of his period.[11] His student, Antonio Stradivari (1644–1737), is the most well-known and often regarded as the finest craftsman of the group. He generally included the date at which he completed an instrument under the left-hand f-hole, but in later years also included his own age at the time it was made. (That fact is the primary argument for assuming 1644 as the year of his birth.) Sacconi's remarks to the contrary, there is no reason to believe Stradivari had any advanced knowledge of mathematics or the science of acoustics.[12] He was adept at laying out the designs for his top plates with a compass and he was a fine, hard-working craftsman. (His last violin was made in his 93rd year!) He had great intuition, but most of his deductions were the outcome of educated trial and error and the heritage of previous instruments created by the early Cremonese makers. (See Hill et al. 1902, Chapter VIII.)

---

[9]Faber (2005, p. 25) suggested that Stradivari never was apprenticed to Nicolo Amati. The usual assumption is based on one violin (the oldest known) made by Stradivari in 1666 in which the label was inscribed "Antonius Stradivarius Cremonensis Alumnus Nicolaii Amati". However, the word Alumnus (meaning "pupil of") was dropped from his labels by the time his next violin was made in 1667 and never appeared on any of his later violins. Faber suggests the wording may have been used to help sell his first violin by using Nicolo Amati's name and that Amati himself may have objected to that practice. Faber goes on to note that Stradivari was already an expert wood carver and inlayer at the age of 13, before coming under the influence of Amati and that the few examples of fancy inlaid work on Amati instruments may actually have been done by Stradivari. Faber suggests that Francesco Ruggieri was a more probable teacher of Stradivari in the art of violin making.

[10]For more exhaustive accounts of the history of the violin-making families, see Heron-Allen (1885), Farga (1950), Silverman (1957), the Hill brothers (1902), and Boyden and Schwarz (1980), Blot (2001), and Faber (2005).

[11]Galileo Galilei is thought to have bought a violin by Nicolò Amati for his nephew. In a letter dated May 28, 1638, Father Micanzio regretted than he had not been able to get it for less than 15 ducats! [One ducat contained ≈ 3.49 g of gold, hence 15 ducats would be worth about $72 in the fall of 2004.] Galileo had been advised by the Musical Director at St. Marks in Venice that the Cremona instruments were incomparably better and much cheaper than those made in Brescia (Hill et al. 1902, pp. 240–243) It has also been suggested that the Cremona violin purchased by Thomas Jefferson for £5 in 1768 was made by Nicolò Amati Salgo (2000, p. 17). Salgo (Salgo 2000, p. 18) went on to suggest that Jefferson probably acquired a Tourte-style bow while serving as a diplomat in Paris, despite his (by then) almost crippled right wrist.

[12]Sacconi's book The Secrets of Stradivari had as a main thesis that there actually were no secrets and that everything Stradivari knew was well-known to other luthiers of the period.

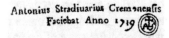

Label inside a violin made by Stradivari in 1719

According to the Hills (1902, pp. 226, 231), Stradivari's lifetime output consisted of 1116 instruments of which 540 are extant. During his prime, he turned out about one violin a week, selling them for some four-to-five pounds each.[13] Stradivari, in addition to making hundreds of violins, also made at least 30 cellos, a few violas and guitars and at least one arpeggione—a guitar-like instrument that was bowed and made famous by the later Schubert sonata with that name. It is said that the cellos he made between 1707 and 1727 (of which only twenty-one have survived) are the finest in the world (Fig. 5.4).

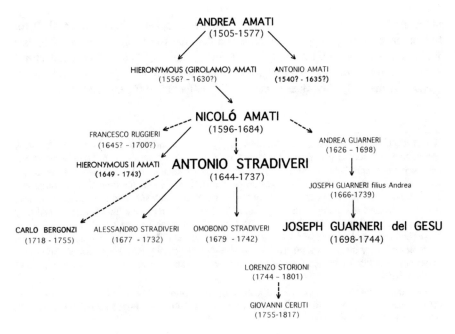

Fig. 5.4 Blood lines (solid) and chain of teaching (dashed lines) in the Cremona school of violin making. Sources: Farga (1950), Heron-Allen (1885), Hill et al. (1902, 1931), Silverman (1957), Sacconi (2000)

---

[13]The most recent, record-breaking sale of a Stradivari violin (the 1707 "Christian Hammer") was for $3.5 million on May 16, 2006. However, Sherlock Holmes only paid 55 shillings for a Strad worth 500 guineas (see Conan-Doyle 1981). Faber (2005, pp. 214–216) gave a careful summary of the sale prices of various instruments by Stradivari; Faber converted the currency used to dollars and corrected the values for inflation as of 2003.

The most famous of the Stradivari cellos is the "Davidov" made in 1712 for the Medicis. That instrument, now named after the Russian virtuoso Karl Davidov (1838–1889) who once owned and played it at the Court at St. Petersberg, has passed through the hands of a large number of famous soloists including Jacqueline du Pré and its present owner, Yo-Yo Ma. I was able to see and hear that instrument first hand in our living room at Silliman College in the early 1980s. Ma brought it with him for a Master's Tea held in his honor and at a request from one of the students, played the Bach unaccompanied cello suite in G major upon it. (She actually had asked merely for the Prelude, but Ma said he hated to play just one small part of the suite and asked if we would mind if he played the whole thing!) Although he played beautifully, the cello did not sound like an unusually brilliant instrument and had a rather subdued tone quality that made me wonder if the spruce top plates on those old Italian instruments wore out from use in the same manner (but at a much slower rate) as clarinet and oboe reeds do.[14] I learned later that Jacqueline du Pré had pronounced the Davidov to be "unplayable" and hadn't used it for several years. Various explanations were given: "You have to draw out the sound on these Stradivari instruments," "the cello probably suffered from changes in humidity in transatlantic flight" and so on.[15] (See, Faber 2005, p. 182.) Possibly the sound post was merely out of adjustment. In any case, Ma seems more recently to have brought it back to life.

In contrast to Nicolò Amati, who had numerous apprentices, Stradivari himself was only willing to teach his two sons and Carlo Bergonzi how to make instruments. An apprenticeship typically lasted at least 6 years and Stradivari clearly did not like to make that commitment of time. Of the two father and son cousins named Joseph Guarneri, the younger one with the appellation "del Gesu" (a name derived from the insignia I H S—with a cross above the H that he inscribed on his instruments) was by far the most famous and the major competitor of Stradivari. Del Gesu was something of a maverick. It has been alleged that he was sentenced to jail as the result of a drunken brawl in which he killed one of his violin-making competitors (Hill et al. 1931, p. 89) Some have claimed that he was able to continue to make violins in jail using tools supplied by the daughter of the jailer. But the Hill brothers (Hill et al. 1931, p. 90) note that except for a "number of more or less commonplace crudities" that have been foisted upon the public, there are no authentic Del Gesu violins known from the jailing period.[16] Storioni made violins in the general Amati style and sold his shop to the last of the Cremonese makers, Giovanni Ceruti.[17] (Ceruti was given the molds and patterns left in Stradivari's house after the great violin maker died.)

---

[14]Some claim that terpenoid resin can rejuvenate the spruce in a sound board.

[15]Those who have seen the 1998 movie Hilary and Jackie might suspect that leaving it out in a snow storm overnight was the trouble.

[16]The Hill et al. (1931) note that the quality of Del Gesu's workmanship deteriorated toward the end of his life. One luthier suggested to me that the unusual brilliance and harmonic content found in the later del Gesu instruments arose because he used extreme quantities of varnish to cover up his mistakes in wood carving.

[17]The Hill et al. (1931) note that J.B. Guadagnini spent a short time in Cremona in 1758, but he was hardly a permanent resident.

Joſeph Guarnerius fecit ✠
Cremonę anno 1735 IHS

Label in a violin made by Guarneri del Gesu

Especially, as modified in the late eighteenth century (by enlarging the bass bar, increasing the length of the strings to permit higher tension, replacing the bridge, and using a modern bow), many Stradivari violins acquired a reputation for being extremely powerful. The Heberlein-Taylor violin made in 1700 (Fig. 5.5.)[18] was no exception, especially in the hands of violinist Joseph Genualdi. I measured sound levels from this instrument of about 95 dB at a distance of 3 ft from that violin and the sound carried exceptionally well in the concert hall. As an example of its power, during the intermission of a concert by The Muir Quartet in Sprague Auditorium at Yale University during the 1980s, Mr. Taylor complained to Genualdi during intermission that he couldn't hear his instrument very well. That turned out to be a mistake because the first violin drowned out the entire quartet during the second half of that concert. (Genualdi, now first violinist of The Chicago String Quartet, is a very sensitive violinist and well-known for achieving excellent balance in chamber music.) Part of Stradivari's secret in making more powerful instruments is thought to be flattening of the belly over the geometry used by Andrea Amati. Other Stradivari violins I checked at 3 ft at the Library of Congress and bowed by violinist Cho-Liang Lin were: the Castelbarco (85 dB), the Ward (85 dB), and the Betts (85 dB). However, Fritz Kreisler's Guarneri del Gesu put out about 90 dB, even with the G-string missing. Faber (p. 53) states that the Strads made after about 1730 were among the most powerful of all, possibly because increasing deafness caused Stradivari to work harder on the problem of tone generation![19]

One of the most remarkable tales dealing with the Cremonese instruments has to do with the way in which many (perhaps, even most) of them survived and were transported to northern Europe. The story goes that an Italian craftsman named Luigi Tarisio (1792–1854) was confronted by one Sister Francesca (grand daughter of Antonio Stradivari who lived in a convent on the outskirts of Cremona) the night before she died. It is said that she made him promise to save as many of the great Italian violins as possible and bring them to artists in Europe. Tarisio then walked all over the hills of Lombardy and Tuscany searching out these instruments, which were often in total disrepair, and obtained them from their owners—sometimes by trading them for far inferior instruments in working condition. In one instance he

---

[18]The Heberlein-Taylor violin was recently donated by Hugh Taylor to the San Francisco Symphony. The Concert Master of that orchestra also has the use of the 1742 Guarneri del Gesu previously owned by Jascha Heifetz. In contrast, Jack Benny bequeathed his 1729 Strad to the Los Angeles Philharmonic.

[19]However, power can be a liability. The recent legal feud between members of the Audubon Quartet may have come about in part because the first violinist acquired a 1735 Carlo Bergonzi instrument capable of drowning out the rest of the quartet. (See Daniel J. Wakin, "The Broken Chord", in the New York Times, December 11, 2005, p. AR 1.) They solved the problem by firing the first violinist.

Fig. 5.5 1700 Stradivari violin known as the Heberlein-Taylor (photo courtesy of Joseph Genualdi)

traded an Amati and some cash to a Florentine noblemen for an entire collection of Stradivari instruments including "The Messiah"—the violin that Stradivari himself is said to have regarded as his finest instrument. According to Silverman (1957), Tarisio carried a bundle of ten violins on his back during his first trip to Paris while walking over the Alps from Milan. In Paris, he first met the violin maker Aldric who only offered him a modest sum which he accepted. On a later trip, dressed more elegantly, he met the famous French violin maker Jean Baptiste Vuillaume (1798–1875) to whom he ultimately sold many such instruments.[20] According to Millant (1972, p. 124), Tarisio brought more than a thousand violins from Italy. Vuillaume

---

[20] According to Silverman (1957, pp. 196–198), Vuillaume was such an expert that (working with physicist Felix Savart) he secretly made a copy of Paganini's Guarneri del Gesu ("The Cannon") while the instrument was on loan for repair that Paganini himself couldn't tell from the original. (However, a later attempt to repeat that experiment failed, probably due to poorer quality wood.)

repaired them, and in some cases copied them, and went on to sell them to famous European artists.[21] These instruments ranged from Paganini's favorite Guarneri del Gesu—"The Cannon" (which is now in a glass bottle in the Genoa Museum) to the "The Messiah" by Stradivari which is locked up in a glass case at the Ashmolean Museum in Oxford. It has been said that the name "Messiah" (or "Le Messie") was dubbed by the French violinist Delphin Alard, who happened to be Vuillaume's son-in-law. It is said that Tarisio kept promising to produce this exquisite instrument during his visits to Vuillaume in Paris, prompting Alard to exclaim, "Ah, therefore, your violin is like the Messiah; one always awaits him, but he never appears." The Hill brothers do not endorse that story (which is one of several different versions) and refer to a violin called "le Messie" as existing during Stradivari's lifetime. Vuillaume actually had to wait until Tarisio died (cradling two violins in his arms), at which point he bought out Tarisio's entire remaining collection of violins, including the elusive "Messiah."[22] After the death of Vuillaume in 1875, Alard decided to buy the instrument himself, although he may never have played it. It eventually passed through the hands of the Hill brothers before it went to the Ashmolean Museum. According to Millant (1972, p. 123), the Messiah was made along slightly more delicate lines than the usual Stradivari violins; the back is in two pieces of flaming maple, the model is flattish and the f-holes are a little more sloping than usual. The ribs are slightly higher than normal, but the length of the body is the standard 35.6 cm. Some maintained that the instrument had never been played. But Milstein said that he had tried out the instrument at the Hill's shop on Bond Street: "An astonishing Stradivarius known as the Messiah—an unforgettable experience!" (Milstein and Solomon 1990, p. 87).

When Vuillaume obtained the Messiah, he took it apart, replaced the bass bar and brought the instrument up to the then current standards for concert hall performance. Physicist Savart encouraged him in this process to see what made it tick. While the Messiah was apart, Vuillaume made enough careful measurements to permit making

---

Faber (2005, p. 116) presents a more prosaic version of this story: The color of the varnish was quite different, as was the tone quality. But Paganini still liked it.

[21]Silverman's account of the life of Luigi Tariso is very engaging, even if you don't play the violin. (His 1957 book also contains numerous color photographs of instruments from the Golden Age of Cremona.)

[22]Skeptics note that Alard's father-in-law (who bought and sold the Messiah several times) was a master at building fake violins and might have been good at concocting fake stories to go with them. At one point, suspicion of the Messiah's authenticity arose from a tree ring analysis suggesting the wood was cut after Stradivari died. However, a recent more thorough study by Grissino-Mayer et al. (2004) implied that the tree rings dated between 1577–1687, well before the completion of the Messiah in 1716. In the account by Silverman (1957, p. 45), Sister Francesca supposedly told Tarisio of her grandfather's "most perfect violin called Le Messie" on her death bed in 1809— long before Tarisio met Vuillaume (1827) and still longer before Alard could have christened it "The Messiah." Another doubter, violinist Joseph Gold (private communication) thought that the purfling on the instrument did not look up to par with the other work of Stradivari.

Fig. 5.6 The 1716 instrument known as "The Messiah" alleged to have been regarded by Stradivari as the finest violin he ever made and by Vuillaume as the most perfect violin he had ever seen. It is claimed that the violin was never played, although Vuillaume replaced the bass bar and it probably was played by Alard and certainly by Milstein (photo courtesy of the Asholean Museum)

accurate copies. It is said that towns like Mirecourt in France and Mittenwald in Germany then became centers of mass-production in turning out replicas of the "Messiah."[23] (Faber 2005, p. 122) (Fig. 5.6).

---

[23] One wag commented that of the original 1116 Stradivari instruments made by the great master, over 4000 still exist. According to the Hills, the Italians themselves began the habit of putting false labels on these instruments, a tradition carried on by Vuillaume. According to Faber, the Messiah is the most-copied violin in the world. In case you want to make one yourself, the Heinrich Dick Company of Metten, Germany sells a kit with precut wood.

Fig. 5.7 The earliest (1838) known cello by Giuseppe Rocca. This cello is one of very few with a one-piece maple back (photograph courtesy of Eric Blot)

The Cremonese tradition did not end with the death of Ceruti in 1817. According to Blot (2001, p. 158), Tarisio met one Giuseppe Antonius Rocca (1807–1865) of Turin circa 1842 and showed him a number of Cremonese instruments, including the famous Stradivari "Messiah" and the "Alard" Guarneri, perhaps for maintenance or small repairs. Rocca was so impressed by these instruments that he went on to spend the rest of his life copying them. In later years, he started adding small innovations of his own. One striking aspect of his early cellos, not to mention his copies of the Stradiveri and Del Gesu violins, was his penchant for one-piece backs of maple with strong flaming. He evidently found sufficiently large pieces of maple even to make cellos with both one-piece backs and top plates! (Fig. 5.7)

5.4   The Bowed String

> "Everything should be made as simple
> as possible, but no simpler."
>
> Albert Einstein

Except when the strings are plucked or hit with the bow, the excitation method used on instruments of the violin family is quite different from those found in the other stringed instruments that we have discussed previously. With bowed

instruments, the string traditionally has been excited by frictional contact with horsehairs fastened on a moving wooden stick. Although friction plays an essential role in exciting the string, it is hard to provide more than an empirical description of its action. The old notion, still taught by some violinists, assumes that the surface of the hair has hooks and barbs spaced along its length that all point in the same direction. The argument then runs that the hairs should be divided into two groups of equal number and arranged in opposite directions before they are attached to the wood of the bow. The supposition is that otherwise, the bow would only work in one direction, producing either an "up-bow" or a "down-bow."

The claim is actually nonsense. There are no significant "barbs" or other asymmetries in the two directions. Microscopic examination shows that any such protuberances are very small in amplitude to the extent they exist at all. (See later discussion of bow hair.) Obviously, the main friction is provided by the rosin violinists apply to their bows. If you don't believe that, try drawing a tone in either direction from an unrosined bow.

The motion of the bowed string consists of a series of "stick-slip" cycles as illustrated in Fig. 5.8. There, we have assumed a steady-state waveform in which the string at the bowing point moves at the constant upward velocity of the bow until it reaches a point where the restoring force from the extended string exceeds the frictional force. At that time, the string rapidly snaps back, launching a pulse (or "kink") that runs along the full round-trip length of the string. The transverse motion of the string at the bowing point continues through zero amplitude until it reaches a negative displacement such that the frictional force from the bow is just able to grab the string once again and reverse the direction of motion. The magnitude of the restoring force at the lower "grab" point is about the same as it was when the

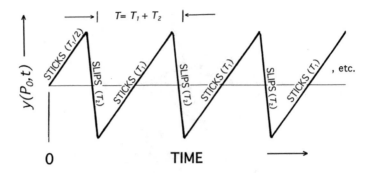

Fig. 5.8 The stick-slip model of the bowing process discussed in the text

string started to slip. In Fig. 5.8, T_1 corresponds to the time out of the cycle that the string is grabbed by the bow, and T_2 represents the time during which the string slips along the bow. The displacements above and below the time axis are about equal on the average. (If they weren't, the amplitude would build up uncontrollably in one direction or the other.) The full period of one cycle is $T = T_1 + T_2$ and in practice, T_1 is much greater than $T2$. The fundamental frequency is $F_0 = 1/(T_1 + T_2)$. The motion of the string thus consists of a continuing series of stick-slip cycles that are locked to the round-trip frequency $(c/2L)$ for a disturbance launched along the string and the motion of the string at the bowing point approximates a sawtooth waveform in time. That locking process transforms the inharmonic nature of the vibrating into one that is precisely harmonic.

As with the harpsichord and the piano, the general solution for the vibrational motion of the string can be written in the form

$$y(x,t) = \sum_{n=1}^{\infty} A_n \sin(n\pi x/L)\sin(2\pi n F t), \qquad (5.1)$$

where $y(x, t)$ is the transverse displacement, L is the distance from the bridge to the opposite point of constraint (the "nut"), and x is the coordinate along the length of the string (here, measured from the bridge). The boundary condition that determines the amplitude coefficients A_n is on the steady-state time-dependent displacement of the string at the bowing point ($x = P_0$ shown in Fig. 5.9). Those coefficients may be obtained by Fourier analysis of the waveform at the bowing point.

5.5 The Helmholtz Model

Helmholtz (1885) gave an approximate solution to the problem by assuming straight-line motion of the string at the bowing point ($x = P_0$) over the two portions of the cycle, an assumption that was compatible with his experimental

Fig. 5.9 Relative coefficients C_n computed for the up-bowed string in Fig. 5.8. As shown in the magnified section, the harmonic amplitudes oscillate in sign

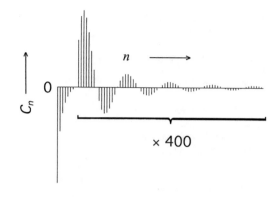

measurements. The sawtooth motion of the string at that point may be written as the Fourier series

$$y(P_0, t) = \sum_{n=1}^{\infty} C_n \sin 2\pi n F_0 t \qquad (5.2)$$

where, as shown in Appendix B,[24]

$$C_n = -\frac{2}{n^2 \pi^2} \left(\frac{T}{T_1}\right) \left(\frac{T}{T_2}\right) \sin\left(n\pi \frac{T_2}{T}\right) . \qquad (5.3)$$

Here, the sawtooth waveform was normalized to the range $-1 \leq y(P_0, t) \leq +1$ and the negative sign corresponds to upward bowing as shown in Fig. 5.8. Relative values of the coefficients C_n are illustrated in Fig. 5.8. By appropriate choice of the time origin, the need for a phase angle disappears, although the amplitudes vary in sign with increasing harmonic number.

Comparing Eqs. (5.1) and (5.2) and substituting C_n from Eq. (5.3), it is seen that

$$A_n = -\frac{2A}{n^2 \pi^2} \left(\frac{T}{T_1}\right) \left(\frac{T}{T_2}\right) \frac{\sin\left(n\pi \frac{T_2}{T}\right)}{\sin\left(n\pi \frac{P_0}{L}\right)} \qquad n \neq L/P_0 . \qquad (5.4)$$

where A is the amplitude of the oscillation at the bowing point. (As with the other vibrating string problems discussed earlier, it is assumed that $A \ll L$.)

The complete motion of the string is then determined by substituting Eq. (5.4) into Eq. (5.1) and is illustrated in Fig. 5.10. As is apparent from Eq. (5.4), the denominator vanishes when the string is bowed at the node for any harmonic. The consequence of that apparent singularity may be avoided in practice by noting that the particular value for An must be zero when the string is bowed at a node for a

---

[24]Helmholtz obtained an equivalent result in a somewhat different manner.

Fig. 5.10 Motion of a violin string at the point of contact with the bow. (The diagram illustrates the case where the bowing is "upward" with $T_1/T = 7/8$ and $T_2/T = 1/8$), and $P_0/L = 1/10$, computed from Eqs. (5.1) and (5.4). For the up-bowed case, time starts from the top. (Time starts at the bottom for the down-bowed case.) The wiggles in the waveforms are due to the omission of the tenth harmonic

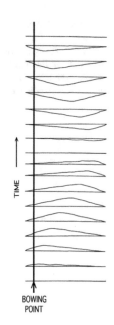

TIME

BOWING
POINT

harmonic.[25] Although the string motion becomes the driving force to produce sound from a violin whose bridge is attached to the string, the spectrum of the vibrating string alone is quite different from that which would characterize the sound of a bowed instrument.

As noted before, one important consequence of the stick-slip mechanism is that it forces the string motion to be precisely periodic at the round-trip frequency $(c/2L)$. Hence, the inharmonicity vanishes that was described in Chap. 4 in connection with piano strings (see Eq. (4.7)) and which is present in any plucked stringed instrument. That is, the waveform is forced to be precisely periodic by the stick-slip process and, hence, is characterized by a pure harmonic series.

As can be seen from Fig. 5.10, the release of the string from the bow, which occurs when the restoring force from the displaced string exceeds the frictional force provided by the bow, results in a "kink" or pulse traveling in a closed loop of total length $2L$.

Because the string is clamped at both the nut and bridge against lateral motion, hard reflections occur at both ends of the string, with the result that the "kink" changes sign at each reflection. As noted by Helmholtz, the peak displacements of the string during one cycle trace out parabolic arcs, as illustrated in Fig. 5.11. The envelope is roughly what one sees by eye in looking down at a bowed string, because the eye integrates the image over many cycles of the fundamental frequency.

---

[25] A more rigorous solution for that case was given by Raman (1918).

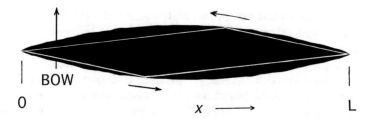

Fig. 5.11 Parabolic arcs swept out by the peaks in the triangular waveform over one bowing cycle. Here, the dark area was computed by superimposing the string trajectories for one cycle over several hundred time increments. The white segments indicate the motion of the "kink" for the up-bow process at one instant in time. Again, up-bowing was assumed and the same parameters were used as in Fig. 5.10. The wiggles in the envelope are due to the fact that the string was bowed at a node for the tenth harmonic, hence that harmonic was suppressed

5.6   The Real Bowed String

Although the Helmholtz model illustrates the main features of the bowing mechanism, there are some aspects of the real problem that it ignores by necessity. These have to do with such things as the string rolling on the bow, the fact that the string motion undergoes acceleration during the bowing cycle and does not move in simple straight lines, and the effect of bowing is strongly pressure-dependent. Further, the variation in the frictional characteristics produced by rosin on the bow also adds to the complexity. For example, the static (sticking) and sliding friction coefficients vary differently with temperature above about 80 °F. The sliding friction coefficient goes down whereas the sticking coefficient goes up. Infrared pictures of the string heating during bowing show that the segment of bow hair in contact with the string is about 25–30 °F warmer than the rest (Pickering 1991, pp. 69,70). Another difference is that the intersection of a real bow with the string does not occur at a point, but can be spread over a centimeter or so. For these various reasons, it is only practical to examine the actual motion of real strings under normal bowing conditions rather than rely on theoretical models. Pickering (1991) has provided the most extensive collection of experimental data on real bowed strings known to the author. Using a bowing machine and a digital measuring technique, he examined the motion of a variety of string materials bowed at different pressures. The materials ranged from gut through various metallic and metal-wound strings to orlon.[26] Although violin strings were often called "catgut" (indeed, there was even a violin journal of that name), the gut strings of choice during Mozart's time were made from sheep intestines. It is said that the luthier's house of that period reeked from the pungent smell of boiling sheep guts.

---

[26]Pickering notes that the word chorde in Greek originally meant the intestines of animals, just as the word gut means today.

Leopold Mozart advised that all four strings on a violin should have nearly the same tension. The only way to do that with gut strings is to increase the diameter as one goes down in pitch (by about a factor of three in going from the E-string to the G-string). Although gut has a large internal damping factor (which is its principal virtue on bowed instruments),[27] the large diameter required on the lower-pitched strings increases the inharmonicity to intolerable levels. (See Eq. (4.5) of Chap. 4 and related discussion.) For that reason, gut strings, if used at all, are now often made with a narrow-diameter gut core and then over-wound with metal— a process that Paganini carried out by using round metal wire that he sanded flat. (For uncertain reasons—except perhaps that he was Italian, he also stored his strings in olive oil.) As noted previously in the case of bass piano strings, the over-winding technique increases the mass per unit length and makes the string more flexible (hence, less prone to inharmonicity). According to Pickering (1993), the most successful strings used on bowed instruments today are of a type using a core of synthetic polymer strands that is often over-wound with aluminum or silver of square cross-section. These polymers (forms of nylon, with trade names ranging from the German Perlon to our Orlon) are roughly twice as strong as gut and can be used with smaller diameter cores. If over-wound with a dense metal, they can be still smaller in diameter. With this technique, rubbing by adjacent coils of the metal helps to increase the damping factor.

Two interesting extremes from Pickering's data are shown in Fig. 5.12. In neither case does the bowed point on the string move in a simple straight line. It is clear that the string is slowing down toward the top of the cycle in each case. In the upper portion of the figure where the bowing pressure is greater by a factor of about 5.5, the ripples in the waveform are much more pronounced because the bow is pushed down in greater contact with the string. As with the ripples in the figures computed earlier from the Helmholtz method, those in Fig. 5.12 are at least partly due to the suppression of a harmonic by the presence of the bow—an effect that is much greater at high bowing pressure. The ripples can also arise from torsional motion of the string. Because the string was bowed at 2 cm out of a length of 32.8 cm, one might expect the 16th harmonic to be suppressed. However, by counting the ripples, it is evident that the suppression occurred at closer to the 11th harmonic. The reason for that seeming discrepancy is that the bow hair had a width of about 1 cm, hence the range of contact was from about 2 to 3 cm from the bridge. (32.8 divided by 3 is close to 11.) It is a straightforward matter to Fourier analyze data of the type shown in Fig. 5.12, once it has been digitized. (See Appendix C.) In doing that and then calculating the motion of the entire string, one would again find ripples in the motion resulting from the missing harmonic. It should be emphasized that, as in the case of the Helmholtz model, the Fourier amplitudes for the data in Fig. 5.12 would have relatively little resemblance to the acoustic harmonics that would be produced

---

[27] In contrast to harpsichord and piano strings, a large damping factor is desirable for the strings on bowed instruments. Here, in contradistinction to the keyboard instruments, one does not want the tone to persist long after the excitation process has stopped.

Fig. 5.12 Waveforms at the bowing point (2 cm from the end) made during a four-second stroke at 15 cm/s on violin A strings of 32.8 cm length. Upper Figure: Light steel core with aluminum winding 0.485 mm in diameter and a tension of 13.1 lbs with bowing force of 110 g. Lower Figure: Heavy steel core with nickel winding 0.488 mm in diameter and a tension of 17.5 lbs with bowing force of 20 g (data reproduced with permission from Pickering 1991, p. 62)

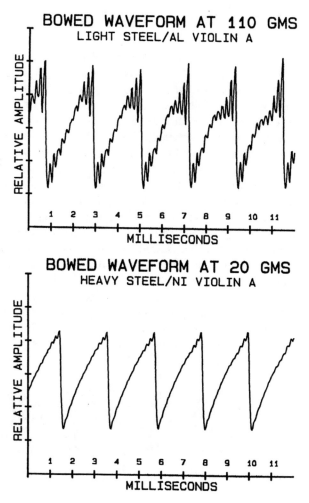

when the string was attached to a violin. Obviously the presence of the vibrating string drives the acoustic resonances and it is important to have lots of harmonics in the waveform. But various air and wood resonances in the instrument predominate, together with the effects of the bridge and bass bar in transmitting vibrations to the violin top plate. Direct radiation from the surface of the violin also changes the tonal color and directionality of the instrument.

5.7   Basic Structure of the Violin

An exploded diagram of a violin (which in reality contains some 70 parts) is shown in Fig. 5.13. The basic elements are essentially the same in the viola, cello and double bass apart from differences in scaling. The main parts consist of a bottom plate made of maple separated by a rib structure (usually maple) from a top plate made of very thin lighter wood (usually spruce with the fiber running in the long dimension of the instrument). Together these pieces comprise a resonant air volume. The top plate, which is hollowed out and has a convex shape in the upward direction, contains two f-holes placed symmetrically about the long dimension of the violin on either side of the bridge which serve to couple sound from inside the violin to the outside world and permit readjustment of the sound post. The exact shape and area of the holes also tune the fundamental air resonance of the instrument in a very approximate version of a Helmholtz resonator. (See Appendix A.) The bottom plate is also hollowed out and is convex (to a lesser extent) facing in the downward direction. The top plate, ribs and bottom plate are glued together with small additional pieces of wood (the top block, bottom block and corner blocks) that serve to strengthen the structure mechanically. Inlaid wood glued along the periphery of the top plate and bottom plate called the purfling prevents the wood of the two plates from splintering at the edges. Stradivari produced some of the most elaborate inlaid carving for the purfling on his instruments of any of his contemporaries.

Section of purfling from a Spanish viola made by Stradivari in 1696

The structure shown above the instrument in Fig. 5.13 holding the strings consists of a tail piece (fastened through a covered loop of wire to a hub imbedded in the bottom block), a bridge which couples the lateral vibration of the strings to vertical motion of the top plate, a finger board above the top plate and neck of the instrument which permits the performer to stop the string vibration at various lengths, a nut which terminates the length of the open vibrating strings, and a hollowed out peg box containing the four pegs (often made from boxwood) which permit tuning the instrument by changing the tension on the open strings. In practice, violin makers often apply graphite lubricant to the groves in the nut and bridge to facilitate the string slipping over these regions during tuning so that equal tension is maintained

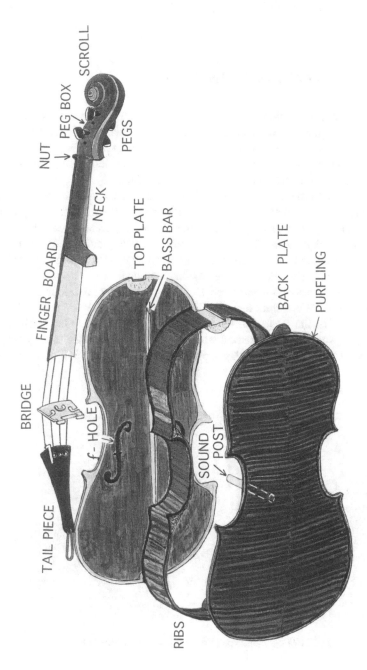

Fig. 5.13 Exploded view of a violin showing the main parts of the violin

throughout the string. The scroll at the end of the instrument is primarily of artistic value but also provides a convenient place to hold the instrument or suspend it from a wire.

5.8   Sound Coupling to the Top Plate

Coupling of the lateral vibration of the strings to the top plate occurs through a rocking motion of the bridge as indicated in Fig. 5.14. The bridge pivots at a point on the top plate near and above something known as the sound post. That post is placed near and below the foot at the treble side of the bridge near the highest string of the instrument. The post is positioned near, but behind the bridge foot, and close to a nodal line for one of the resonant modes.[28] Both the bridge and sound post are held in place only by friction and the component of force from the string tension pushing downward. The vibrational amplitude of the top plate increases with the torque applied to the bridge by the transversely vibrating string, and the torque applied to the bridge from the lateral force exerted by the bow increases with the height of the bridge. Hence, one would expect the loudness of the instrument to increase with bridge height for constant bowing force. However, increasing the bridge height also means increasing the vibrating length of the string between the bridge and the nut, hence requires that the tension in the string be increased to keep the pitch of the open string the same. Such increased tension may detract from the tone quality of the instrument. As well as providing a pivotal point for the rocking of the bridge,

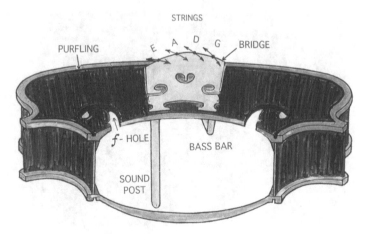

Fig. 5.14 Positions of the bridge, sound post and bass bar in the assembled violin

---

[28]See Jansson et al. (1997).

the sound post plays an important role in transmitting vibrations to the back plate and in determining the fundamental resonant modes of the assembled instrument. Known by the French as the "soul" of the instrument, the sound post requires critical adjustment to get the best tone quality from the instrument. That adjustment, done by listening to the sound of the instrument, involves locating the post at the right node for vibrational resonances in the two plates. (See later discussion of the top plate modes.)[29]

The other leg of the bridge rests at a point on the top plate that is above the bass bar, a long piece of wood made of spruce glued to the underside of the top plate that transmits vibrations from the bridge to the full length of the top plate. Although the bass bar strongly couples the lower frequency vibrations to the top plate, it also plays an important role in transferring energy from the higher harmonics of all the bowed strings.[30]

5.9   The Bridge

The bridge is one important part of the violin that has changed markedly since the time of Stradivari. His bridges were also made of hard maple, but with significant differences in shape from the modern ones.[31] (See Fig. 5.15.) A hole (either circular or heart-shaped) introduced in the center of the Stradivari bridge was an aid in alignment. One could run a string through this hole that attached to centered positions at the ends of the violin. However, a large hole in that location attenuates the high-frequency response of the bridge. As shown in Fig. 5.15, Paganini used a substantially different bridge shape than other violinists. His bridge was flattened enough so that he could easily sound three strings at once. According to Giordano and Dilworth (2004), the bridge on Paganini's Guarneri del Gesu "Cannon" was very narrow—38 mm across the feet as opposed to 42 mm for a modern bridge and the feet themselves were only 7.8 mm by 3.6 mm thick, as opposed to a thickness

---

[29] Another kind of "sound post" was invented by the Belgian cello virtuoso Adrien-François Servais (1807–1866) who became so fat that he could no longer support the instrument between his knees (the traditional method used since the time of the viola da gamba.) He designed an end-pin to support the instrument on the floor, an innovation that has since been adopted by most modern cellists. Faber (2005, p. 131) suggests that vibrations from the cello may be enhanced by coupling through the end pin to resonances in the wooden floor of a stage.

[30] Because the main body of the violin is symmetric about the long dimension, it is relatively easy to adapt the instrument to left-handed players merely by interchanging the bass bar and sound post locations and reversing the bridge. Thus, for example, the left-handed Austrian violinist Rudolph Kolisch (1896–1978), founder of the Kolisch Quartet, and Charlie Chaplin could easily be accommodated. (Accommodating left-handed pianists would be much more of a problem.)

[31] Although modern bridges are generally made of hard maple, they sometimes contain a V-shaped insert of still-harder ebony at the location of the E string to minimize the danger of that thin wire cutting into the bridge.

Fig. 5.15 Different bridges
(as seen from the fingerboard)

STRADIVARI   PAGANINI   MODERN

of 4.5 mm at that point on a modern bridge. There is evidence of the same type of
alignment pinhole used by Stradivari in the Guarneri del Gesu bridge on Paganini's
violin. Although its location is marked on both sides, it does not go through the
wood and may have been filled in after the violin was made.[32]

Although Helmholtz (1885, p. 86) suggested that all of the sound was coupled to
the top plate through the same, single foot of the bridge above the bass bar, that idea
was wrong.[33] With the help of luthier Hiroshi Iizuka, I investigated this question
further in the case of both a violin bridge and a viola bridge on instruments made
by him. Two identical piezoelectric pickups[34] were mounted in various places. In
the preferred location in which the violin data shown below were taken, they were
mounted in the slots on each side of the bridge. In other instances, we tried mounting
the elements on the outer, flat sides of the bridge and then finally on the top plate
near the 2 ft of the bridge. The output voltages were fed into the two input channels
of a digital recorder while different notes were bowed on each of the strings of the
instruments. In none of these cases was there a significant difference encountered
between the magnitude of the voltages from the two piezoelectric elements.

With piezoelectric elements mounted in the slots on each side of the bridge, the
signal was characterized by sharp spikes at a repetition frequency characteristic of
the pitch of the note and the two pulse signals were clearly 180° out of phase.
(See Fig. 5.16.). This phase relationship should tend to excite modes that have
odd symmetry about the long axis of the instrument. The sharp spikes result from
Newton's "Law of action and reaction" when the "kink" in the bowed waveform is
reflected by the bridge. (The bridge receives an impulse in an opposite direction to
that in which the kink is reflected.) However, the two vibrational pulses running
through the bridge to the feet occur in opposite directions, producing pulses of
opposite polarity. The relative polarity of the two pulses changes when the bowing
direction changes. Although the extreme low-frequency content was somewhat
higher from the side of the bridge above the bass bar, there was not a great deal
of difference between the spectra obtained in those two locations in either a violin
or viola bridge. (See Fig. 5.17).

---

[32] Acoustic data for some twenty-one different modern violin bridges was given by Atwood (1997).

[33] Cremer (1981), p. 206) suggested that the 2 ft of the bridge transmitted sound very differently
and that the bridge acts like an electric circuit with four input terminals and two output terminals.

[34] Model SH SV2 made in Erlangen, Germany and distributed by Shadow Electronics. These
elements are said to have nearly constant response within the audio band up to 20 kHz.

Fig. 5.16 Diagram showing the locations used for the piezo electric elements on the bridge

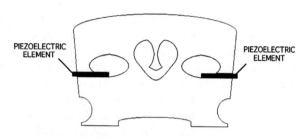

PIEZOELECTRIC ELEMENT   PIEZOELECTRIC ELEMENT

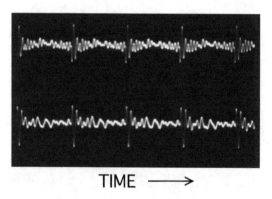

TIME ⟶

Fig. 5.17 Trace taken with a double beam oscilloscope showing the output voltages from piezoelectric elements placed in the slots on each side of the bridge as shown in Fig. 5.16 when an Iizuka violin was bowed on the open A string. The upper trace is from the element above the bass bar. Note that the two sets of periodic pulses (resulting from the "kink" round trip cycle) are approximately 180° degrees out of phase. Similar results were obtained on different notes on the other strings

In addition to the rocking motion of the bridge, the bridge can also bend back and forth slightly about an axis perpendicular to the strings. The side of the bridge toward the nut is curved ("like someone with a puffed-out chest," in the words of Iizuka.) This effect (which can be minimized by skillful violinists) was observed by placing the piezoelectric elements on the face of the bridge. The signal amplitude was typically smaller in this case than that produced by the rocking motion by about 5–10 dB and the pulses from the two piezoelectric elements placed on opposite sides of the bridge face were approximately in phase in this case.

One thing that strongly affects the overall bass response of an instrument is the relative position of the treble foot of the bridge (on the right side from the player's point of view) with respect to the sound post. If this point is positioned directly above the sound post (a location preferred by Isaac Stern in his later years[35]), the higher frequencies are emphasized at the expense of the low-frequency response.

---

[35]Paul Arnold, private communication.

That position may help the violin tone to be heard over a large orchestra in the performance of a concerto. However, a more balanced result occurs when the treble foot of the bridge is located about 1–1.5 sound-post diameters away from the center of the post. In that position, the top plate is freer to vibrate up and down at low frequencies. Another thing that affects the relative high and low frequency response of the instrument is the amount of wood (or thickness) in the bridge. Low-frequency coupling of the instrument to the air requires a relatively large motion of the top plate up and down. In contrast, high-frequency coupling involves a smaller amplitude excursion (Fig. 5.18).

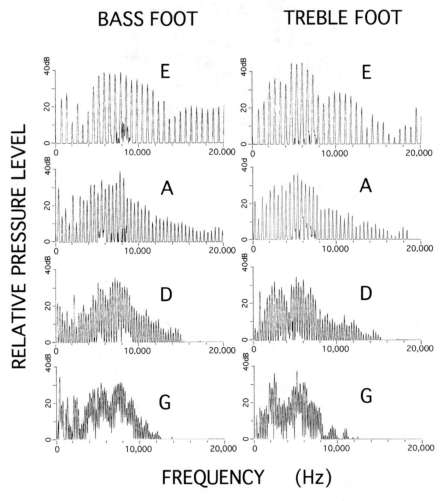

Fig. 5.18 Spectra from the two piezoelectric elements mounted at opposite sides of the bridge on an Iizuka violin bowed on the open G, D, A, and E strings. The bridge used was the lighter version whose acoustic spectra are shown in Fig. 5.19 (the same Iizuka violin was used in each case)

The presence of a large mass of wood in the bridge impedes the motion of the bridge, hence, the less wood, the more bass response. (Placing a mute on top of the bridge increases the mass and thus attenuates the sound.) In contrast, a smaller mass of wood in the bridge does not impede the high frequency response as much. For similar reasons, violins made with thick wood throughout tend to have very poor bass response. The exact shape of the bridge (especially the height) is often cut to match the tastes of individual performers. Relatively small changes in the mass and shape of the bridge can have large effects on the sound produced. For example, the two sets of acoustic spectra shown in Fig. 5.19 were obtained from bowed open strings on the same Iizuka violin played by the same violinist (Paul Arnold) before and after about 0.2 g were removed from sides of the bridge. The effect was not huge, but was certainly noticeable in the spectra and sound of the instrument.

5.10 Bass Bar

Sacconi remarked that every Stradivari instrument he had ever worked on had had the bass bar replaced. He went on to say that the bass bar actually wears out after some 5–20 years of playing and has to be replaced for that reason. Surprised by that comment and being aware that removal of the top plate to get at the bass bar was an inherently risky and delicate operation, I sampled the opinion of two other contemporary experts on that matter:

According to one New York luthier, the main thing that needs adjustment from time-to-time is the sound post, whose position depends critically on the position and size of the bridge and its location with respect to the bass bar. He felt that many instruments, including very old ones, never need to have the bass bars replaced. Sometimes they are replaced when the instrument doesn't sound as good as it might to the owner or when there is some warpage of the top plate. At other times it might be replaced when it is found to be improperly positioned. For example, some instruments became severely stressed in changing from gut to modern higher tension strings and may have required a new or stronger bass bar for that reason (Fig. 5.20).

Instrument maker Hiroshi Iizuka agrees with the notion of replacing the bass bar under some circumstances. In new instruments (as opposed to the repair of old violins) he has replaced the bass bar after it had been in for a while with an improvement in sound quality. Similarly, it makes sense to replace the bass bar in old instruments that have too large or too small a bass bar. He notes that there is also controversy regarding whether the bass bar should be installed under tension. In an old instrument he would not fit a new bass bar under tension unless the arching is sagging and seems to require additional support. He does install the bass bar with a little tension in new instruments in order to counteract the downward pressure from the string tension through the bridge against the arching of the top plate and to give some extra "punch" to the sound.

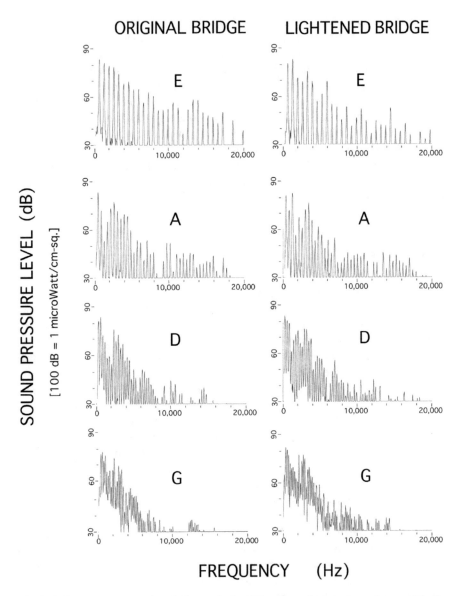

Fig. 5.19 Spectra on open strings before and after 0.2 g of wood were removed symmetrically from the two sides of the bridge. Note that the acoustic spectra shown here for the lightened bridge are quite different from the vibration spectra for the same bridge and violin shown in Fig. 5.14

Fig. 5.20 Under-side of a
1709 Stradivari top plate
showing the location of the
bass bar (long piece of wood
at the bottom) and of the
sound post. (black circle at
the top) (after Sacconi 2000)

5.11 Violin Construction

The comment is occasionally made that physicists tend to over-simplify the design
of instruments such as the violin in order to provide models that are mathematically
tractable—a process that at least dates to Felix Savart of the Biot-Savart Law in
electromagnetic theory. (Savart worked closely with Vuillaume and supposedly took
apart several dozen Stradivari instruments to see what made them work.) At the
other extreme, the makers of fine violins are generally much too busy to bother
explaining what they are really doing and primarily employ empirical methods
to improve the quality of their instruments. The philosophy of the present author
is to try to describe the actual physical properties of the best instruments, rather
than to emphasize very approximate physical models. For that reason, the main
emphasis here will be on instruments of the Cremona School, or at least those
modeled after the Cremonese instruments because they seem universally regarded
by musicians as the best ones made. That is not to say that there weren't great
violin makers that arose in other countries—for example, the Austrian genius Jacob
Steiner (1621–1683), a contemporary of Nicolò Amati who created the Austro-
German School of violin making, Matthias Klotz (1656–1743) who founded the
violin-making industry in Mittenwald, and Jean Baptiste Vuillaume (1798–1875)
of Paris, the most prolific violin maker of all time. It is estimated that Vuillaume
made about 5000 violins in his lifetime. (Of course, he had a large team of helpers.)
Fortunately, I developed a very helpful friendship with an expert violin and viola
maker from Narberth, Pennsylvania named Hiroshi Iizuka,[36] who apprenticed with

---

[36] After an initial apprenticeship in Tokyo during the 1970s, Hiroshi Iizuka spent some 4 years as
an apprentice to a well-known Yugoslavian expert named Josef Kantuscher, who had settled in
Germany. After his years as an apprentice, Iizuka moved to Narberth, PA where he set up his shop.
He brought with him a lifetime supply of Balkan maple and Dolomite spruce and has made some
265 instruments over the past 25 years. Michael Tree told me that he thought Iizuka's violas were
just as good as any of the famous Italian instruments. He feels that young musicians make a bad
mistake by going into debt to buy over-priced old instruments when much cheaper and equally
good ones are being made by contemporary craftsmen.

Fig. 5.21 Luthier Hiroshi Iizuka in his workshop

Josef Kantuscher in Mittenwald, Germany and was very familiar with the traditional techniques of the great Italian masters (Fig. 5.21).

Iizuka, himself, has supplied his instruments to such people as violists Michael Tree of the Guarneri Quartet and Steve Ansell Principal Violist of the Boston Symphony and of the Muir Quartet, not to mention members of the Philadelphia Orchestra and the Berlin Philharmonic. He is perhaps best known for his original design of a very powerful viola of especially rich and warm tone quality. His violas are of a symmetric shape (about the long axis of the instrument) and are designed to make it easier for the player to reach the fingerboard. They contain an indentation where the instrument rests near the neck of the performer and have a rounded trapezoidal shape where the fingerboard begins. Nevertheless they contain about the same air volume (hence primary air resonance) as a more conventional large viola. His design is slightly similar in those features to the Italian viola da braccio of the baroque period, but is of a more modern appearance. He also makes violins and an occasional cello. Among the musicians from the Philadelphia Orchestra who use his instruments is violinist Paul Arnold, who demonstrated both an Iizuka violin and a viola for me during research in preparation for the present chapter. In Iizuka's shop, Michael Tree of the Guarneri Quartet was able to produce an astounding sound level of 85–90 dB on the C string of one of Iizuka's violas! (Normally, the viola is the weakest sounding instrument of a string quartet.) (Figs. 5.22 and 5.23).

Fig. 5.22 An especially powerful viola designed and made by Hiroshi Iizuka

Fig. 5.23 Spectrum from open C on one of Iizuka's violas played by Michael Tree. Going up the scale produced about 90 dB on the resonance near F#

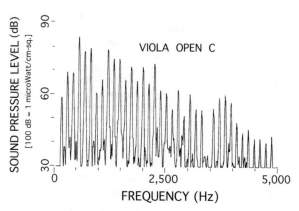

5.12 The Role of Wood

The first requirement for a good instrument is carefully chosen and well-seasoned wood. Stradivari, like most of the other makers from Cremona, had an open shed to permit drying the wood (generally for periods of at least 3 years and in some cases over 20 years) on top of his house. Later experiments using artificially applied heat to accelerate drying carried out by Vuillaume turned out disastrously: the wood continued to change dimensions long after the violin had been made. According to Heron-Allen, others who tried steaming, steeping, stoving, boiling and baking the

Fig. 5.24 Supply of Dolomite spruce and Balkan maple Iizuka brought with him from Europe some 25 years ago. The well-seasoned wood he uses is similar to that employed by the great Cremonese luthiers. (Photograph by the author.) Grissino-Mayer and Burckle (2003) recently suggested that the Cremonese makers might have been assisted by the "Little Ice Age" that gripped Europe from the mid-1400s for the next 400 years and slowed tree growth generally, producing spruce with unusually dense spacing. This "mini ice age" reached its coldest point from 1645 to 1715, an era known as the "Maunder Minimum" (after the astronomer E.W. Maunder, who documented the lack of solar activity during that period). That period was coincident with the lifetimes of Andrea Amati and Antonio Stradivari

timber, in addition to treating it with spirits, caustics and acids also experienced poor results compared to the natural aging approach. Iizuka, in the fashion of the famous makers from Cremona, stores quarter-cut samples of wood in wedge-shape form of the dimensions required to economize on space. (See Fig. 5.24.) No basic improvements in the material for the violin body have been discovered since the time of Andrea Amati, although different makers have tried an astounding variety of things (e.g., leather, paper mâché, copper, silver, steel, glass, fiberglass and even concrete!)—generally with very poor results. Of the various shapes tried, ranging from pear-shaped violins through violins with a trumpet-like horn protruding from the air chamber to the flat trapezoidal model with straight "f-holes" designed by physicist Félix Savart,[37] the basic original design by Andrea Amati has prevailed.

---

[37]Plans of the Savart violin are given by Heron-Allen (1855), pp. 117–120. The instrument looks more like a balakaika than a violin. Savart's reasoning was that a flat top-plate would radiate sound

As discovered by Andrea Amati, wood with fairly uniform, closely-spaced and symmetric grain spacing running in the long dimension is best for the top plate. As with many of the Cremona masters such as Andrea and Nicolò Amati, Stradivari and Guarneri del Gesu, Iizuka usually uses quarter-cut (radial-cut) spruce for the top plate (or belly) which gives a highly symmetric grain pattern when opened up like a butterfly's wings and Balkan maple for the bottom plate. The Hill brothers (1902, p. 161) note that neither Nicolò Amati nor Stradivari were able to afford the more exotic Balkan maple until late in their careers. However, according to Farga, Andrea Amati was well-off enough financially to permit using Balkan maple, which was supplied by the Turks and available on the Venice market, early in his career. It is said that the Turks unloaded a lot of this maple on that market with the malicious hope that it would be used to make oars for warships; the strong "flaming" resulted in weak pieces of wood that would easily break under stress. Luthier Iizuka (private communication) has found that violas and cellos made with backs having strong flaming tend to be slightly deficient in low-frequency response. Maple weakened in this way probably is less resilient in reflecting low-frequency sound waves. Although the flaming marks in maple are often described as the result of variable light reflection, there are palpable indentations in the wood where the pattern occurs, and the flaming bands often produce bumps in the bark (The phenomenon is not confined to maple, but also is found occasionally in other woods such as ash and oak.)

Flaming (alias "curly," "tiger-striped," or "wavy-grain" figure) results from waves in the grain direction at nearly right angles to the longitudinal axis of the board. Split faces will show these waves on the radial face and usually on the tangential face as well. When the grain corrugations are close and abrupt, the resulting pattern is called "fiddleback figure," because of the common use of such figured maple in violin backs. "Flaming maple" is of the species known as pseudo plantanus. Most experts I consulted were unable to explain why the phenomenon occurs. The most believable suggestion is that these marks are induced by strain when a tree bends in the wind. Leaf growth results in increased wind resistance and would provide an annual periodicity in the strain. Voichita Bucur (C. Hutchins, private communication) has found a genetic aspect to the effect: seeds from maple trees showing the effect result in trees that also have a flaming pattern.

Although Loen[38] is skeptical of this argument, one would nevertheless expect that slower growth of the wood would produce spruce similar to that only found

---

more efficiently than a curved one. Further, with the flat geometry, the "f-holes" no longer need to be curved and could be made without cutting the long fibers in the wood that he thought tended to vibrate more efficiently at low frequencies. The violin was also cheaper and easier to make. Although a panel of "experts" (including Savart's colleague Biot) thought it compared favorably with a Cremona masterpiece, most audiences disagreed.

[38]Loen (2004). Also see Schneider 2004, and Kolbert 2005.

Fig. 5.25 Close grain structure in the top plate of a viola made by Hiroshi Iizuka of 30-year-old quarter-cut dolomite spruce from the Italian Alps. See Figs. 5.26 and 5.27 for the method used to cut the wood (photograph by the author)

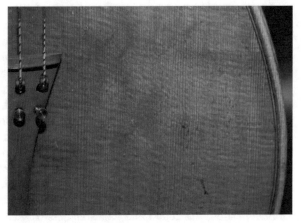

Fig. 5.26 Quarter-cut Dolomite spruce and Balkan maple and slab-cut Balkan maple. The grain is vertical in each case, although the "flaming" in the maple samples is nearly horizontal (photograph by the author)

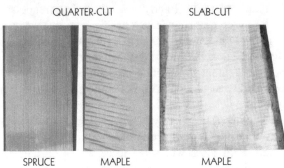

now in the mountainous areas. (See Fig. 5.25 for an illustration of the fine grain structure in such a sample of dolomite spruce.) Sunspot activity (indicative of solar emission) was typically under 20 per year in the fifteenth century, whereas it went up to about 75 per year during the twentieth century. Of course, the more recent global warming is also dependent on complex interactions of solar radiation with gases in the atmosphere (Fig. 5.26).

The advantages of quarter-cut (or wedge-shaped segments) in providing symmetric grain patterns parallel to the long dimension of the violin will be evident from the left side of Fig. 5.27. As shown at the left, quarter-cut (radial, wedge-shaped segments) are cut from the tree and then sliced in half radially to provide the top and (often) the bottom plate. The outer (bark) sections are then planed flat and perpendicular to the radius, and the two pieces are opened as a butterfly would open its wings. The outer (thicker) portions are then glued together, providing a highly symmetric grain pattern about the radial mid-line and running parallel to the long dimension of the instrument. With slab-cut maple, shown at the right, the entire bottom plate is laid out on a single original piece of wood, again with the grain running vertically. The "flaming" from annual growth in tree height, which

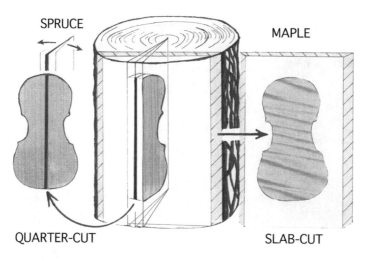

SPRUCE

MAPLE

QUARTER-CUT

SLAB-CUT

Fig. 5.27 Portions of a tree used to provide quarter-cut and slab-cut wood

makes the maple bottom plate so attractive, is in a slanting horizontal direction. The old Italian masters were fond of using slab-cut maple for the backs of violins and violas when suitably large pieces of wood could be found. As will be evident from the geometry involved in Fig. 5.26, it is much easier to produce a bottom plate of large width by joining two pieces of wood back-to-back, a technique often used by the old masters.

Annual tree growth occurs in a series of conical shells that are coaxial in the vertical direction. (See Fig. 5.28.) The outer edge of each shell intersects a tree ring if a lateral section is cut through the tree near the base. These shells mark the outer boundary for each year's successive growth of the tree. Sawing vertically in the radial direction exposes wood in which the grain runs in nearly straight vertical lines from the ring pattern produced along the edges of the growth cones. However, a slab cut perpendicularly to the radius near the bottom of the tree and slanting away from the vertical axis intersects the growth-cone surfaces in quasi-horizontal lines called "flaming"—the colorful horizontal bands that cross the conic sections and are related to vertical tree growth. In the case of contemporary Balkan maple, these bands can be ≈ 3/8 in. wide and separated by ≈ 3/4 in. The flaming pattern is most striking in red maple.[39]

---

[39]Illustrations of grain patterns obtained by numerous different saw cuts are given by Berlyn and Richardson (2001).

Fig. 5.28 Growth patterns of
a tree. The numbers represent
successive years of growth
(after Robbins and Weier
1950)

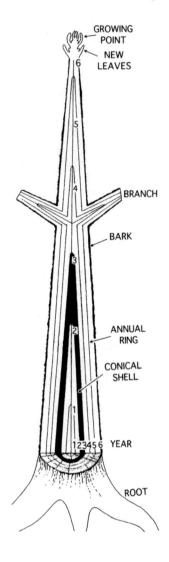

5.13   The Top and Bottom Plates

As with harpsichord and piano sound boards, spruce provides a suitably flexible
membrane to transform the string vibrations into motion of the top plate perpendicular to its surface. The close-spacing of the fiber also results in vibrational mode
patterns in the top plate that are elongated in the fiber direction. (As discussed
before, the velocity of sound in the wood is faster by a factor of about three in
the fiber direction than perpendicular to it; see Table 1.2 of Chap. 1.) It is probable
that this results in more efficient coupling of plate vibrations to the air because of
the narrow, elongated nature of the instrument. Maple was generally chosen for the

back plate in violins and violas because it is a hard wood, reflecting sound back into the violin cavity efficiently, and because it provides decorative beauty through the flaming marks in its texture. (Vibrations in the back plate are also coupled from the top plate through the sound post.[40]) The great Cremonese makers preferred maple from the Balkans (near Sarajevo) because it was harder, lighter, and more resilient in reflecting sound than that grown in western Europe. [See Sacconi 2000, p. 56.]

In making the top and bottom plates, only hand tools (chisels, gouges, planes, knives, and scrapers) are used for the initial shaping (Fig. 5.29). Stradivari used old sabers to cut with because of the high quality of their steel. For finer shaping, thin sharp-edged blades are employed that are bent at about 45° at the honed edges to provide a burr used to cut the wood fibers. The object is to cut the fibers in the direction of the grain without injuring them laterally. (The technique is very similar to that used by oboists in shaping their reeds.) Although the task of reducing a nearly half-inch thick slab of maple to a curved surface about 5 mm thick sounds Herculean, Iizuka says that the rough shaping can be done with sharp gouges in about 2 h. But smoothing out the surface takes another 6 h. (The top plate made of spruce is much softer and takes far less time.) Finer finish of the top and bottom plates was obtained by Stradivari and the other Cremonese craftsmen by using the sharp ridges in dogfish skin and a kind of horsetail "grass" (asprella or equisetum) that grew along the rivers and canals in the outskirts of Cremona. Botanists note that Horsetail is closely related to ferns.[41] (It reproduces with spores rather than seeds.) The samples I've seen consist of segmented hollow stems 1–3 in. long between joints and about 0.35 in. in diameter with ridges running lengthwise that have abrasive qualities. Slitting the horsetail lengthwise permits unrolling it into flat strips that may then be used like fine emery paper in the fiber direction without the danger of bruising the wood fibers laterally. It's scientific name is Equisetum hyemale, but is sometimes called "scouring rush." (Early American settlers used it to clean their cooking utensils.) Horsetail looks identical to the plant material that some clarinetists have erroneously called "rush"—an entity that could be picked along the railroad tracks bordering swampy areas in the New Jersey meadows. ("Rush" used to be sold at Linx and Long's Woodwind Store on 48th Street in New York City for about 25 cents a piece.) But real rush ("Juncus") does not have joints or a rough exterior and is a seed plant. Horsetail is an evergreen and, ironically, is poisonous to horses. The plant is easy to grow in almost any shady, moist environment. To use it for polishing violins, Iizuka soaks the horsetail in water and then slits it lengthwise and unrolls the plant into flat strips. After drying, he glues the back to paper.

According to Sacconi (2000), Stradivari also immersed both the dogfish skin and the horsetail in water and then dried the pieces in cotton cloth before use. Iizuka uses

---

[40]This interpretation, given by William Huggins in his 1883 paper (see Helmholtz 1885, p. 86 footnote), is a slight oversimplification. As shown by Savart, a major role of the sound post is in defining the nodes for top plate resonances. (See later discussion of top-plate modes.).

[41]I am indebted to botanist Dr. Frone Eisenstadt for a helpful discussion of the properties of "horsetail."

Fig. 5.29 Top (upper) and bottom (lower) plates for a violin being carved by Hiroshi Iizuka. Note the flaming in the bottom plate (photograph courtesy of Iizuka)

shark skin as well as "horsetail" for the same purpose. The expert makers never use sandpaper because the random orientation of the cutting particles in that medium would tear the wood fibers laterally rather than cutting them cleanly in the fiber direction.

Sample of horsetail, alias "scouring rush"

The polished surfaces of curvature in the top and bottom plates are obtained entirely by hand and the final adjustments of each plate are made by listening to the "tap tones" produced when the plate is rapped by the knuckles or tapped by the fingers of the craftsman. According to Faber (p. 114), Felix Savart concluded from the Vuillaume dissections that the dominant tap tone on the back plate of a Strad was always a semitone higher than the one on the top plate. One way the resonant frequencies of a plate were determined was by supporting it at the intersection of two nodal lines—for example, by resting the plate on a piece of cork at that point and pressing down on it with a finger while tapping or bowing at the edges. The resonant frequencies were then adjusted by varying the thickness of the plate. Those of a free plate will be rather different from those from a plate that is clamped along the edges. However, the rib section is generally made of very thin wood that probably does not serve as a very effective clamp for the plate vibrations. Nevertheless, violin makers also note that very poor tone quality results when the rib section is made of thick wood.

It is natural to suspect that the shape and thickness of the top and bottom plates have a large effect on the tone quality of an instrument. According to Sacconi (2000), who had measured numerous instruments by Stradivari, the belly (top plate) thicknesses were typically about 2.3–2.5 mm and did not vary appreciably over the curved surface. The center portion between the f-holes is flattened where the feet of the bridge are positioned. As discussed before, the bass bar is glued to the underside of the top plate passing below the bass foot of the bridge. Animal glue (or "hide glue") is used on all glued joints to permit taking the instrument apart for later repair.

5.14   Thickness Graduation Maps

Jeffrey Loen (2003, 2005) has determined thickness maps for the top and bottom plates of a very large number of violins by Stradivari and other makers. Although Sacconi presented some data of this nature, it was seldom clear to what instruments they referred. The thickness graduation measurements reported by ? represent the most comprehensive results currently available. This tome contains data taken by Loen and his colleagues over a period of 5 years. It is displayed in a series of computer-generated contour maps in color that are particularly easy to read. The measurements were made using a magnetic gauge (Hacklinger Gauge) in which a small counter magnet is moved about inside the instrument and attracted by a magnet on a spring with a calibrated scale. Although not as accurate in principle as a mechanical caliper, it is more practical to use with instruments that cannot be taken apart. Loen estimated his errors in measurement were about 0.1 mm over the range from 1.0 to 4.0 mm and about 0.3 mm for greater than 4.0 mm thicknesses. He checked the accuracy of his gauges occasionally using wood strips of known thickness. His measurements were with a contour interval of 0.25 mm and the plates were displayed as viewed from outside the surface. An example of his technique is shown in Fig. 5.30 for both a violin and a cello made by Stradivari in 1697.

Stradivari's early violins had spruce top plates with thicknesses in the order of 2.4 mm (about a tenth of an inch!), with relatively little variation over the surface. In his early period, he stuck to a symmetric arrangement of the thickness in the top and

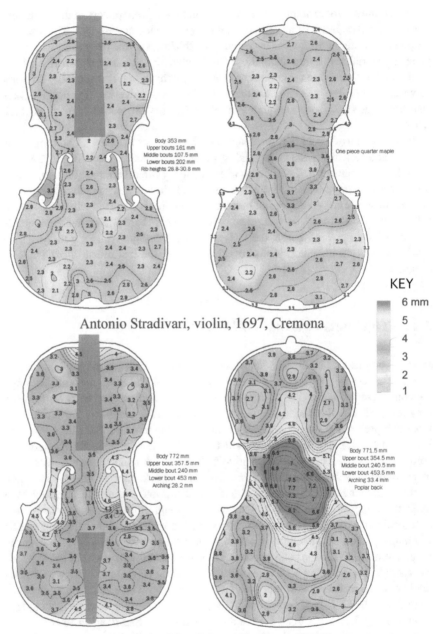

Antonio Stradivari, violin, 1697, Cremona

Antonio Stradivari, the "Castlebarco" cello, 1697, Cremona

Fig. 5.30 Thickness maps for a violin and cello made by Stradivari. Reproduced from Loen (2005, p. 28) by permission

bottom plates of the violin and made violins in the style of Andrea Amati.[42] The maple back plates in early Stradivari instruments were thickest in the center region between the two f -holes (typically, in the order of 4 to 5 mm), but in the "Golden Period" (starting after 1700) there usually was a strong asymmetry. Measurements of later "Golden Age" Stradivari violins by Loen (2002, 2003) showed a distinctive "bull's eye" pattern on the back, with the thickest portion off-axis (under the bass bar) in contrast to the longitudinal "back-bone" structure of many other makers. In the period of the "Messiah," small regions of the top plate were thinned out deliberately as if Stradivari were trying to enhance particular vibrational modes.

Other parts of the violin are made from different woods. Generally, Stradivari carved the scroll, neck, and peg box (which require unusual strength) out of one piece of quarter-cut maple. The ribs on violins were always of maple (often slab-cut showing flaming lines), but for violas and cellos Stradivari also used poplar and willow. The ribs had a thickness of about 1 mm and were bent to shape using dry heat (almost to the point of burning) with a curved iron. Stradivari placed a leaf of lead on top of the wood and bent it together with the heated wood to protect it. (He didn't have to worry about the Environmental Protection Agency.) The ribs were then glued to the six (willow or spruce) blocks of wood, using a mold to position them accurately. (Rib liners made of 2×8 mm strips of willow or spruce were glued to the ribs after the shaping of the belly and back had been completed.) The bass bar was made of spruce with an annual ring spacing of about 1 mm and positioned to pass under the left foot of the bridge (made from maple) with its external edge about 1 mm inside the foot of the bridge. In order to keep the instrument light and reduce damping of the resonances Stradivari used willow for the fingerboard, although he covered it with a veneer of ebony. Of course, as the Hill brothers have pointed out, not all Stradivari instruments are good.

5.15 Vibrational Modes and Resonances

5.15.1 Modes of a Thin Membrane

As a first approximation to the vibrational modes of a violin top plate, one might simply regard the plate as a thin membrane. The simplest case to treat analytically would be a rectangular shape clamped along the edges. As shown in Appendix B, the resonant modes in that case take the form

$$Z(x, y, t) = \sin(n_x \pi x / L_x) \sin(n_y \pi y / L_y) \cos(\omega_{n_x, n_y} t) , \qquad (5.5)$$

---

[42]Other measurements for an enormous number of Stradivari instruments are contained in Appendices III–V of the Hill brothers (1902) biography.

where $Z(x, y, t)$ is the amplitude of the vibration at the point x, y at time t. The basic shape of the modes is seen by letting $t = 0$ in Eq. (5.5). The expression gives rise to a whole family of vibrational modes that are dependent on the integers n_x, n_y. Several of these are shown in Fig. 5.31. Here, the angular frequency ω ($= 2\pi f$, where f is the cyclic frequency) of a particular mode is determined by the integers n_x, n_y which take on the values, 1,2,3,..., and by the tensions (or stiffness) and lengths of the membrane in the x and y direction, L_x, L_y. Even for equal tension in the two directions, the frequency has a somewhat complex dependence on the above quantities given by

$$\omega = \pi c \sqrt{\left(\frac{n_x}{L_x}\right)^2 + \left(\frac{n_y}{L_y}\right)^2} \qquad (5.6)$$

where c is the surface wave velocity (not to be confused the velocity of sound in air). Each mode has a different frequency dependent on the two integers n_x, n_y, the two lengths L_x, L_y, and the running wave velocity, c. (See Appendix B for the derivation of these equations.) It is easily seen from Eq. (5.6) that these modes are not harmonically related unless very special values of the parameters are chosen.

The modes shown in Fig. 5.31 would be similar to those in the top plate vibrations of a cigar-box instrument—say, one in which a piece of paper was glued to the top edges under tension. In a square cigar box, the 1,1 mode (having equal integers in the x and y directions) would tend to radiate equally in all directions. However, in an actual violin that mode would be substantially distorted by the effects of the off-center bass bar and sound post, not to mention the differences in wave velocity in the membrane between the two directions. With a closed cigar box, the 2,1 mode would tend to be dominant because of its odd symmetry and because it would produce the least resistance against compressing air in the volume below the top plate. As

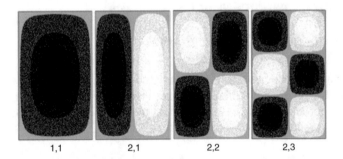

1,1   2,1   2,2   2,3

Fig. 5.31 Representative modes of vibration in a thin rectangular membrane somewhat related to those of a violin top plate. The numbers below each mode pattern correspond to values of the integers n_x, n_y, in the x and y directions. The black and white portions of the figures correspond to out-of-phase vibration. For example, if black stood for a positive amplitude up out of the page, white would represent a negative amplitude going down into the page. The 1,1 mode amplitude alternates between vibrating upward above the plate and downward below the plate

one side went up, the other would go down, resulting in no net air compression. It is analogous to the sloshing mode for liquid in a teacup. As well as being the dominant mode in the system, the 2,1 mode would give rise to radiation approximating that from a dipole and, hence, would tend to carry farther in directions perpendicular to the long axis of the instrument. The 2,2 mode would give rise to quadrupole radiation, and so on. (See the discussion of radiation patterns in Chap. 1 and the section with that title in this chapter.)

5.15.2 Modes in Violin Top and Bottom Plates

The modes in a real violin top plate are substantially different due to the curved boundaries, the difference in wave velocity in the two orthogonal directions, the outward arching, and the thickness variation in the plate. The higher wave velocity in the fiber direction adds further to the complexity and the "thin" membrane approximation would also tend to fail. It is obviously more practical to measure the modes in a real violin top plate than to try to calculate them.

E.F.F. Chladni (1809) devised a technique for studying vibrational modes in plates. By sprinkling sand on them and then bowing the plate at the edges, the sand would bounce around on the plate until it landed on non-vibrating nodes. In that way, the nodal lines surrounding the amplitude maxima or minima in a vibrating mode were outlined. Studies of vibrational modes in violin top plates were made by a large number of physicists, ranging form the early work of Felix Savart on Stradivari instruments using sand to illustrate Chladni patterns in the vibrating wood (see Savart 1819) through that of Carleen Hutchins (1981) in which she suspended the top or bottom plate over a loudspeaker hooked to an audio oscillator and sprinkled the plate surface with small aluminum particles ("sprinkle") so that the nodal lines could be easily seen and photographed. However, this approach is only practical for studying the inside patterns on these plates. (Because the top and bottom plates are convex outwards, the particles would roll off the outside surfaces during excitation of the normal modes.)

Experimental studies of surface vibration modes generally tend to map out the nodal lines between maxima and minima in the vibrations, rather than the peaks and valleys. Hence, they do not usually show the phase relations between different portions of the vibrating surface contained in Fig. 5.31. One, of course, can infer what those relations must be in many cases. Most studies have been made using freely suspended top and bottom plates. (Saunders (1998) and Hutchins 1981.)[43] Except for the early work of Savart and Vuillaume, really great violins

---

[43] See Hutchins (1983) for a review of early research on violins. A massive review of Research Papers on Violin Acoustics has been given in two volumes, compiled and edited by Hutchins and Benade (1997).

Fig. 5.32 Nodal lines for the first three violin top plate modes (left) and first three bottom plate modes observed by Carleen Hutchins (reproduced from Hutchins (1981), by permission of the author)

were not generally studied. Several representative violin top-plate and bottom-plate resonances measured by Hutchins are shown (Fig. 5.32).

The boundary conditions in a real violin are quite different from those for a freely suspended plate. The top plate is glued to the rib structure and that in turn is glued to a relatively stiff bottom plate. In addition to the bass bar, the assembled violin has a sound post that forces a node in the surface vibrations where it is in contact with the top plate and bottom plate of the instrument. (Savart concluded that the main acoustic role of the sound post was to force a vibrational node at its location—hence to select particular modes of vibration—rather than just to transmit vibrations from the top plate to the back plate as had previously been assumed.)

There are numerous resonant modes in the wood of the assembled violin, viola and cello in addition to those in the top and bottom plates and rib structure. Some are a little surprising to the novice, such as the bending mode of the tail piece and fingerboard about an axis through the end of the main body of the instrument. Since there is a tendency for the fingerboard and neck to bend back under tension from the strings, stability of the instrument is improved by making the neck surface a bit convex (instead of flat) and to make the fingerboard gluing surface to fit. Of course, the use of well-seasoned wood is extremely important.

5.16 Air Cavity Resonances

5.16.1 The A_0 Resonance

It used to be thought that there was only one important air resonance in the violin (often labelled "A_0") and that this so-called "breathing mode" resembled a Helmholtz volume resonance coupled to the outside world through the f-holes. Sacconi (2000) reported that this air resonance was tuned to B below middle C (about 247 Hz) in all 350 instruments by Stradivari that he examined. (Of course, one should note that A was lower in pitch than 440 Hz by about 5 Hz in the time of Sacconi and by nearly 30 Hz in the time of Stradivari.) Some modern instrument makers monitor that mode by blowing across an f-hole as they would across the mouth of a flute or by humming into the instrument.[44] If one is to treat the A_0 air resonance of the violin or cello as a Helmholtz resonator, it is necessary to know the volume of the instrument and the area of the f-holes in order to calculate the resonant frequency. Although it is a straightforward matter to determine the area numerically, how do you determine the volume without destroying the instrument? Certainly filling it with water is out of the question So I asked my luthier friend Hiroshi Iizuka the question. His reply was Confucius says, "Use our precious rice!" He went on to say that violin makers actually do use rice to clean the insides of their violins.[45] Of course, the other difficulty is in determining the length L of the effective cylindrical neck of the equivalent Helmholtz resonator. The thickness of the top plate around the f-holes of a Stradivari violin is typically only about 2.5 mm. As a very rough approximation, we could simply take the entrance cylinder of the equivalent Helmholtz resonator to have a length $L \approx 2.5$ mm, and an area equal to that of the two f-holes combined. I calculated the area of the f-holes and of the top and bottom plates for a typical 1700-vintage Stradivari violin by weighing pieces of heavy cardboard that had been cut out from enlarged tracings of the scale drawings given by Sacconi (2000). This procedure gave a combined area for the two f-holes of $A \approx 9.2$ cm^2 and a total air volume in the cavity of $V_0 \approx 2320$ cm^3. (Using the "rice method," Iizuka found a volume of about 2 L in one of his violins.) From Eq. (A.68) of Appendix A, we might then expect the frequency of dominant A_0 air resonance to be

$$f_0 = \frac{c}{2\pi} \sqrt{\frac{A}{L V_0}}$$

where c is the velocity of sound. Substituting my estimated values for A, L, and V_0 with $c \approx 34,400$ cm/s (from Table 1.2 of Chap. 1) gives a resonant frequency of about 689 Hz. That result is too high by a factor of about three from the Sacconi's measured values of the dominant air resonance in Stradivari violins. The approximation for the effective length of the entrance cylinder of the Helmholtz resonator is probably the main source of error, however the complex shape of the volume is no doubt also involved. The length $L \approx 2.5$ mm (based on the thickness of the top plate) is seemingly too small by a factor of nearly eight! The formula for the Helmholtz resonance is apparently only of semi-quantitative value in the case of the violin. Increasing the volume should bring the resonance down and increasing

---

[44] See Carolyn Wilson Field, "Tuning the B0 Mode in Four New Violins," in the Journal of the Catgut Acoustical Society Retrospective Issue (Op. Cit., 2001), pp. 60–61. In that same issue (pp. 57–58), Carleen Hutchins describes the advantage of adjusting the A_0 air and B_0 wood modes to the same frequency.

[45] He added that you wouldn't believe how dirty the inside of a violin can sometimes be. If you're a good detective, you can tell the lifestyle of the owner, or even tell where the musician has traveled. Upon taking old instruments apart, you sometimes find large balls of fuzz that had rolled around inside the instrument gathering dust and dirt.

the area of the f-holes should bring it up. However, the frequency varies very slowly with these quantities. The A_0 mode corresponds to one in which the top and bottom plates move in opposite directions, resulting from the expansion and contraction of the air volume. Hutchins (1962) found empirically that a 20% decrease in air volume or a 59% increase in f-hole area would raise the actual enclosed air volume resonant frequency by a whole tone. Hence, a pragmatic procedure in tuning a violin seems to be to measure the actual resonant frequency and then change the parameters as necessary in accordance with Hutchins' prescription. In practice, you probably wouldn't want to alter the delicate shape of the f-holes, so one would be left with volume-altering adjustments.

5.16.2 *A₁ Resonance*

In contrast to the Helmholtz resonator model, one can treat the violin cavity as an organ pipe, closed at both ends. Its inside length ($L \approx 31.73$ cm) in the long direction between the flattened end supports gives an answer that is fairly close to measured values for the second air resonance found in a violin, often known as the "A_1 resonance." From the discussion in Chap. 1, this fundamental resonance should be given by[46]

$$f_1 = c/2L \approx 542\,\text{Hz}\,,$$

for the 1700-vintage Stradivari violin assumed above. I measured the first two air resonances on several violins supplied by luthier Hiroshi Iizuka. These values were determined by coupling the output of a small loudspeaker driven from a variable frequency oscillator through a very short (about 3 in.) narrow diameter rubber tube into one f-hole of the violin while another short rubber tube went from above the opposite f-hole to a condensor microphone. (It is rather important to keep the rubber tubing short in order to avoid the effects of organ pipe resonances in the tubing itself at the frequencies involved.) In two cases, the sound was introduced through a small hole at the front end of the violin. The results in Table 5.1 represent the average of several measurements with variations in the peak locations of about 2 or 3 Hz. All the instruments used were strung with sound posts installed and with the strings under normal tension. The Q's were usually much higher for the A_1 resonance than for the lower one. However, the A_0 resonance was generally the strongest and in the case of the Iizuka viola was some 55 dB louder than the A_1 resonance when the sound was introduced through an f-hole. The relative intensities varied with the method of excitation. Large differences in relative excitation were noticed when the tube from the loudspeaker was simply placed against the bridge of the instrument rather than introduced inside the cavity. By placing one microphone above each of

---

[46]Using the method of phase shift analysis discussed in Chap. 1 and noting that "hard reflections" are present at both ends of the doubly-closed pipe, it is seen that the resonance in this case is of exactly the same form as that for an open pipe of the same length.

Table 5.1   Air resonances measured by the author

Instrument	A_0 (Hz)	Q	A_1 (Hz)	Q
Eastern European Violin (14.25 in.)	265	≈ 7	475	≈ 16
Chinese Factory Violin	273	≈ 15	478	≈ 16
Sound fed into f-hole				
Sound fed into end hole (13.9 in.)	270	≈ 13	472	≈ 25
Iizuka Amati-model	266	≈ 9	488	≈ 30
(13.9 in., high arch)				
Iizuka Guarneri-model (13.9 in.)	260	≈ 16	461	≈ 58
Iizuka Viola (16.75 in.)	211	≈ 18	476	≈ 48
(Sound fed in end hole)				

the two f-holes and looking at their output voltages on a two-beam oscilloscope, it was clear that the phase of the sound was the same from each f-hole. In contrast, an out-of-phase relationship should exist for some of the higher-order modes reported by Jansson. (See below.) Measurements made on several instruments that were not strung, were not under tension, and did not have sound posts, exhibited numerous resonances that probably arose from top plate vibrations.

5.16.3   Higher-Order Air Resonances

Studies by Jansson (1973) found the existence of an entire family of high-Q, higher-frequency air modes.[47] He examined these air resonances in a violin-like wooden cavity that was encased in plaster to suppress vibrations of the wooden structure. (For practical simplicity, the cavity had a flat top and back.) He further measured these air resonances with the f-holes open and then closed. Although the f-holes have a very small area (about 1%) compared to the entire surface area of the instrument, he found that they played an important role in suppressing some of the higher frequency modes in the same way that drilling a hole in the middle of an open organ pipe suppresses the fundamental mode in a harmonic flute. The main coupling of energy from the higher-order air modes to the outside world in a real violin with open f-holes occurs through vibration of the violin surfaces. Of course, these vibrating surfaces are what excite those air resonances in the first place when the violin is played. The f-holes seem either to suppress the higher frequency resonances altogether or not to affect them very much. Jansson's results (summarized in Table 5.2) were computed from Figure III-C-2 of his paper.[48]

---

[47] See, for example, "A Retrospective on Air and Wood Modes," edited by Jeffrey S. Loen, Catgut Acoustical Society Journal, vol. 4, No. 3 (Series II), May, 2001.

[48] See the reproduction of Janson's article " On Higher Air Modes in the Violin" in the Catgut Journal retrospective issue, May, 2001 (op. cit.)

Table 5.2 Data for air resonances from Erik Jansson (1973)

Air mode	f-holes open		f-holes closed	
	Frequency (Hz)	Q	Frequency (Hz)	Q
0	294	15	–	–
1	487	55	488	75
2	1088	75	1050	106
3	1196	17	1112	81
4	1285	115	1312	130
5	1603	79	1562	121
6	?	?	1767	125
7	1906	106	1899	160

Comparing Tables 5.1 and 5.2, it is seen that the results for A_0 and A_1 are reasonably consistent. From Table 5.2, it is seen that opening the f-holes has relatively little effect on most of the higher frequency air modes with the exception of Jansson's modes No. 3 (where the Q dropped substantially) and No. 6 (which was extinguished completely). As he noted, those two modes are probably suppressed because the f-holes fall on pressure maxima in the mode standing wave patterns. Jansson suggested that the higher frequency air resonances he found in a violin are similar to those that would occur in an organ pipe that is closed at both ends.

As with the vibrating membrane modes, it is useful to examine what happens inside a rectangular enclosure—say, a cigar box with rigid walls. (For ease in later comparison, we will take one that has about the same length-to-width ratio as a fine Stradivarius.) Because the thickness of the "instrument" is small compared to its length and width, we will ignore variation in the mode structure in that direction and just consider two-dimensional modes formed by the walls. The basic boundary condition for determining the resonant modes in a rigid box is that the pressure waves be a maximum at the walls. Hence, the pressure modes in the cigar box should be of the form

$$P(x, y, t) \propto \cos(n_x \pi x / L_x) \cos(n_y \pi y / L_y) \cos(\omega_{n_x, n_y} t) \qquad (5.7)$$

where $P(x, y, t)$ represents the pressure difference from the ambient atmospheric pressure, $0 \leq x \leq L_x$ and $0 \leq y \leq L_y$ as in the thin membrane case, and the frequencies are determined by Eq. (5.6) except that here c is the velocity of sound in air (or other gas contained in the box). Again, $n_x, n_y = 1, 2, 3, \ldots$ The modes for $n_x = 1$ all have nodes running vertically down the middle of the box and maximum values of the pressure of opposite phase at the walls. Hence, for example, the 1,1 cigar box mode would couple efficiently to f-holes placed in the lower half of the top lid. For a perfectly rigid box (say one made of plaster or concrete), the sound coming out of the two holes in this case would be distributed approximately in a dipole radiation pattern. More complex distributions in various directions would occur for sound coupled through a top plate that was allowed to vibrate (Fig. 5.33).

In the case of a real violin with curved side walls or "bouts"[49] and a constriction at the middle (formed between the "C-bouts"), much more complex two-dimensional modes exist beyond the Helmholtz volume mode. Here, Jansson argued that the dominant ones corresponded to closed pipe resonances propagating in the long dimension of the violin. (See the data in Table 5.2. and Fig. 5.34.)

The situation may be loosely analogous to that in a generalized confocal laser cavity—one having curved mirrors at each end. In that case, low loss modes propagate in the long dimension normal to the mirror surface and (at least for wavelengths short compared to the length of the cavity—a condition that is not well-

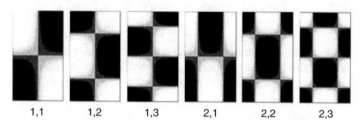

| 1,1 | 1,2 | 1,3 | 2,1 | 2,2 | 2,3 |

Fig. 5.33 The first few two-dimensional air resonance modes for a rigid cigar box. The numbers below each mode pattern represent values of n_x, n_y. As in previous illustrations of this type, the black and white areas represent those of maximum pressure magnitude, but opposite phase

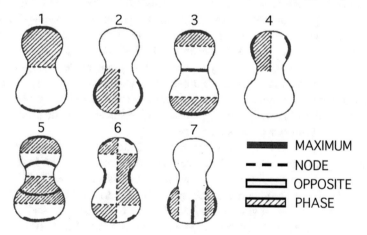

MAXIMUM
NODE
OPPOSITE
PHASE

Fig. 5.34 Jansson's determination of the higher-frequency air-mode distributions in a violin-shaped cavity. Source: Jansson (1973)

---

[49] According to the 1989 Oxford English Dictionary vol. II, p. 449, 5. A "bout" is the inward curve of a rib in a violin or similar instrument, by which the waist is formed. From the 1893 Fiddler's Handbook, 4. Bouts are the sides of the fiddle, divided into the lower, middle, and upper bouts. The term apparently comes from the obsolete word "bought" for "bending." In German, the word for "bout" is "bogen" meaning "arch," "vault," or "bow."

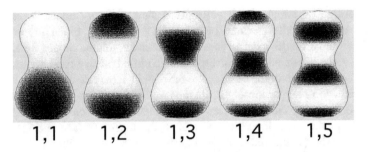

Fig. 5.35 Higher order air pressure modes in a violin assuming closed pipe propagation in the vertical direction and that the modes are independent of the side wall spacing

satisfied here), the mode distribution is independent of the sidewalls.[50] Hence, there is some justification for a closed organ pipe approximation in which one assumes longitudinal propagation in the long dimension of the violin with a distribution that merely falls off symmetrically in the lateral direction. Calculations for such mode distributions for longitudinal propagation using the proportions of a Stradivarius violin cavity are shown in Fig. 5.35. Here, the boundary condition assumed was that a pressure maximum occurred at the two ends of the violin and that a cosinusoidal sound pressure distribution occurred about the long vertical axis. The pressure distribution for Mode 1,1 matches Jansson's Mode No. 1 and Mode 1,3 matches his Mode No. 5 (apart from trivial changes in phase). However, his Mode No. 3 looks closest to the present Mode 1,4, and his mode No. 6 has a standing wave pattern similar to the 1,2 cigar box mode shown in Fig. 5.33. His remaining three higher-order modes (Nos. 2, 4, and 7) may represent closed pipe modes with propagation in the lateral direction in the separate upper and lower halves of the violin. Jansson concluded that the air modes in the upper and lower cavities of the violin are only loosely coupled, and several of the higher-order air resonances would produce sound that would be out-of-phase between the two f-holes.

However, the actual frequencies are another matter. Although Jansson does not give the length of his test model, the resonant frequencies for the higher modes should simply be of the form $n_y c/2L$ in his model, where $n_y = 1, 2, 3, \ldots$, and $c/2L \approx 487\,\mathrm{Hz}$ corresponding to his value for the A_1 mode. According to his model, the next several higher modes should then have resonances at 974, 1461, 1948, and 2435 Hz. With the possible exception of Mode 1,2 at 1088 Hz, these numbers simply do not agree very well with his measured frequencies. (Compare these numbers with those in Table 5.2).[51] In another study of a violin, Marshall (1985) found the first higher Air Modes to be at 478 and 839 Hz. Hence, there is good agreement on the location of the A_0 and A_1 mode, but not on the still higher

---

[50]See Bennett (1977, pp. 21–32).

[51]Effective values for a 1700 Stradivarius of $L_x \approx 7.59\,\mathrm{in.}$ and $L_y \approx 14\,\mathrm{in.}$ were used for the equivalent cigar box calculation. See *Appendix III* of Hill brothers (1902).

modes. It would be nice to have a more rigorous (but simple) mathematical solution to the air mode problem; however, such a calculation would involve the complex boundary shape in a real violin and probably have to be done numerically for each violin geometry. Although the closed pipe model is of doubtful quantitative utility beyond determining the first of the higher air resonances, Jansson's experimental determinations of frequency and Q for the higher air modes shown in Table 5.2 are quite useful in themselves.

5.17 Modes of the Assembled Violin

As noted before, an enormous list of assembled-violin resonances was given by Marshall (1985). Marshall found some 34 modes (including both air and wood resonances) in a violin made by Carleen Hutchins (her SUS #295), ranging from 119.5 to 1228.5 Hz. Marshall concluded that the A_1 and A2 modes had somewhat different nodal distributions than those found by Jansson. (Compare Marshall's Fig. 9 with Fig. 5.34 above.)

There are two main types of resonance present in the assembled violin: the wood resonances that result when the whole body is assembled and air resonances from the resultant enclosed cavity. As may be deduced from the changes in top plate modes as the violin is assembled, the resonances in the complete violin are very complex indeed and result from dozens of both wood and air resonances.

Frederick A. Saunders (at the Croft Acoustical Laboratory at Harvard) started a practice in the early 1930s in which electronic equipment was used to measure loudness curves from a violin. He would play the chromatic scale, bowing each semitone for four seconds, recording the amplitudes from the first ten harmonics and then combining the results to get the overall response curves. (See Hutchins 2004.) This technique was carried out further by Hutchins (1962) to investigate a variety of different violins, with results such as those shown in Fig. 5.36.

Hutchins concluded that the best violins, as well as having many strong resonances, were those that had the first air resonance (A_0) just above the open D string and the second wood resonance (W2) just below the open A string. Hence those two resonances should be about a fifth apart. In some poor instruments the main wood and air resonances were as much as 12 semitones apart. Another interesting result from this work was that the finest old instruments could produce very loud sound levels—for example, over 95 dB at one meter when bowed on the open strings. (In contrast, a poor instrument might produce sound levels 20 dB less.) The aspiring violinist should be warned that the levels produced by a fine Stradivarius are intense enough to cause substantial hearing loss according to OSHA (Occupational Science and Health Administration) standards when the exposure amounts to several hours a day. (Paganini seems to have anticipated this danger and always practiced with a mute on his Guarneri del Gesu "Cannon.")

Marshall (1985) determined some 34 separate modes in one assembled violin ranging from about 119 to 1228 Hz and was able to identify about two dozen of

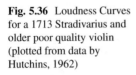

Fig. 5.36 Loudness Curves for a 1713 Stradivarius and older poor quality violin (plotted from data by Hutchins, 1962)

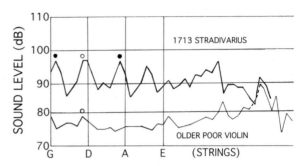

them. He detected vibrational resonances with a small light-weight accelerometer while driving the vibrational modes with a small, light-weight impact hammer. His test violin displayed many modes that are best described as "bending modes." He noted that there were typically about three of those per 100 Hz in the violin. Hence, very few modes were adequately separated from others to be dominant. The motion generally involved a mixture of different ones. Of course, these will all vibrate at the driving frequency, rather than their own resonant frequencies when the violin is played. But the different vibrating areas and phase relations result in complex radiation patterns. Some of these modes are ones in which the neck joint acts as a torsional spring between the body and the fingerboard. Several others (at about 303, 666 and 1173 Hz) represented bending modes of the neck and fingerboard of the violin. The neck joint itself had three rotational modes clustered about the frequency of the open G string (196 Hz).

These resonances, together with the variation of coupling of sound from the vibrating instrument with frequency, make up what has been called the "formant" for the instrument. As in the case of the human voice, the formant is a frequency-dependent envelope that effectively multiplies the spectrum of the vibrating string. Thus, as one goes up the scale, the relative distribution of harmonics from the string changes substantially with the result that no two notes sound precisely the same. In addition, the distribution of harmonics produced also depends on the relative position of the bow with respect to the bridge as well as upon the pressure applied to the bow. These various effects produce an enormous variety of sound on different notes within one particular instrument. As discussed below, the direction in which the sound is emitted also varies substantially with the note on the scale. For these reasons, synthesizing the sound by digitizing the waveform on one particular note and then changing the fundamental frequency for different notes (as is done with some commercial electronic equipment) is doomed to failure.

A major problem in making a violin is deciding how the separate plate resonances (or "tap tones") should be adjusted so that the assembled violin has the final resonances in the right place. This had to be done by trial and error in the early days of violin making and is pretty hard to accomplish even with modern electronic instruments to measure the resonances. Obviously, a certain amount of adjustment is needed after the instrument is assembled. (See, for example, Firth 1976/77a, 1976/77b.)

5.18 Holographic Studies

An elegant method for studying plate resonances incorporating laser interferometry was devised by Stetson and Powell (1965, 1966). This "holographic" technique only requires displacements of the violin surface of as little as one wavelength of visible light to portray the mode patterns. (Of course, that also means that the violin also has to be held in a position stable to within a wavelength of light during the measurement.) Another advantage is that one does not have to worry about the particles used in the Chladni approach rolling off the convex violin surfaces. Hence, the method can be used to study the assembled violin. It is interesting to compare patterns determined for a real violin top plate with the case of the simple thin membrane example.

One of the most interesting mode studies was conducted by Jansson et al. (1970) in which holographic techniques were used to examine the variation of resonant modes in the top plate as the components of a violin were assembled. The plate was excited by a mechanical vibrator from underneath and the interference pattern was produced by a $15\,\mu W$ red ($\lambda = 0.6328\,\mu m$) helium-neon laser. They photographed successive time-averaged interference fringes that corresponded to differences in vibration amplitude of about $0.3\,\lambda$ from one fringe to the next. The individual fringes thus represented equal amplitude contours of the vibration pattern on the top plate surface.

In order to illustrate the large changes in some of these modes and the relatively small changes in others, I assembled the illustrations in Fig. 5.37 from the data of Jansson et al. (1970). As can be seen from the figure, the even-symmetric 1,1 mode takes on more and more odd-symmetric character as the bass bar and sound post are added. Surprisingly, its resonant frequency remains little changed from its initial value after addition of the sound post. On the other hand, the 2,2 mode retains its even-symmetric character until the sound post is added, but undergoes large shifts in resonant frequency. In contrast the hybrid 3,2* mode changes relatively little in frequency in the assembly after the f-holes are incorporated.[52] (The particular mode designations used here were chosen to facilitate comparison with the computed modes for a thin rectangular membrane in Fig. 5.31.)

---

[52] Jansson et al. (1970), also reported interferograms for the bottom plate and for the completely assembled violin. A later interferometric study by Jon Luke (1971) of the body vibrations of a violin also indicated the relative phases for many modes.

MODE

TOP PLATE + f-HOLES + BASS BAR + POST

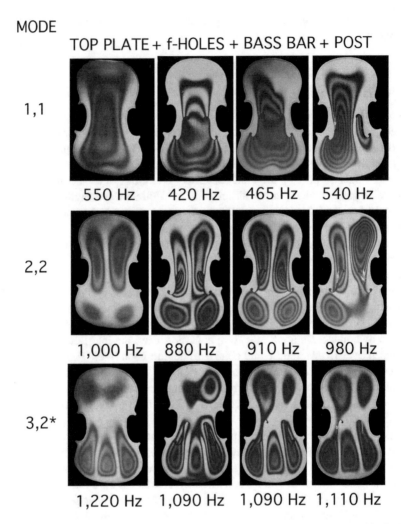

Fig. 5.37 Mode patterns based on data by Jansson et al. (1970) measured holographically. The original data has been rearranged here so as to show the progressive changes in the three modes illustrated as the violin was assembled

5.19 Radiation Patterns

It has often been noticed that some violins and cellos "project" very much better than others. That seems to be especially true of those made by the old Italian masters such as Stradivari and Guarneri del Gesu. For example, Sacconi cited the case of a cellist getting ready for a concert at Carnegie Hall who was trying out a modern cello in addition to his Stradivari instrument in a hotel room before the concert. He thought they both sounded equally loud in the hotel room and was all set to leave the Strad behind during the recital. Sacconi warned him that he had better try them both out in the concert hall before the concert. He did so and found that the Strad carried very well far back into the hall, whereas the other instrument could scarcely be heard beyond the first few rows. Similar stories have been reported by others for violins. For example, Paul Arnold mentioned that the sound of the Guarneri Del Gesu played by the concert master in one of his orchestras seemed quite loud at the middle of the concert hall.[53] After he was promoted to sit next to the concert master, he was surprised to find that the del Gesu did not sound unusually loud right next to the instrument. Similarly, a Joseph Gagliano instrument played by another violinist friend only produced a sound level of about 70 dB at 3 ft, yet carried extremely well in the concert hall. Violins with flatter (less arched) top plates project farther, but have darker tone quality. (See Hutchins and Schelleng 1967.)

It is clear that the differences in carrying power between these instruments must lie in their far-field radiation patterns—that is, the distribution of sound radiated at distances large compared to a wavelength from the source. The wavelength of sound in air at 1 kHz (near the frequency where the ear is most sensitive) is about 12.4 in. and is comparable to the length of a typical violin body (about 14 in. for most violins made by Stradivari.) Hence, diffraction effects resulting from the different surface modes produced on a violin are bound to be important in the region where hearing is most sensitive.[54]

To do an accurate calculation of the radiation pattern from a real instrument, one would have to integrate the contributions to the sound (including the relative phases at different parts of the vibrating surface) over the entire instrument—a task for which precise data are not available. However, it is instructive to see what might happen using an approximate calculation of radiation patterns based on the data for the 1.11 kHz 3,2* mode shown in Fig. 5.37. Figure 5.43 shows a result from a numerical calculation in the plane of the top plate for that particular mode. The method used was to replace the full surface mode by point sources located at the maximum vibration spots with intensities roughly equal to the central area in each case. The openings for the f-holes were each replaced by three circular sources as shown in the diagram at the left of the figure. The two main surface maxima at the left were assumed to represent negative deflections (downward on the plate) and the remaining three (two at the right and one at the bottom middle) were taken to be positive (upward from the plate). The radiation from the f-holes was assumed to be of opposite phase from the nearest top-plate maxima. (That is, a downward motion of the top plate would compress the air in the violin cavity locally, resulting in a positive air pressure at the nearest f-hole.) The convention used here to display the results is the same as that employed in the discussion of multipole radiation in Chap. 1. The gray background represents the ambient pressure in the room; the white patterns represent local air-pressure minima and black represents local air-pressure

---

[53] Paul Arnold, private communication.

[54] For a review of the properties of vibrational modes that are controllable by the violin maker, see Woodhouse (2002).

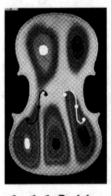

1,110 Hz

Fig. 5.38 Approximate computed radiation pattern in the plane of the top plate from the 1.11 kHz 3,2* mode of an assembled violin top plate (The bass bar, f-holes, sound post and rib structure are all in place)

maxima.[55] (The white regions in the diagram at the left are assumed to generate air-pressure minima, and vice versa.) Thus, for example, the wavelength in air at one instant in time corresponds to the distance between successive black bands on the gray background. As shown in Fig. 5.35, the radiation pattern is roughly that of a quadrupole source distribution and has four directions of maximum intensity. Other resonant modes at different frequencies would generate quite different radiation patterns and directions of maximum intensity.[56]

Now suppose that you were Mr. Stradivari, seated at your work bench and wanting to enhance the radiation from your violin in the direction of the audience (i.e., to the right in Fig. 5.38.) After the basic shape is cut out and the bass bar is glued in place, about all you could do (other than move the sound post) would be to change the relative thickness of the wood and the width of the f-holes.

---

[55]The interference patterns were computed by the author using a Monte Carlo calculation in which sums were evaluated of the type

$$\left[\sum_i A_i \frac{\sin(2\pi R_i/\lambda)}{R_i}\right]^2$$

where R_i is the distance from the ith source to the point of observation, A_i is the source amplitude (proportional to the area of the source), and λ is the wavelength of the sound in air. The white versus black color was determined by the sign of the total amplitude at each observation point, while the intensity was computed from the square of the total amplitude. The intensity was displayed as a "probability cloud" using random fluctuations in the vicinity of each observation point. (See Bennett 1976, Chapter 3.)

[56]For a review of measurements of sound radiation patterns from the violin, see Weinreich (2002).

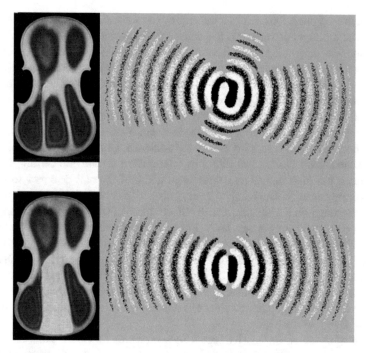

Fig. 5.39 Modification of the radiation pattern computed for the 1.11 kHz mode in this figure were obtained by suppressing the f-holes (top) and eliminating the central maximum between the lower bouts (bottom). Note that the radiation pattern at the bottom is aimed preferentially at the audience

Presumably, making the wood thinner in one spot would increase the flexibility of the top plate and enhance a surface vibration maximum or minimum at that point. Conversely, increasing the thickness at another spot would decrease the amplitude of the vibration locally. Figure 5.39 was computed to illustrate how such changes might affect the radiation pattern. In the top portion of the figure, sound from the f-holes has been suppressed (say, by narrowing their width) and the radiation lobes parallel to the long dimension of the violin have been decreased. In the bottom part of the figure, the central maximum between the lower bouts has been eliminated (for example, by increasing the relative thickness of the wood at that point). In that instance, radiation directed toward the audience has been selectively increased. (Suppressing radiation in the direction away from the audience is harder; a good practical solution to that problem is simply to add an acoustic reflector to the left of the player.) Although it is relatively easy to make these changes in a computer calculation, Stradivari would have had to scrape away with his sharpened sabers for quite a while to achieve the same effect. He clearly must have worked out empirical solutions to the radiation problem through years of judicious trial and error.

The above discussion is meant merely by way of qualitative illustration and should not be interpreted to imply that all, or even most, of the radiation occurs in the plane of the top plate. Direct measurements by the author and others have shown that substantial amounts of radiation (especially at frequencies above about 1 kHz for which the wavelength is shorter than the length of the top plate) also occur in many different directions. (See, e.g., Weinreich 1997.) Although the radiation well-below 1000 Hz is largely isotropic, much of the sound above that frequency is radiated into the hemisphere above the top plate. The higher frequencies that produce the much-desired feeling of "presence" are usually lost in large concert halls that have no shell or acoustic panels above and behind the violin and viola sections of an orchestra so as to reflect that sound outward toward the audience. Under these conditions much of the sound from violins and violas travels upward toward the ceiling, which is often much too high due to an overly large width of the hall designed to provide great seating capacity. The result is a very "distant" character to the sound, if not its complete elimination. The lack of such reflections is one of the worst short-comings of many contemporary concert halls. For example, there are locations toward the back of Philadelphia's new Verizon Hall where one can hardly hear the violins at all, even though the cellos and double bass come through quite strongly. Often the only practical way to overcome that problem after a hall has been constructed is through the suspension of Lucite or wooden panels ("acoustic clouds") from wires attached to the ceiling. Their presence is a sure sign of inadequacy in the original acoustic design of the hall.

Some of the most impressive recent measurements of violin radiation patterns have been made by Lily Wang (1999) using a linear array of 15 microphones that could be scanned at right angles to the array with a stepping motor. She used a bowing machine in an anechoic chamber to study the sound from three different instruments. (See Fig. 5.40.)

Fig. 5.40 A slice through the NAH reconstructed intensity vector field for the Hutchins mezzo violin at 1320 Hz showing nearfield effects (reproduced from Wang 1999 with permission of the author)

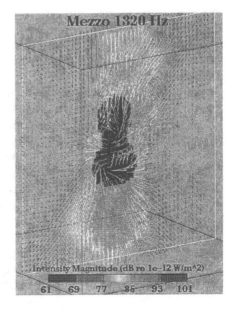

Her method is related to one suggested by Weinreich (1997) in which near-field measurements of the sound were to be expanded in terms of a complete three-dimensional set of Hankel functions and spherical harmonics. Solving for the expansion coefficients then effectively determines the whole radiation field. Wang's results were analyzed using a technique of multi-planar "Near-field Acoustic Holography" (NAH for short). The basic idea is that near-field measurements of amplitude and phase made with the microphone array can be used to construct intensity vectors in three dimensions. The directions of these vectors depend on phase differences of the sound field at close points. An example of her technique is shown in Fig. 5.40 as applied to the partial at 1328 Hz on the Hutchins mezzo violin. (Also see Wang and Burroughs 2001.) Two of the violins Wang studied had been made by Carleen Hutchins, including the over-sized "mezzo violin" used to generate Fig. 5.40, an instrument similar to that described originally by Hutchins and Schelleng (1967).

5.20 Additional Phenomena

5.20.1 Wolf Tones

The phenomenon of the "wolf tone" often plagues very good violins and, especially, fine cellos with very thin ribs.[57] These instruments often have many more strong resonances than do mediocre ones. The phenomenon involves a near-coincidence of a strong high-Q resonance in the body or air cavity of the instrument with a note on the musical scale. When two such oscillations are strongly coupled, two important things happen. First, energy is transferred back and forth between a body or air resonance and the string resonance at the difference frequency between those two resonances with the result that the bow pressure necessary to maintain the stick-slip process varies strongly with time. This required time-varying bow pressure makes it hard for the musician to control the note.

The other difficulty is that interaction between the two modes results in a "repulsion" of the two closely spaced resonant frequencies in the coupled mechanical system. That is, the original two resonances move in opposite directions in the coupled system. (See the equivalent circuit derivation in Appendix A.)[58] Hence, when one resonance is the fundamental mode (i.e., the first harmonic) of a vibrating string and the other is a strong air or wood resonance, the fundamental mode is shifted out from underneath the harmonic series of overtones in the string. As

---

[57]This phenomenon has been studied by numerous people, ranging from the early work of C.V. Raman (see his collected Scientific Papers, vol. II, 1988) and John Schelleng (1963) to Arthur Benade (1975).

[58]A mathematical derivation of this effect for a string resonance in strong coupling with another resonance has also been given by Gough (1981; see Fig. 10).

a result, the sound tends to jump up an octave and the note becomes unstable. (Two "kinks" in the string motion may develop during the slip-stick process at the fundamental frequency.) Schelleng (1963) concluded that the even harmonics tended to remain stable while the odd harmonics (including the fundamental) grew and shrank at the pulsation rate.

Several things may be done to minimize the wolf-tone effect. Putting more pressure on the bow usually suppresses it and expert players learn to do that automatically.[59] Shifting the frequency of the air or wood resonance can also reduce the effect significantly. That sometimes may be accomplished by adding extra mass to the system—e.g., by a placing a small weight on the string between the bridge and the end point, or adding more wood to the back plate or top plate, or somehow changing the dimensions of the body of the instrument—for example, by shortening the rib height. Adding some dissipative mechanism to the vibrating system can sometimes reduce the Q of one of the modes enough to suppress the effect. In any case, it is a curse that often accompanies a highly resonant instrument and is apt to be much more of a problem with a priceless Stradivari or Guarneri del Gesu violin than a garden variety "cigar box." (Fig. 5.41).

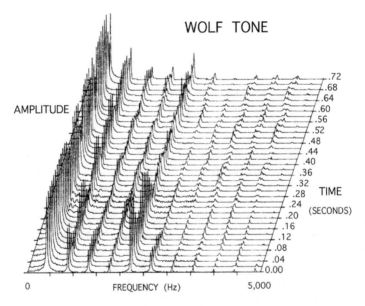

Fig. 5.41 Unstable behavior of a resonant note near a "Wolf Tone" on a 1700 Stradivari violin bowed by Joseph Genualdi

---

[59]David Oistrakh's advice: "I do three things to avoid a wolf. I press my left shoulder tightly against the violin, I vibrate more, and I pray." (Steinhardt 2006, p. 78.).

5.20.2   Scaling and the Hutchins Octet

One early experiment in scaling resulted in the elephantine three-stringed Octobasse designed by Savart and Vuillaume for Berlioz. The Octobass had a total height of 3.45 m ($\approx$ 11.3 ft)! (See Fig. 5.39.)

According to Berlioz[60] the octobass "produced a sound of rare beauty, full and strong without any roughness... every orchestra of any importance needs two of them." (But only two of them are extant—one in the Conservatory of Music in Paris and the other in the State Museum in Vienna.) Alas, neither Berlioz nor Wagner (who also liked the concept) ever used them. One problem was that the performer had to stand on a platform and change the active string lengths using a system of levers actuating pads to contact frets on the "fingerboard." (The left hand pulls levers up and down at the back of the neck and the player uses seven foot pedals for the other notes.) In a normal-sized room, one needed a hole in the ceiling so that the pegs could be adjusted by a tuner upstairs (Fig. 5.42).

Various resonances in bowed string instruments were studied in some detail for a Stradivarius violin and several other instruments by Carleen Hutchins (1962, 1981, 1983). Of the several wood resonances, the one just below the open A string on a good instrument seemed to be most important in evaluating an instrument. On bad instruments, that resonance either didn't exist, or was above the A string. The other main wood resonance on the Stradivarius fell just above the open G string. A strong (Helmholtz) air resonance is also important and typically occurs just below the open D string on a Stradivarius violin. Sacconi's (2000) observations on air resonances in violins by Stradivari discussed above were in reasonable agreement with Hutchins' conclusions.

In doing this work, Hutchins speculated that to make a powerful viola, cello, or double bass, one ought to reduce the arching of the top plate and increase its area and to scale each instrument in such a way as to make the principal wood and air resonances fall below and above the two middle strings as in the better Stradivari violins. The reduced arching of the top plate was to increase the projection of sound from the instrument. To get the right volume with reduced rib height, greater length was required for the sound box. For violas and cellos this scaling process meant enlarging the dimensions of the new instruments. One had to change the length of the wood box so that the first excited air mode and a wood resonance fell above the open-string resonance for the third string and to adjust the volume (shape and rib height) so that the "Helmholtz" air resonance fell below that for the open second string. The wood resonances of course depend on the thickness and size of the top and bottom plates—not to mention the acoustic properties of the wood itself.

One problem with the Hutchins approach is that her new viola is about the same size as a cello, and her new cello is about the size of a normal double bass. By the

---

[60] See Hector Berlioz's report "Of the Great Exhibition in London (1851) to the French Commission of the International Jury" and his Traité d'Orchestration et d'Instrumentation Modernes; also see, Millant (1972, p. 116).

Fig. 5.42 The Vuillaume
Octobass, played here by Paul
Badura-Skoda (photo
courtesy of Heming
associates)

time you get to the new double bass, it takes on proportions mildly reminiscent of
the Vuillaume octobass. Hutchins designed and made an entire octet of such scaled
instruments that had remarkable power and great beauty. (See Fig. 5.43.) I measured
the sound output of several of these instruments. Most were well-above 95 dB at 1-
m and the new double bass peaked at over 105 dB at a distance of 3 ft. That is nearly
the sound level obtained from the entire double bass section of a large symphony
orchestra. Hutchins' double bass was at least playable by a tall person standing on
the floor.

However, there are several practical problems with the instruments in her "octet."
First, anything larger than her violin (which is about the same size as a normal
Stradivari) is difficult to play in a conventional manner. One's arms simply aren't
long enough for comfort. Ironically, the idea (although quite clever and very fine
from the standpoint of acoustical design) goes in the opposite direction of the
evolution of violas and cellos from the Amatis through Stradivari and Guarneri del
Gesu. The initial instruments designed by Andrea Amati, although very mellow,
were cumbersome to play. Hence, there was a deliberate effort by his descendants
to make them smaller and more agile. In the case of the cello, this effort was fought
for many years by the Catholic Church (a major patron of the instrument makers)

Fig. 5.43 The Hutchins Octet (photographed by John Castronovo in the living room of Carleen Hutchins and reproduced here by her permission)

because they wanted deep, booming violoncellos in their religious processions. (Holes were drilled in the back of the instruments into which pegs could be inserted to support the instruments from a chain around the neck of a walking performer.) For many years, there was no thought given toward use of the cello or viola as a solo instrument; they were exclusively employed to carry the bass line in church music. That situation obtained until the death of Nicolò Amati in 1684. After that, Stradivari and other descendants of Nicolò Amati began making cellos of smaller size. The Hill brothers concluded that no further improvement was desirable beyond the later design by Stradivari.

In spite of being Bach's favorite stringed instruments, large violas had a special problem unto themselves. First, they were not used very much in chamber music until the time of Haydn. As a consequence, Stradivari, who made instruments during the Baroque Period, only made a few violas in his lifetime. (One was the famous viola that was eventually passed on to Paganini, who refused to use it to play the solo part in Berlioz's Harold in Italy.) Consequently, little effort was applied to making them more powerful, as was done with violins and cellos toward the end of the eighteenth century. Various solutions to that problem have been tried in recent years in addition to that of Hutchins. Generally, the approach has been to produce violas with much greater air volume, sometimes by attaching tumorous-looking protrusions to the instrument. At least the Hutchins viola looks like a cello and not a viola with the mumps. One of the most successful modern viola designs I have seen is that by Iizuka. (See Fig. 5.22.) His approach provides a powerful, symmetric instrument that is easily accessible to the fingers of the player.

Another problem with the concept of the Hutchins Octet is that by scaling the instruments so radically, she has given all of them a fairly uniform tonal character. It had been a basic tenet of Stradivari and his later contemporaries that the timbre of the different instruments should be kept distinctive. The instruments in the Hutchins Octet, although very powerful, all sound very similar. The resonances in the viola and cello aren't where you are used to hearing them in a conventional instrument. For example, Mozart clearly made use of the resonances in the viola near F# on the C string in his viola quintets. When played on a much larger instrument, the viola parts sound out of character as Hutchins had pointed out herself. (See, Hutchins 1983, p. 1429)

Finally, there isn't much music written for her ensemble, which includes a double bass, cello, tenor viola, viola, violin and piccolo violin in this new scaling. (Hutchins hopes that contemporary composers will get interested.) Although her basic idea is indeed a very clever and ambitious application of acoustical science, it is doubtful that her scaling concept will catch on and that contemporary composers will write very much for the new octet. Composers with the genius of Mozart, Beethoven, Schubert, and Brahms are in very short supply these days.

5.20.3   Vibrato

Continuous vibrato on the violin did not become the norm until the time of Fritz Kreisler according to violinist Sandor Salgo (2000). Interestingly, Mark Katz (2005) argued that the development of the phonograph played an important role in the adoption of continuous vibrato by violinists such as Kreisler and Eugéne Ysaÿe. He notes that the timing was right (circa 1910) and that the technical problem of capturing the sound of the violin on the cylinder phonograph was reduced by the presence of vibrato. He felt that a greater feeling of "presence" was produced with vibrato and that the sound could be captured without danger of the instrument or bow bumping into the large horns used in the Edison recording process. He also noted that the frictional sounds of the bow on the string that were picked up by later more sensitive microphone technology were masked by vibrato. One further possible argument that Katz did not mention is that the presence of "wow" and "flutter" (i.e., periodic speed variation from the eccentricity in moving parts of the machine) in a recording is much less conspicuous when the source is not a pure tone of constant frequency and amplitude.

Of course as Katz acknowledged, there were other reasons for adopting vibrato: It covered up the harshness of the metal strings that were used after World War I and at least gave the impression that the performer played with better intonation. As one study concluded, vibrato gives the violinist more time to adjust the pitch before the listener notices he is out-of-tune. Katz also thought the presence of vibrato compensated somewhat for the absence of the normal visual element of enjoyment in a performance which was missing in audio recordings. He also noted some listeners feel that the particular form of vibrato used permits identification of the performing style of different soloists.

Donnington (1980) suggests that a slight and inconspicuous vibrato was probably in normal use since the invention of the bow. But Louis Spohr in his *Violinschule* of 1832 regarded it as an ornament for special expressive purposes only, a belief evidently adhered to by Brahms' violinist Joseph Joachim and the legendary teacher, Leopold Auer. (One wonders how Joachim felt about the vibrato used by Brahms' clarinetist friend, Richard Mühlfeld.) Even today, some violinists (for example, Anne-Sophie Mutter, in her recordings of the Beethoven Violin Sonatas) often use vibrato as a form of expression coupled to the crescendo. They sometimes start a note pianissimo without any vibrato at all and then increase the vibrato with the loudness of the note—a technique also used by oboists to very good effect. The difference in sound produced by vibrato also can compensate for the limited dynamic range of an instrument.

Salgo (2000, p. 41) goes on to explain that "two types of vibrato existed in the eighteenth century violinist's repertoire: (1) the 'true' vibrato in which one finger undulates the pitch above and below the given note; and (2) the two-finger method ...[called] the 'close shake.' There, one finger is pressed firmly on the string and a second finger makes a rapid beating or shaking very close to the first...actually, not a true vibrato at all." Salgo's second example is really more of a trill than a vibrato. Yale violinist Broadus Erle had a still different theory of the vibrato—one that seems to be shared by a number of current professional violinists. He felt that the vibrato should primarily go below the note in an asymmetric way, spending more time in the vibrato cycle on the flat side. Around 1970, Erle went so far as to ask Professor Kindlmann of the Yale Electrical Engineering Department to design and build an electronic circuit so that he could analyze the vibrato produced by his students quantitatively. Kindlmann built the circuit and tried it out on one of Erle's students from the Yale School of Music. The measurement showed that the student's vibrato was symmetric about the note, much to Erle's annoyance. Professor Erle concluded that the student must have been playing out of tune and lost interest in the investigation.[61]

If it were just a matter of modulating the frequency, one would not expect to find much amplitude variation. However, in most cases I have examined, there is also substantial amplitude modulation on the note. That suggests that the process produces significant bending motion of the bridge, which in turn affects the coupling of sound from the string to the violin top plate.

Vibrato is so built into the technique of modern violinists that they sometimes use it unconsciously. I once asked Joseph Genualdi (then first violinist in the Muir Quartet) to play an open G on his Heberlein-Taylor Stradivarius, thinking that it would surely be without vibrato and therefore easy to analyze. Much to my astonishment there appeared an "Open G" on my spectrum analyzer that was loaded with vibrato! It turned out that he was applying vibrato to the D string an octave above the G, while bowing the open G string. On seeing that, my first thought was that the effect represented a resonant interaction through the bridge of the frequency on the D-string with the second harmonic of the open G. However, the vibrato was actually strongest at the fundamental of the open G where it showed up primarily as amplitude modulation, rather than at its second harmonic. What may have happened is that modulation of tension on the D string was moving the bridge back and forth, thus producing vibrato on the open G. Violinist Paul Arnold demonstrated to me that he could produce a similar vibrato by modulating almost any note on any other string. He explained that that vibrato resulted merely from shaking the instrument and thus affecting the bow velocity—an effect probably similar to the "fast and narrow wrist vibrato" of Toscha Seidel described by Steinhardt (2006, p. 78).

There has been some difference of opinion about the most desirable vibrato frequency. Some violinists (for example, Joseph Zigetti) used an extremely slow rate whereas others produced what might be best described as a nervous tremor (as one English critic put it, "like jello on the plate of a nervous waiter.") Broadus Erle felt that an optimum frequency for vibrato was about 6 Hz. There probably are neurophysiological reasons why different vibrato speeds are preferred by different people. At very slow vibrato rates (often used on the tremolo engines in pipe organs), the ear hears the frequency swinging back and forth over a wide range much too clearly, sometimes introducing a sense of nausea in the listener. The comment by Sir Thomas Beecham comes to mind: When an English oboist gave the A for the orchestra Beecham said, "Gentlemen, take your pick!"[62] At 6 Hz, the perceived frequency seems to be the average pitch. But at much higher frequencies the vibrato sounds more like gargling.

---

[61]P.J. Kindlmann, private communication.

[62]Atkins and Newman (1978, p. 20).

5.20.4   The Effect of Varnish

Violins are often treated with sizing to seal the wood, leaving it in a state called "The White." The wood actually gets darker through gentle exposure to ultraviolet radiation in sunlight. (See Fig. 5.44.) After perhaps three to 6 months of this treatment, it is ready for varnishing, coating, which provides the rich, warm color associated with the finished violin. (See Fig. 5.45.)

The varnish is normally either oil-based or spirit-based. The mystique associated with the varnish used by Stradivari was partly created by the secrecy with which he was said to have guarded the formula himself. It was said that he wrote down the formula on the inside cover of the family Bible. One of his sons claimed to have copied the formula in 1704 and then destroyed the heavy and cumbersome Bible after his father died. But having destroyed the evidence of its authenticity, no one was convinced that he had the real formula. (Hill brothers, 1902, p. 170.) According to Faber (2005, p. 141), this "preposterous" story was soon proved to be a fabrication. That Stradivari used solely a pure oil varnish consisting of a gum soluble in oil and possessing good drying properties is incontrovertible according to

Fig. 5.44 A group of violins and violas "in the white" basking in the morning sunlight on Iizuka's porch (photo by the author)

Fig. 5.45 A group of varnished instruments languishing in Iizuka's shop

the Hills.[63] They also put much faith in the notion of gradual exposure to subdued sunlight and the influence of time. As Galileo was told when impatient for the delivery of his Amati, "the violin cannot be brought to perfection without the strong heat of the sun." According to Faber (p. 63), Galileo's nephew ended up with an old violin because Father Micanzio was too impatient to wait for "the strong heat of the sun to bring a new one to perfection."

Part of the varnish mystique was reinforced by the Hill brothers themselves. In their opinion the most important things were: (1) the varnish, (2) the workmanship, and (3) the quality of the wood in that order. Sacconi (2000) states in several places in his book that he thought the varnish produced "ossification" of the wood. However, other sources maintain that there really was nothing very special about the varnish used by Stradivari as it was commonly available to the furniture makers of Cremona at that time (See Farga). There clearly were some changes in the varnish Stradivari used over his lifetime. Early in his career, it gave a yellowish tinge to the wood similar to that found in the Amati instruments, but later in his life (and

---

[63]Sacconi (2000, Chapter XII) gives numerous prescriptions for varnish that may or may not have been used by Stradivari.

probably to escape comparison with Amati's work), he added coloring to produce a more reddish glow. It is suggested by Millant (1972) that Vuillaume came the closest to reproducing the varnish of Stradivari.[64]

Although it seemed clear that the varnish results in protection of the wood over long periods of time and enhancement of the beauty of the natural wood grain, it was not obvious what quantitative effect it really had on the acoustic properties of the instrument.

Narberth luthier Hiroshi Iizuka feels that the initial solutions he uses actually change the acoustic properties of the wood somewhat. He thinks that the pitch of the tap tones is raised by small amounts and that the timbre is altered. He also feels that the treatment of the wood is one of the most important aspects of making a good violin. His formula for the varnish is probably as complex as that used by Stradivari. I won't try to give a detailed description of the procedure here, but merely summarize its major features.

After hanging the instrument outside in the sun for about 4–6 months, Iizuka puts a very light coat of oil (he is still experimenting with linseed, walnut, or rosin oil) over the entire surface, making sure not to saturate the wood. The top is then sealed with a casein solution. After that is dry, he puts two coats of an emulsion of egg tempura plus oil varnish on the instrument. He then follows that with three-to-four coats of colored oil varnish. [The violin we tested had initially been coated with walnut oil. (See below.)]

In order to clarify my own understanding of the role of varnish, I undertook an experiment with Iizuka and Philadelphia Orchestra violinist Paul Arnold. Iizuka strung up one of the violins he was making during the spring of 2004 while it was in the "white" and before it had received any coating, although it had had some exposure to sunlight. Arnold then played the violin on open strings with a fine Peccatte bow in Iizuka's shop, and I made a digital recording of the sound using a high-quality Sennheiser MKH 104 condensor microphone placed about 1 meter laterally from the violin.[65] The recording was then analyzed for spectral content. After a long exposure to the sun during the summer, Iizuka applied his varnish to the instrument, he subjected the violin to more sunlight, and we made another digital recording in the same room, with the same equipment and violinist (Fig. 5.46).

---

[64]In attempts to understand why the old Italian instruments sounded so great, all kinds of suggestions have arisen: floating the logs down the Po river from the Alps, soaking them in salt water [no doubt mixed with sewage] in the canals of Venice, and coating the wood with volcanic ash before applying the varnish. The most bizarre suggestion was provided by the movie industry in which varnish for The Red Violin (supposedly inspired by 1720 Red Mendelssohn Strad) was mixed with human blood.

[65]This microphone (Serial No. 19689) was the same one used to study the spectra from different bows and had a frequency response curve provided by the manufacturer that was flat within about ±1 dB over the range from 50 to 20,000 Hz. (Two similar microphones were used to make stereo recordings, the second placed about 1-m laterally on the right side. But the spectra shown here were taken from the microphone on the left side facing the instrument—i.e., in the usual direction of an audience.)

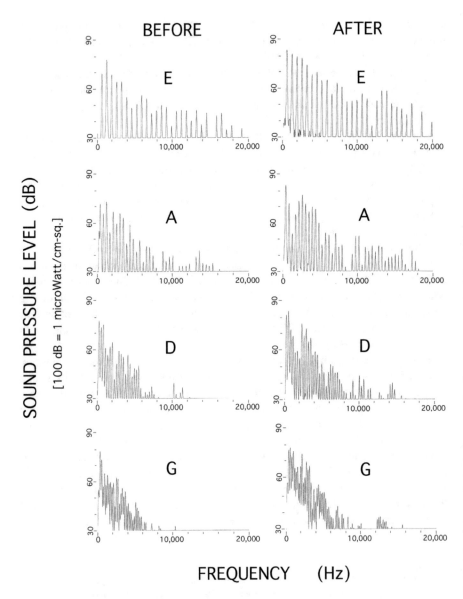

Fig. 5.46 Spectra obtained from an Iizuka violin before and after applying varnish

The "before" and "after" results of our experiment are shown in Fig. 5.43. Here, the sound pressure levels on a logarithmic scale are shown for each of the four strings. (On this scale, 100 dB corresponds to the standard pressure level adopted for such measurements of 1 μW/cm^2.) Somewhat to my surprise, there was a fairly substantial difference in the two sets of spectra obtained from the same microphone. The intensity of the harmonics has gone up by about 5 to 10 dB with the biggest increase being at higher frequencies and on the higher pitched strings. Hence, the results are in agreement with the notion that the varnish increases the stiffness of the wood. (The reason why some older instruments with thin top plates sound very strong is probably due to the use of oil-based rather than spirit-based varnish.) Iizuka felt that there also were noticeable differences in the directional properties of the radiated sound in the room. The maximum sound level on any harmonic at a distance of one meter was about 80 dB on all of the open strings when played at a normal forte level and my meter readings did not vary much over the dimensions of the room. Finally, Iizuka replaced the bridge on the varnished instrument with a somewhat lighter one. That modification resulted in a further increase in higher harmonic content of a few dB. There, even on the G string, significant energy was emitted above 10 kHz. (See Fig. 5.19. Also see Schleske 1998).

5.20.5   Evolution of the Bow

As with piano hammers, the modern violin bow has evolved through a remarkable series of forms.[66] It is said that the low-tensioned bow of Bach's time, which was shaped like an archer's bow, was capable of sounding all four strings of the violin simultaneously. (See Fig. 5.47.) Thus, the opening chords of the famous Bach Chaconne could be played without rolling the bow across the strings.[67] In a desire to reconstruct the bow that Bach might have used, something known as the "Vega Bach Bow" was made by Knud Vestergaard in the 1950s and demonstrated by the Hungarian violinist Emil Telmányi (1892–1988). A CD version of a 1954 recording made by this violinist was released by Testament Records (SBT2 1957) in 2003. Although the bow was pretty bizarre looking, the organ-like chords produced with it by Telmányi are both beautiful sounding and perfectly amazing to anyone used

Fig. 5.47 Artist's conception of a Bach bow in action

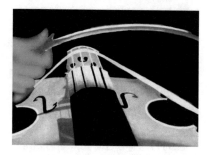

---

[66]Heron-Allen (1885, Chapter IV) gives an exhaustive discussion of the history of the violin bow.

[67]Steinhardt (2006, p. 227) notes that the "chaconne" derives from a dance form known as the chacona—an invention frequently attributed to the devil and outlawed in Spain in 1615. He also implies it might have originated with the natives of South America and brought to Europe by the Spanish conquistadors.

Fig. 5.48 The double bow
technique of Frances-Marie
Uitti

to hearing the Chaconne played with broken chords. The New York Times review
hardly did justice to the elegant sound produced in that recording.[68] However,
mastering the Vega bow must be incredibly demanding on the part of the performer.
(Vestergaard had built into the frog of the bow a mechanism that permitted the
performer to tighten and loosen the horsehair during a performance.)

Another approach to playing chordal and polyphonic music on a single bowed
instrument was developed more recently by cellist Frances-Marie Uitti. She uses two
bows: one above the strings and the other, below (See Fig. 5.48). She apparently can
control the two bows so as to get different tonal color from each (by moving the
two at separate distances from the bridge) and play two voices at once with cross-
rhythms (e.g., three against four, etc.) She says the underbow tends to have a softer
sound resembling the sweetness of a viola da gamba. Her double bow technique
was evidently well-suited to performing the music of Giacinto Scelsi, especially the
one-note pieces from his "Quattro Pezzi," in which each movement is based on just
one note—but in which the tuning gets "bent" and the tonal color is altered.[69,70]

As shown in Fig. 5.49, the shape of the bow gradually evolved during the
seventeenth and eighteenth centuries from a solid wooden stick holding the stretched
horsehair (as in the case of the Mersenne bow and those made by Stradivari himself)
through the shorter version used by Corelli, to a longer, lighter and springier bow
with higher tension of the type designed by the Italian virtuoso Giovanni Battista
Viotti (1755–1824). Viotti was the first to introduce the power of a Stradivari
instrument to French musical audiences and then later to those in London. While in
Paris, Viotti advised the legendary bow maker François-Xavier Tourte (1747–1835),
alias "the Stradivari of the bow," of his own requirements for a new type of bow in
which the Pernambuco wood curved inward toward the bow hair in contrast to the

---

[68] Jeremy Eichler, "The Bow of Bach's Dreams? Not Quite." The New York Times, August 10,
2003, p. AR22.

[69] See Paul Griffiths, "Bringing a Reclusive Composer to Light," The New York Times, Art Section,
February 16, 1997.

[70] Still another method of bowing the cello was discovered by Piatigorsky. While rehearsing with
Stravinsky, he was so nervous that the bow jumped out of his hand and slid behind the bridge
producing a strange whistling sound. "Marvelous!" said Stravinsky. "How did you do it?" (Milstein
and Volkov, 1990, p. 143.)

Fig. 5.49 The evolution of
the violin bow from
Mersenne to Viotti (after
Abele 1905)

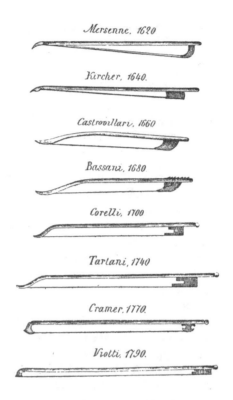

previous convex design. This change gave the bow greater strength and elasticity
and permitted larger tension on the horsehair.[71] In this design the wood was bent
into a convex shape facing the string using dry heat. A heavier head section than
had been normal was used in order to equalize "down and "up bows." The horsehair
was held away from the stick with a metal clip to hold the hair in a flat band. The
bow was balanced by adding weight to the "frog." Tension was applied through a
screw at the frog end where the bow was held. It has been said that Tourte did not
add many new concepts to bow making himself, but largely just combined the best
features previously discovered by others (especially those by Viotti.) In so doing,
he produced bows that surpassed those by anyone who had gone before him.[72]
(Fig. 5.50).

The requirements for the wood included springiness, strength and high density,
in addition to being workable[73] and capable of retaining a curved shape produced

---

[71] See discussion in Faber (2005, p. 75).

[72] The author is indebted to cellist Frances-Marie Uitti and her bow maker, Andreas Gruter, for
helpful comments on the art of bow making.

[73] A experimental study of the bouncing bow has been given by Askenfelt and Guettler in the CAS
Journal, vol. 3, No. 6 (November 1998), pp. 3–8. Also see the second article in that same journal
by these authors on spiccato and ricochet bowing; Askenfelt and Guettler, CAS Journal, pp. 9–15,

Fig. 5.50 (Top) A piece of pernambuco cut to the rough shape from which a bow can be made. (Bottom) A modern bow fashioned from pernambuco wood by Iizuka (photograph by the author)

after the wood had been heated. Although various hardwoods (among them "snakewood") had been used previously to make bows, Tourte discovered that a particular form of wood known as pernambuco imported from Brazil satisfied these requirements best. It has the highest density and least dissipation rate of any of the woods commonly used in violins. (See Bucur 1988.) By fortunate coincidence, large quantities of this furnace-red wood (or "pau-brasil" as it was called in Portuguese) were available in Paris because a red pigment extracted from it was used to dye the robes of the nobility. Even so, it has been said that one might have to go through eight-to-ten tons of this material in order to find a satisfactory piece to fabricate a single 70–80 g bow, 75 cm long. In a way, the problem of finding a suitable piece of wood for a fine bow is similar to that encountered by woodwind players in locating good reeds—with the exception that the wood in the bow seems to retain its springiness forever. Vast quantities of the wood were used in the dye industry until the mid-1800s (when aniline dyes came into use), but the trees that produced this wood and which were located on the coastal regions of Brazil were savagely cut down as the forests were cleared to produce sugar plantations and the wood was burned to produce charcoal for use in the steel industry. Not surprisingly, the trees from which this wood is obtained have ended up on the environmental protection list. As with cane reeds used in woodwind instruments, attempts to replace the wood with plastics and other synthetic materials (e.g., fiberglass, composite fiber, carbon graphite and steel) have proved to be unsatisfactory to professional musicians.[74] (See Rymer 2004.) (Fig. 5.51).

Environmentalists and violinists alike will be glad to know that an International Pernambuco Conservation Initiative was officially approved by the Brazilian Government in 2004 and is dedicated to the preservation of pernambuco wood in Brazil.[75] (Also see Hannings and Chin 2006.) (Fig. 5.52).

Horsehair has proved to be the only satisfactory material for exciting a violin string. Three commonly used sources of bow hair are shown in microscopic views

---

and the earlier study by George Bissinger of bounce tests and modal analysis of the violin bow in the CAS Journal, vol. 2, No. 5 (1995), pp. 17–22.

[74]It is said that manikara kauki wood grown in Asia is similar to pernambuco and is suitable for making good bows (Dick 2003, p. 14). However, the violinists I've talked to who tried it remain unconvinced.

[75]The Violin Society of America Newsletter, March 2004, p. 6.

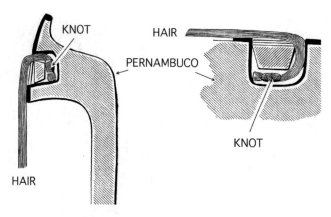

Fig. 5.51 Method used by Tourte and others to fasten the horse hair to the tip of the bow

Fig. 5.52 The "frog" from a modern bow made by Iizuka containing a screw with which the tension in the horsehair may be adjusted. The frogs are made from a variety of materials: ebony (as in the figure), 50,000 year-old mammoth tusks from Siberia, and horn. Ivory from elephants is now forbidden (photograph by the author)

of single strands in Fig. 5.53. The top photograph is Chinese black (listed by the supplier as being "very strong"), the middle is Chinese light grey ("strong and elastic") and the bottom is Siberian horsehair ("uniform in color and texture").[76] Single strands of these three types examined by the author varied in diameter as follows: Chinese black: 0.18–0.25 mm; Chinese light grey: 0.18–0.28 mm; and Siberian: 0.15–0.20 mm. The top two photomicrographs in Fig. 5.50 of Chinese horsehair were taken with reflected light, whereas the bottom photomicrograph of a Siberian horsehair was taken using transmitted light.

In none of these cases were any significant "barbs" encountered at the walls of the hair of the type described in the old violin-teacher's lore. However, there were long groves running lengthwise along the hairs surrounded by ridges which are best seen stereoscopically. These ridges should increase the strength of the hair against lateral shearing over that which would be obtained from the same amount of material concentrated in a perfectly circular cross section. One rough way of understanding this is to note that resistance (hence, strength) of a beam against bending goes as the third power of its thickness.

---

[76]The author is much indebted to the Heinrich Dick Company of Metten, Germany for supplying these samples. He is also indebted to Dr. Jean Bennett Maguire for taking the photomicrographs shown.

Fig. 5.53 Photomicrographs
of individual strands from
three different kinds of
horsehair commonly used n
the making of violin bows.
Top: Chinese black; Middle:
Chinese light grey; Bottom:
Siberian (photographed by
Dr. Jean Bennett)

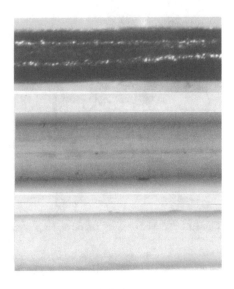

These ridges probably arose through years of evolution in the swatting of horseflies and may be the reason why horsehair is so much better than other material for violin bows. As well as being substantially thinner, human hair does not exhibit such structure. (Rapunzel would not have been of much use in providing hair for violin bows.) In addition to questions of strength, the groves in between the ridges may also facilitate the adhesion of rosin. The ridges show up most clearly as shiny reflections in the case of the Chinese black hair at the top of the figure. (Rosin, of course, was not applied to the hair in any of the photographs shown here.) Photographs taken by others under extreme magnification using electron microscopy have also shown the presence of a scalely structure on the surface of the horsehair that may also promote adhesion of rosin.

One occasionally hears the mysterious (to the uninitiated) comment that "it's better to have a fine bow and a mediocre violin than a fine violin and a mediocre bow." In addition, the notion that a fine bow should cost as much as (or even more than) a good violin seems incredible to non-violinists. Part of the reason, quite apart from the artistry necessary to fabricate a good bow, seems to be the scarcity of wood with suitable elasticity. However, there is much more to the question than that. The words of violinist Arnold Steinhardt (1998, p. 263) are instructive:

"The uninitiated might ask why the fuss over a long curved stick with horsehair attached to both ends. Made by artisans from Brazilian pernambuco wood, each bow varies in thickness, weight, and elasticity. A bow that works for one violinist may be out of the question for another. The bow has to match not just the player but his instrument as well. A lightweight, supple bow fits one instrument; a heavy, stiff one, another. String players can talk for hours about bows, whose effect is so intangible—their playing qualities, the difference between the great makers, what is for sale, where, and for how much. At this writing [1998], some bows sell for upwards of $100,000. Could I hear the difference between the Mendelssohn Violin Concerto played on a Dominique Peccatte or a Xavier Tourte bow? Probably not, but in a way that doesn't really matter. If the performer finds a bow comfortable and responsive, then that's the one for him."

From the scientist's point of view, it is of interest to see if there is any major difference in the spectra of sound produced on a good violin by bows of widely different quality. For that purpose, a professional violinist from the Philadelphia Orchestra (Paul Arnold) played on a very fine violin made in 2003 by Hiroshi Iizuka, using three different bows of vastly different cost: a bow appraised at about $65,000 made by Peccatte,[77] one made by Eugène Sartory[78] in 1905 valued at about $15,000 to $18,000 and a contemporary Chinese bow only costing about $1000. (The reader should be warned that there is considerable subjectivity involved in determining the monetary value of any particular bow; some of the most expensive ones are largely valued on the fame of their previous owners.) The Sartory bow used in the present experiment had a replacement frog made from the tusk of a woolly mammoth about 50,000 years old. (According to the violinist, the presence of this mammoth frog on the end of the bow actually detracted from its balance.) All three bows were made from pernambuco and had been strung with Siberian horsehair.

Figure 5.54 shows the acoustic spectra obtained from these three different bows on each of the four strings of the same violin. These data were all taken in the shop of violin maker, Hiroshi Iizuka, using the same high quality Sennheiser Type MKH 104 condensor microphones used in the "varnish study," again placed about 1 meter laterally from the violin. Although definitely not an anechoic chamber, the room had a reverberation time of less than 1 s and did not appear to have any strong resonances in the range studied. The relative intensities are shown in decibels as a function of frequency on a linear scale ranging from 0 to 20 kHz in each case. The violinist was asked not to use vibrato; however, there was still some variation in the spectra with time due to the fact that the bow did not remain at precisely the same distance from the bridge while the data were being taken. One of the things that makes the sound from a real violin different from the more mechanical sound which may be produced by synthesizing the tone from the harmonic amplitudes and phases of the spectral components is the fact that the tone quality does normally vary in this way. (Of course, the presence of vibrato would produce still more variation in the tone quality in normal playing of the instrument.)

As may be seen from the figure, there is not a great deal of difference among the spectral samples from these three bows. The differences between the spectra were comparable to the variations in harmonic content found as a function of time when one bow was used by itself. As is apparent from the spectra, and as we found previously with the waveform from an Amati violin, the sound is loaded with harmonic content. With the Peccatte bow, frequencies up to about 12 kHz were quite pronounced on the G string. With the other strings, the harmonic content increased to about 15 kHz on the D and A strings and exceeded 20 kHz on the E string. The

---

[77]Dominique Peccatte (1810–1874) worked for J.B. Vuillaume and was an acquaintance of François Tourte. He is generally regarded as the second best maker after Tourte and apparently started in making bows where Tourte left off. (Beare 1980a). He was probably a great disappointment to his father, who intended him to be a barber. (Millant 1972, p. 112.)

[78]Eugène Sartory (1871–1946) studied bow making with Charles Peccatte (son of Dominique Peccatte) and made several bows for the famous violinist Ysaÿe. (Beare 1980b).

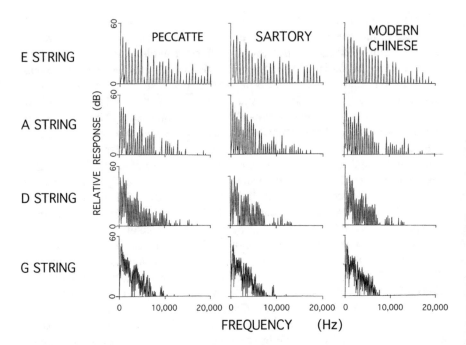

Fig. 5.54 Spectra obtained using three different bows to play the same violin. The instrument was played by professional violinist Paul Arnold. The three bows were by Peccatte, Sartory and a contemporary Chinese maker. The individual harmonics of the string are clearly resolved and result in the peaks shown in each case in the relative response

latter range is most readily perceived by small children, dogs and bats. (Typically, the hearing in adults falls off markedly above about 10 kHz.[79]) The energy in the harmonics is indicated logarithmically in the figure over a range of about 60 dB. (A 60 dB change in intensity corresponds to a change in pressure amplitude by a factor of one thousand.) As discussed in books on hearing, the sensation of loudness in the human ear is roughly logarithmic—something known as "Fechtner's Law."

According to violinists, the difference between various bows has more to do with the ease of playing than with the tone quality. For example, the elasticity of the bow has much to do with the ability to play rapid spiccato passages. A heavier bow may be able to produce louder tones, but that probably is at the expense of a decrease in the player's technical agility. As violinist Arnold put it, of the three studied, the Pecatte bow adheres best to the string in rapid passages without a lot of "chatter."

---

[79] Aging tends to reduce the upper limit of hearing in people by about 300 Hz per year after the age of 20—hence, often leading to near-total deafness by the age of 100.

5.21 Special Effects

5.21.1 Pizzicato

Because the science of plucked strings was discussed in detail in Chap. 3 in connection with harpsichords and guitars, as well as in the mathematical development in Appendix B, pizzicato (plucked string) playing has not been emphasized in the present chapter. It is, of course, an extremely important technique used in compositions ranging from the string quartets of Beethoven and Debussy to large orchestral works such as the third movement of Tchaikowsky's 4th *Symphony*.

As discussed in Chap. 3, the motion of the plucked string may be understood through a simple method first described by d'Alembert (see the discussion associated with Figs. 3.4 and 3.5) and detailed computations of the motion of the plucked string based on analysis of the wave equation were displayed in Figs. 3.7 and 3.8.

There are some important differences that arise in the motion of plucked violin strings as compared to those of guitars and harpsichords. First, the large damping factors in violin strings that are so important in permitting fast bowed passages result in a very rapid decay of single notes when the instrument is played in the pizzicato mode. In addition, the higher frequencies damp out very much faster than the low ones. As a result, the tone quality changes markedly even within the short time of the note's decay. (The decay times of individual notes on the violin, viola, and cello are very fast in comparison to those found in the guitar, harpsichord, and piano.) In addition, with pizzicato excitation the inharmonic nature of the string discussed in Chap. 4 in connection with Eq. (4.5) is not suppressed as it is when the instrument is bowed. The overtones of the plucked string are no longer precisely harmonics of the fundamental pitch. (As discussed earlier in this chapter, the slip-stick process in the bowed string forces the waveform to be periodic and locks the overtones in phase so that they are perfect harmonics of the fundamental tone.)

As with the bowed string, the plucked string has an overtone spectrum strongly dependent on the point of excitation. Plucking the string at the mid-point results in only odd harmonics and produces a clarinet-like sound. Plucking at a point one seventh of the way between the bridge and the nut results in both even and odd harmonics, but with the suppression of the "clashing" seventh harmonic. In general plucking the string at a point $P_0 = L/M$ away from the bridge, where L is the length of the string and M is an integer, results in suppression of the Mth harmonic and all of its multiples. Examples of this effect were given in Figs. 3.6 and 3.10 of Chap. 3. (Just as in the case of the bowed string, the spectrum of the radiated sound will differ from that of the vibrating string alone due to the frequency response of the bridge and top plate of the instrument.)

5.21.2   Harmonics

Nicolò Paganini (1782–1840) had much the same effect on violin playing that Liszt later had on that for the piano. One of his friends, violinist and conductor Guhr 1831, spent much of his professional life studying Paganini's technique and published a treatise on the virtuoso's playing in 1831. In addition to the unusual tunings that Paganini employed,[80] he was a master at the use of harmonics. According to Guhr, he used very thin strings to enhance harmonic production and played in high positions with the bow very near the bridge. The bridge, being flatter than usual, rendered the higher positions easier to handle. He played without a chin rest and kept his left hand with which he supported the violin in a fixed position; double-jointed fingers made it easier for him to reach the different locations on the fingerboard than would be the case with most people. According to Joseph Gold, who played on Paganini's Gaurneri del Gesu "Cannon," Paganini had worn an indentation in the side of the instrument where his thumb constantly rested.[81] Much of Guhr's treatise is devoted to the notation and technique the legendary master used to produce harmonics and double stops with harmonics. The latter, of course, were achieved by playing harmonics simultaneously on two different strings.

One probable source of confusion should be clarified at the start of our discussion: Paganini and most other musicians count harmonics differently than do physicists. Violinists usually think of the first overtone on a string as the "first harmonic," whereas the physicist starts counting with $n = 1$ corresponding to the fundamental frequency of the open string; $n = 2$ is then the first overtone or second harmonic, and so on. The frequencies of the various harmonics of the string are then simply given by

$$f_n = n(c/2L),   \tag{5.8}$$

where $n = 1, 2, 3, \ldots$ as described in earlier sections of this book. (Here again, frequency is in Hz, L is the length of the string, and c is the wave velocity for transverse vibration along the string.)

In practice, the performer plays a harmonic by lightly touching the string on a node for the harmonic desired while bowing near the bridge. For example, to excite the second harmonic (first overtone), one touches the string in the middle.[82] This

---

[80]For example, instead of playing his well-known first concerto as originally written in E-flat major, he tuned all his strings a semitone higher so that he could read it off in D-major. He never tuned in front of an audience and did not even let the conductor see the part he played from. (His first concerto is generally played in D major currently.)

[81]Joseph Gold (private communication) learned much about Paganini's technique from discussions with a person who actually knew H. W. Ernst, the famous nineteenth century violinist and friend of Paganini's.

[82]As demonstrated to me by Syoko Aki, some violinists are able to sustain the harmonic after removing the finger from the node merely by bowing in the right position.

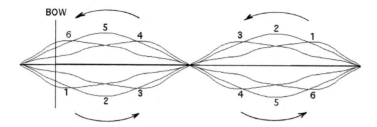

Fig. 5.55 Motion of the "kinks" (or pulses) during second-harmonic (first overtone) generation in an up-bow cycle. A node is created at the middle of the string by the violinist's finger that in effect reflects the kinks in the two halves of the string which circulate out-of-phase by 180°. The running waves actually go right through that node, but the net string motion behaves as if there were a "hard" reflection at the node. The string slips from the bow at point 1 and successive points in the cycle are indicated by the numbers 1 through 6. At point 6, the string sticks on the bow again and the cycle repeats (computed by the author using the Helmholtz model of the bowing process discussed earlier)

process forces a node in the middle that kills the fundamental frequency ($n = 1$) in much the same way that drilling a hole halfway down an organ pipe produces the first overtone ($n = 2$) in a harmonic flute. In both cases the frequency jumps up by a factor of two from the open string (or pipe). The motion of the string is now altered from that shown in Figs. 5.10 and 5.11. It is still described approximately by Eq. (5.4) except that the Fourier coefficients in Eq. (5.3) are now forced to be zero for all of the odd harmonics in order to produce a node in the middle of the string.

The result for the string motion is shown for second harmonic generation is shown in Fig. 5.55, where time increases in the steps 1,2,3,...,6 over one cycle of the note. At each instant of time, there is one continuous waveform over the full length of the string, but by choice of the coefficients, C_n, the waveform always has zero amplitude in the middle. The summation of these terms at different times throughout the cycle gives rise to two separate "kinks" in each half of the string rotating in the same direction, but out-of-phase by 180°. The system behaves as if there were two identical halves of the string vibrating separately at exactly the same frequency. Similar constructions can be made for each of the higher harmonics. Producing the higher harmonics on a violin proceeds in an analogous manner, except that the location of the right nodes to use is more complex. To get the third harmonic, two nodes exist at one-third the length of the string between the bridge and the nut; hence, there are two possible locations for the finger. Which is chosen might depend on the previous finger position. However, the fourth harmonic is trickier to excite. Although it has three nodes at intervals of one-quarter of the active string length, only the outer two of these will work. That is, placing the finger at 2/4 = 1/2 of the length just excites the second harmonic as discussed above.[83]

---

[83] Knut Guettler (2002) has discussed the mechanics of exciting such harmonics in more detail.

5.21.3   Tastiera Bowing and "Ghostly Tones"

Violinist Mari Kimura has developed an unusual version of "tastiera" bowing technique that produces a delicate "ghostly" sound.[84] The technique is somewhat related to the mechanics of playing normal harmonics in that the transverse excitation of the string must pass under one finger lightly pressed downward on the string. However, in this case the bow is placed far down over the fingerboard of the instrument where the neck of the violin emerges from the belly and the single finger that determines the vibrating length of string is placed between the bow and the bridge. The pitch of the note is determined by the length of string between that finger and the nut; however, the short length of string between the bridge and the finger is then automatically tuned (from the geometry involved) to a harmonic of that same pitch. (Only one finger can be put on the string at a time and that finger must move toward the nut when the player goes up a scale.) The bowing location at the end of the belly of the violin permits up and down motion of the bow without striking the body of the instrument. However, the technique is difficult in that one must avoid hitting an adjacent string with the bow. A photograph of her producing this effect was shown in the glowing review by New York Times music critic Edward Rothstein of her 1994 debut recital at Merkin Hall. (See Fig. 5.56.) The acoustic spectra obtained in this way for the note D bowed on the G string is shown at the right in Fig. 5.57 together with the normal spectra obtained for that same note when the string is bowed near the bridge. For the "ghostly" example shown, the active length of the bowed string between the finger and the nut was tuned to D at about 294 Hz, whereas the length of string between the bridge and the finger was (automatically) tuned to the second harmonic of that note at about 588 Hz. This resulted in an enhancement of the second harmonic (by about 5 dB) in the bowed acoustic spectra, whereas the fundamental at 294 Hz was largely attenuated (by over 30 dB) since the running wave must pass by the finger on the string to get to the bridge. Nevertheless, the various frequencies present in the "ghostly tone" spectra are all precise harmonics of the normal D fundamental at 294 Hz. (The D at 294 Hz is an octave below the third harmonic of the open G string, which itself is tuned to 196 Hz; hence all of these frequencies in the case shown are harmonically related.) It was verified that the "ghostly" sound was indeed coupled through the bridge to the top plate by studying the sound generated by piezoelectric elements placed in the bridge.

---

[84]The word "tastiera" means "keyboard" in Italian. Bowing over the keyboard is often described by that same term in the world of violinists.

Fig. 5.56 Position of the
bow and finger for the tastiera
bowing technique developed
by violinist Kimura (from a
photograph taken by David
Corio at Kimura's 1994 debut
recital in Merkin Hall.
Reproduced by permission of
Mari Kimura)

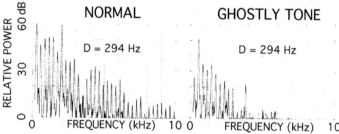

Fig. 5.57 Normal spectrum for D (left) bowed on the G string compared with that for the "Ghostly Tone" (right) achieved by bowing over the finger board on the same note. (See text.) The data were taken with the same microphone in the same relative positions with respect to the instrument. As with all studies of this type, there is some fluctuation in the relative harmonic distribution from one instant to the next due to changes in bow pressure and position. However, the main features (enhanced second harmonic and attenuated fundamental) were consistent

5.21.4 Tartini Tones

As is well known to accomplished violinists, playing two notes that are in a precise harmonic relationship results in the seeming appearance of a third, subtone lower than either of the two notes played. Indeed, the occurrence of the subtone is used by violinists to tune the top two notes forming a double third, fourth or fifth. If you don't hear the subtone, the top two notes are probably not harmonically in tune.[85]

---

[85] Meyer and Wit in the CAS Journal vol. 3, No. 5, (May 1998), pp. 22, 23 claim to have detected the emission of difference tones produced by a violin. However, their resultant tones were at least 70 dB

   Although they surely must have been known to violinists prior to his time, Giuseppe Tartini (1692–1770) claimed to have discovered these subtones, which he dubbed terzo sono in 1714. His main claim to credit for the discovery is based on the fact that he incorporated the phenomenon in his theory of harmony.[86] Although Tartini realized the terzi soni represented difference frequencies, it is not clear that he really understood how the third sounds were produced. (His derivations are based on tedious geometrical models that contain some mistakes.) Many contemporary violinists do not understand the source either and think that the subtones are actually physically present and come out of their violins! I first encountered that phenomenon when one of Broadus Erle's students was doing a senior research project with me on the violin spectrum in the Yale physics department. I had to take out a high-quality condensor microphone and spectrum analyzer to convince him that the subtones did not exist, except in his head.

   The main source of these subtones is nonlinearity in the human ear. (One of the reasons that the ear is usefully sensitive over such an enormous range of sound intensity is that it is nonlinear.[87]) By "nonlinearity" in a device is meant that the output signal (e.g., in the case of the ear, the rate of neurons firing in the cochlea) does not fall on a straight line as a function of the input amplitude (here, the pressure wave amplitude striking the ear drum). As a very rudimentary example, consider the presence of a quadratic nonlinearity in an otherwise linear system whose characteristic response is of the form,

$$\text{Output} = \text{Input} + (\text{Input})^2 . \tag{5.9}$$

If the *Input* signal consists of two different sound waves of the form

$$\text{Input} = a \sin A + b \sin B , \tag{5.10}$$

the *Output* signal will be

$$\text{Output} = a \sin A + b \sin B + a^2 \sin^2 A + 2ab \sin A \sin B + b^2 \cos^2 B . \tag{5.11}$$

   Here, A and B represent periodic, time-varying angles proportional to the frequencies involved (i.e., $A = \omega_A t$ and $B = \omega_B t$). Equation (5.10) may be

---

lower than the primary tones and it is not obvious that they were not produced by nonlinearities in their instrumentation. The psychoacoustic tones heard in the violinist's ear are extremely loud.

[86]Published as Trattato di Musica Secondo la Vera Scienza Dell' Armonia in Padova in 1754. This treatise contains a table of intervals such as fifths, fourths, thirds, and their difference-frequency subtones.

[87]This logarithmic nonlinearity is known as "Fechtner's Law" in psychology. The actual nonlinearity in the ear is a good deal more complex than the example used in the text to illustrate the generation of difference-frequency tones by a nonlinearity.

Frequency term	Output amplitude	Output intensity
A	a	a^2
B	b	b^2
$(A - B)$	ab	$a^2 b^2$
$(A + B)$	ab	$a^2 b^2$
$2A$	$a^2/2$	$a^4/4$
$2B$	$b^2/2$	$b^4/4$

Table 5.3 Frequencies, Output amplitudes, and Output intensities produced by the quadratic nonlinear expression in Eq. (5.10)

rewritten using simple trigonometric identities[88] to obtain the results in Table 5.3. The main point is that the quadratic nonlinearity generates output signals involving the sine or cosine of $(A-B)$, $(A+B)$, $2A$, and $2B$, hence at frequencies of $\omega_A \pm \omega_B$, $2\omega_A$, and $2\omega_B$ that are quite different from the input frequencies, ω_A and ω_B. That comes about from simple trigonometric identities for $\sin A$, $\sin B$, $\sin 2A$, and $\sin 2B$. The "Tartini Tones" are at the frequencies $\omega_A - \omega_B$, which are below either ω_A or ω_B and therefore quite conspicuous. Note that the nonlinear terms produce intensities effectively proportional to the fourth power of the original sound-wave amplitudes. Hence, if the original sound wave intensities fall off with distance r as $1/r^2$, the nonlinear terms will seem to fall of as $1/r^4$. Those extra signals will seem to fall off much more rapidly with distance than the original sound. Thus, a subtone that seems loud to the performer may not even be heard by the audience.

Now suppose that each of the tones A and B is represented by a Fourier series whose fundamental frequencies ("pitches") are harmonics of some lower frequency, ω, that will be the "subtone." Let's say A and B correspond to the second and third harmonics of ω. Then $A = 2\omega t$ and $B = 3\omega t$, and the fundamental frequency of the difference term is represented by $A - B = 1\omega t$. Now when we multiply the two Fourier series representing the input signals together, we will get a result of the form

$$\left(\sum_n = 1 a_n \sin 3n\omega t \right) \times \left(\sum_m = 1 a_m \sin 2m\omega t \right),$$

that will generate difference frequencies of the sort $(3n - 2m)\omega$ which are written specifically in Table 5.4 in units of ω.

Note from Table 5.4 that the difference frequencies $(3n - 2m)\omega$ themselves represent a harmonic (Fourier) series in the fundamental frequency ω. Hence, since the ear has a quadratic term in its nonlinear response, playing two notes a fifth apart (say C and G on the scale) whose pitches are at the frequencies 3ω and 2ω (which are the third and second harmonics of 1ω, hence in the ratio of 3:2) will generate in the brain the illusion of a Fourier series about the subtone one octave

---

[88]The trigonometric identities, which are among the most useful things taught in high school, are $2ab \sin A \sin B$ $ab \cos(A - B) - ab \cos(AB)$, $a^2 \sin^2 A = \frac{a^2}{2} - \frac{a^2}{2} \cos 2A$ and $b^2 \sin^2 B = \frac{b^2}{2} - \frac{b^2}{2} \cos 2B$.

Table 5.4 Difference
Frequencies in units of ω for
the example in the text for
integers $n, m = 1, 2, 3,\ldots$

n or m	$3n$	$2m$	$3n-2m$
1	3	2	1
2	6	4	2
3	9	6	3
4	12	8	4
5	15	10	5
, etc.			

below the frequency 2ω (i.e., an octave below the first C in the example) that is not really present in the incident sound wave. Listening for that subtone is the way violinists play double stops in tune. Of course, violins are not the only instruments that can create such subtones. If, say, a flutist and an oboist play a separate G and C that are harmonically in tune, one will also hear the subtone an octave below the C. (Alas, most orchestras do not play in tune that well. However, I have often noticed the subtone effect in performances by the wind section of the San Francisco Symphony.)

5.21.5   *Bowed Octave Subtones*

In April 1994, the young Japanese violinist named Mari Kimura astonished the audience at Merkin Auditorium in New York by performing several original compositions featuring what she referred to as subharmonics.[89] (See Rothstein 1994.) She had extended the violin's range by a full octave below the open G string without changing the tuning. Since then she has extended the technique further to include nearly all the chromatic intervals for one octave below the fundamental open string note by applying different amounts of bow pressure and different bow positions (See Kimura 2001). From Fig. 5.58, the bend in the string produced by heavy bowing with a string tension of 10.4 lbs (= 4.7 kg) was about 3°. Hence, the lateral force ($\approx T \sin 3$) was about 247 g. In her description of the technique, she said she was applying unusually heavy bowing force. The extreme bowing force can be estimated from the bending of the G string from the photograph in Fig. 5.58. The technique (sometimes known as "Russian bowing") had been suggested to her by her first teacher, Armand Weisbord (former concertmaster of the Ottowa Philharmonic) as a means of "developing smoothness and purity of sound."

Other violinists (for example, Norman Pickering and Michael Tree) have told me that their teachers suggested the same bowing technique for developing tone quality. While practicing this technique at slow bowing speeds, Kimura noticed the presence of various subtones at musical intervals in the otherwise scratchy sound.

---

[89] When Kimura demonstrated that she could play an entire scale below open G for a master class at Juilliard, one of the students exclaimed, "HOLY SHIT!".

Fig. 5.58 Mari Kimura playing an octave subtone on the G string (photograph by the author)

These subtones apparently are well-known to some violinists, who learned to avoid them in normal playing and never regarded them as a possible source of musical sound. Violinist Gold (1995, p. 83) believes that Paganini discovered a similar technique for playing octave subtones on the G string and used it in his concerts. Indeed, Gold suspects that Tartini was probably able to do it, also.[90] However, Kimura seems to be the only contemporary violinist who makes use of this effect in performance. She found a reliable and systematic way to produce a whole series of such tones with remarkably consistent intonation. As she put it. "they were supposed to be faint or ghostly, but I proved that wrong."[91] Direct measurements by the author showed that these subtones were typically produced at a total sound level of about 70 dB at a distance of about 4 ft from her violin, or at about the level of "loud speech." One musical problem Kimura explained is that they always come out "forte." She is able to produce these tones on almost any violin, although her own instrument is a fine Italian one. But it does help to use a bow that has strong weighting in the middle. The tones are best produced with a down bow near the frog for one can then obtain the most bowing force (Fig. 5.59).

The relative spectra on different subtones produced by Kimura observed in a series of measurements by the author were roughly similar. The waveform over one cycle and its corresponding spectral distribution are shown in Fig. 5.56 for the G one octave below the normal open G-string pitch. The periodic waveform at the subtone frequency seems phase-locked to the normal open string frequency. The relative harmonic content is shown on a linear scale for six of these notes in Fig. 5.57. However, the fundamental frequency components of these subtones was nearly non-

---

[90] Gold's source of information about Paganini was in private discussion with a friend who actually knew Mme H.W. Ernst—wife of the famous nineteenth-century violinist who was also a close friend of Paganini.

[91] See Neuwirth (1994) for a more extended background discussion.

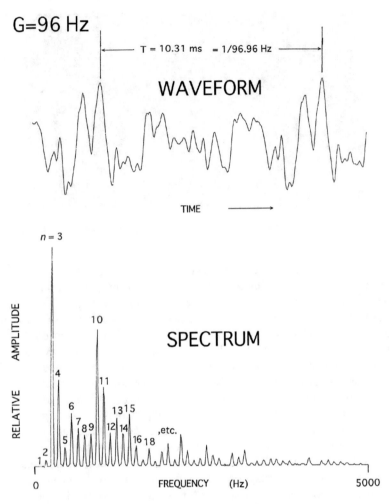

Fig. 5.59 Waveform and spectrum for the octave subtone below open G produced by violinist Mari Kimura

existent acoustically, being down typically by more than 25 dB from the stronger harmonics present. The fundamental is not of key importance in determining the apparent pitch of the note, although its absence does affect the tone quality. As is well known from various psychoacoustic experiments, the ear generates the illusion of the fundamental pitch from the presence of several harmonics, even when the fundamental is totally absent. The data in Figs. 5.56 and 5.57 were taken in a large room ($\approx$18,000 cubic-ft volume) with a 1-s reverberation time constant using a highly linear Sennheiser MKH 104 condensor microphone at about 1-m and a professional model Sony DAT recorder. (Recording was necessitated by noise from the spectrum analyzer. However, other measurements have shown that negligible distortion was introduced by the DAT recorder used.) (Fig. 5.60).

A very important question for the physical understanding of the production mechanism is whether or not the fundamental tone is actually produced in the string by the bowing process. Because it seemed probable that the acoustic response of the violin fell off so rapidly that the suboctave frequency of open G would be inaudible, we investigated whether that frequency was actually present in the vibrations of the bridge of the instrument. The technique used was the same as that discussed earlier in this chapter to investigate the bridge coupling in a violin bowed in the normal manner. Piezo electric transducers were fitted in the bridge of Kimura's violin for this purpose. Some data from this experiment are shown in Fig. 5.61. These results clearly demonstrate that the fundamental octave subtone is definitely contained in the vibration of the bridge. The reason it is not more prominent in the acoustic spectrum from the violin is simply that the coupling efficiency at the suboctave frequency is very low due to the small size of the violin top plate compared to the wavelength of the sound. Typically, the length of the top plate is about one tenth the wavelength of the subtone fundamental and it is expected that the coupling efficiency would vary roughly as the square of that ratio. Hence, it is not surprising to find that the fundamental component of the subtone is down by about 40 dB from the larger harmonics in the acoustic spectrum.

Kimura's approach to producing these remarkable subtones is best described in her own words (Kimura 2001). The bowing arm creates tones and the left fingers are kept in normal position pressing down on the fingerboard. (The subtones can also be produced with vibrato.) The bow pressure must be heavy, slow and uniform throughout the stroke and the bow hair kept flat. The higher the finger position, the closer the bow should be toward the bridge to maintain a subtone octave. The approximate relationships between finger positions and resultant subtones are shown in Fig. 5.60 and Table 5.5. In Table 5.6, the bow positions are shown for a new G string. As the string ages, Kimura found that the ease of producing the subtones increases, but the bow position must be moved farther away from the bridge. One important discovery she made (stimulated by a suggestion from Shigeru Yoshikawa) is that one can enhance the production of the subtones by twisting a rigid new string counterclockwise (as seen by the player looking down the string from the bridge). Twisting a new string in this way makes it work as effectively in producing subtones as an old worn string. Further, twisting the string one revolution counterclockwise shifts the subtone down one half step on the musical scale for the same finger and bow positions.[92] Some examples of the sound obtained are included in her web site article.[93] Her interest in producing such new tones was, of course, from a musical

---

[92]Measurements by the author showed that twisting the string counterclockwise in Kimura's sense slightly increases the torsion constant (K), presumably by tightening the metal winding on the string. Typically $K \approx 1135$ ergs/radian when a cylindrical weight is suspended by the 33 cm string and adjusted to produce the normal tension (10.4 lbs).

[93]Mari Kimura, http://pages.nyu.edu/~mk4.

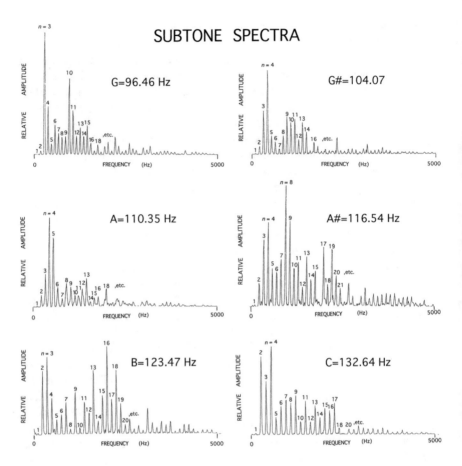

Fig. 5.60 Summary of spectra observed for six subtones produced below open G by violinist Mari Kimura. The data are all presented on the same linear amplitude scale over the range from 0 to 5000 Hz. The results were recorded digitally using a Sennheiser MKH-104 microphone and a Sony TCD-D10 PRO 2-channel DAT recorder and then analyzed with the Wavetek spectrum analyzer shown in Fig. 2.16. This microphone is one of few tested by the author that does NOT produce discernible difference frequencies from nonlinearities in its circuitry

point of view with the hope that they would enhance the range of the violin and be used in future compositions. Anyone who is skeptical of the musical applicability of this result should listen to the pieces she has composed using this technique.[94]

A variety of different explanations has been offered for the generation mechanism of these bowed violin string subtones. According to Lord Rayleigh (1877,

---

[94]Six Caprices for Subharmonics (1997); Gemini for Solo Violin (1995); ALT in Three Movements (1992) produced by Mari Kimura (private publication) and available from her privately by Email: mari.kimura@nyu.edu.

Fig. 5.61 Oscillograms taken from a piezo-electric element in the bridge of Kimura's violin. The top figure is the result of playing open G normally. The lower one, displayed with the same time scale, shows the pulses produced by the subtone an octave below open G. The result proves that the fundamental frequency was actually present in the bridge, hence, coupled from the string vibrations

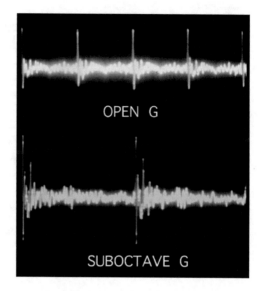

Table 5.5 G-string bow positions for subtone production used by Mari Kimura (Dimensions in cm)

	Sub-octave		Minor third below sub-octave	
Fingered note	Bridge to bow edge	Bow edge to stop	Bridge to bow edge	Bow edge to stop
C	2.70	19.68	3.33	19.53
B	3.02	20.40	3.25	21.11
A#	2.86	21.67	3.17	22.01
A	3.02	23.34	3.65	23.81
G#	3.65	25.00	3.81	24.45
G	4.37	26.67	4.13	26.75

Notes: The active G string length was 32.38 cm from bridge to nut and the string diameter 0.79 mm. (Metal-wrapped nylon, No. 123 by Thomastik-Intels, of Vienna.) End of fingerboard was 27.00 cm from the nut. As the string ages (or if the string is twisted), the optimum bow positions move away from the bridge (measurements made by the author)

p. 253), Savart's experiments on longitudinal vibrations occasionally showed a tone whose pitch was an octave below that of the normal longitudinal vibration frequency. Rayleigh stated that Terquem had concluded the cause was a transverse vibration whose own frequency was approximately the same as the sub-octave longitudinal vibration. However, it is unlikely that longitudinal resonances could be excited by Kimura's particular bowing technique.

Kimura herself thinks of her subtones as subharmonics, an interpretation that is certainly tempting from a musical point of view and that might be justified on the basis of nonlinearity in the restoring force of the violin string due to the large bowing force and extreme deflection of the string. (Hooke's Law probably breaks down to some extent in the bowing process.) Interestingly, McIntyre et al. (1982) discussed the production of subharmonic perturbations in the normal periodic waveform.

Table 5.6 Alternative Bow Locations for Octave subtone production (distances are in cm)

Note	Distance to bridge	Distance to nut	Fingered note
G	5	26.8	
	26.8	5	
	4.7	27.4	
	27.4	4.7	
D	23	8.5	B
	8.5	23	
F♯	20	12.8	D♯
	12.8	20	
D♯	8	24.3	F♯
	24	8.4	
B♭	16.4	16.4	C

Source: Unpublished research by Julie Haas and the author at Yale University. Note that Julie Haas' violin and G string were quite different from those used by Mari Kamura to produce the data in Table 5.5

They concluded that if the dissipation in the bowing process is sufficiently low and reflection from the bow during sticking is sufficiently strong, "negative resistance" at the bow during slipping causes subharmonics to become self-excited. They concluded that this process was the physical mechanism underlying a mathematical instability in the oscillation of bowed strings discovered earlier by Friedlander (1953). They referred to these sounds as "Ghostly Subharmonics" and noted that they were irregular, very faint, seldom persisted for more than a few subharmonic cycles, and were "a bit like the response of a cavity to random-noise excitation." These descriptions definitely do not fit the characteristics of the subtones produced by Kimura. Her subtones are quite loud (typically, about 70 dB at about 4 ft from the violin), unmistakable in their presence, and quite coherent.[95]

Although it is tempting from a musical point of view to refer to Kimura's subtones as "subharmonics," there are several reasons why they are not likely to be true subharmonics of a nonlinear oscillator in the conventional physics sense. As discussed in Appendix A, the conditions for producing such subharmonics usually involve substantial instability. The potential well from the quadratic force term needed to produce a subharmonic tends to make the oscillator producing the result very unstable. But Kimura is able to sustain her subtones for many seconds using a downbow.

---

[95]The subharmonics described by McIntyre et al. (1982) were incoherent. The difference between a coherent audio signal and an incoherent one is similar to that between the filtered output from a light bulb and the output amplitude from a laser. In the former case, the amplitude fluctuates wildly at low intensities, as well as the phase. In the latter case, both the amplitude and phase are constant within normal statistical fluctuation limits. (e.g., fluctuations in the photon count over a prescribed time vary as the square root of the number of photons detected.)

The subtones might result from nonlinear mixing with two modes of oscillation in the string by the bow motion under heavy pressure. That type of process should generate acoustic difference frequencies in the sound from the violin similar to those created inside the ear by the Tartini effect discussed in the previous section. Such difference frequencies could occur at the frequencies produced by Kimura. However, it is not so obvious how they would be locked to fractions of the normally stopped G-string frequency.

The bow positions relative to the bridge used by Kimura (see Fig. 5.55) are clearly desirable for rapid switching between normal playing and subtone production—especially in the manner employed by Kimura in her own compositions. However, there is one important aspect of the production mechanism that is obscured by these data. Namely, many of the subtones can be produced by bowing at symmetrically opposite positions from the midpoint of the active length of string as illustrated in Table 5.5. This aspect of the problem was discovered by the author and one of his special students at Yale University, a violinist named Julie Haas. We had initially misinterpreted the photograph reproduced here in Fig. 5.53, thinking that the bowing position shown was being used by Kimura to produce subtones. We then realized that there were two stable positions for the bow with respect to the bridge that worked and that were located symmetrically about the midpoint of the active length of string. Our measurements were made sometime after the publication of the Rothstein (1994) review and are reproduced in Table 5.6. The importance of this observation is that it rules out several cleverly designed mechanisms to explain how pulses propagating between the bow and the nut could be transmitted to the bridge at the subtone frequency. The point here is that the subtone production mechanism works even when the long dimension of the string is always in direct contact with the bridge.

5.21.6 Torsional Waves

The involvement of torsional waves has been suggested by several authors as a possible explanation of these subtones,[96] and certainly the effect on the subtones from twisting the string described above tends to reinforce that suspicion. Unfortunately, the previous studies have been inconclusive—largely due to the absence of any quantitative determination of the torsional wave velocity in the bowed string. As shown in Appendix B, torsional waves should obey a wave equation similar to that for lateral vibrations and the boundary conditions for reflection of these waves are also much the same. Determining that velocity is not trivial since exciting the torsional waves takes a great deal of torque applied very quickly and the waves damp out very fast. The most effective way to produce them indeed seems to be

---

[96]See, e.g., Hanson et al. (1994) and Shigeru Yoshikawa (1997).

Fig. 5.62 Torsional deflection of a 0.033 in. silver-wound G string under 10.4 lbs of tension at a bow weight of 130 g applied 2 cm from the bridge. The 1-cm wide bow hair is moving to the left horizontally and the rotation of the string downward from the horizontal is about 40° as indicated by a cat whisker glued to the string. Under these conditions Pickering estimated the coefficient of friction was about 58% (figure reproduced from Pickering 1991 with permission)

with a violin bow. (Several attempts by the author at pulsed excitation of torsional waves on a violin G string under normal tension with motors did not work.)

Torsional waves in a bowed cello string were observed as early as 1896 by Cornu,[97] and the torque on violin strings produced under continuous heavy bowing pressure was studied quantitatively by Pickering (1991). Pickering's conditions (see Fig. 5.62) were roughly similar to those used by Kimura, except that the bow weight was about half that implied by Fig. 5.58.

Stroboscopic illumination of the bowed string on Kimura's violin was viewed with a digital video camera and showed two important features of the motion: First, the open G-string waveform from normal bowing consisted entirely of lateral oscillation of the string. Second, the string motion during the subtone production had an extremely strong torsional component in addition to a lateral component of comparable amplitude. But the torsional mode frequency was at least 50 percent higher than that of the normal lateral vibration mode. To see these effects, a small strip of stiff paper was glued to the midpoint and other points on the string and illuminated with a stroboscope operated at the normal open G-string frequency. (See Fig. 5.63.) Strong torsional motion was found for all of the subtone frequencies produced by Kimura. Clearly, both vibrational and torsional standing waves must play some role in the production of these unusual sounds.

As shown in Appendix B, torsional waves satisfy the same type of wave equation as the more common lateral string vibrations discussed before, and consequently

---

[97] M.A. Cornu, Journal de Physique vol. 3, p. 5.1 (1896).

Fig. 5.63 One frame from a video camera recording using stroboscopic illumination near the normal open-G frequency. Here, a single small piece of paper was glued to the middle of the string. The figure illustrates the maximum rotational deflection when the subtone is produced. The fact that there are at least two images of the single piece of paper means that the torsional frequency was much greater than 196 Hz

the boundary conditions for the reflection of torsional waves at the fixed ends of the string are similar to those for lateral vibrations. Since the torsional wave velocity is significantly higher than the normal lateral wave velocity, difference frequencies involving submultiples of the normal transverse string resonances could well be produced through non-linear mixing by the bow of these two forms of resonance. Such an interpretation could explain why the subtone spectrum is phase-locked to the normal harmonics (and subharmonic) of the G string.

5.22   Possible Models of Subtone Production

Although it is hard to be sure precisely what mechanism is used by Kimura to produce octave subtones, there are several ways in which this production might come about.

5.22.1   Subharmonic Generation

It is appropriate to start the discussion with the mechanism that Kimura herself has tacitly suggested. A nonlinear oscillator with a strong quadratic term in the spring constant could certainly produce an octave subtone, or a true subharmonic. One might argue that the heavy bowing pressure applied to produce the violin subtones would tend to produce such a nonlinearity in the restoring spring constant. This case is examined in some detail in Appendix A. Instead of the normal form of Hooke's

Law, one assumes there that the restoring force from the spring can be expanded in a power series in the displacement from equilibrium as done in Eq. (A.48) of Appendix A. One then solves the differential equation for the oscillator with that nonlinear force term present. An immediate problem arises in the shallowness of the potential well for the oscillator as shown in Fig. A.7 of Appendix A. Although stable solutions may be obtained at very low excitation levels (see Fig. A.8 and, especially, Fig. A.10), relatively little force is required to cause the oscillator to fly apart—which in the present case would correspond to the bow slipping on the string. This difficulty might be overcome through the addition of a cubic term to the restoring force as shown at the right side of Fig. A.7. Although that assumption may actually provide a realistic model of the bowing interaction with the string and does stabilize the oscillator, solutions of the equations of motion at high restoring force in that case show signs of chaotic motion in which the waveform is no longer precisely periodic. (See Fig. A.9.) The waveform does have a basic fundamental repetition rate at the driving frequency, but the shape of the waveform varies from cycle to cycle. In contrast the waveforms for the actual subtones produced by Kimura are precisely periodic over long periods of time and have precisely defined harmonics. (See Fig. 5.60.) For these reasons, it seems unlikely that the stable octave subtones obtained by Kimura are the result of a subharmonic solution to a nonlinear oscillator.

5.22.2 Coincidence Model

The stick-slip mechanism of the bow moving on the string results in the generation of pulses traveling between the bow and the nut (the stop at the tail end of the violin). It can therefore be argued that the heavy bowing pressure results in such pulses being reflected back and forth between the bow and the nut. Some of these pulses will result from lateral vibration of the string and some from the torsional waves generated on the string. Since the torsional waves are expected to have a higher velocity of propagation than the lateral waves, the successive pulses from these two different wave motions will generally not arrive at the bow at the same time. One then argues that when the two types of pulse arrive in coincidence, the combined pulse (or "kink") is large enough to slip through the bow and get to the bridge. This model is illustrated schematically in Fig. 5.64.

The basic idea is that two pulses are launched simultaneously in the two separate loops when the bow slips on the string. The two loops have different delay periods. The long period is associated with lateral vibration waves and the short period with torsional wave propagation. The pulses in each loop are recirculated by the bowing process which provides some amplification to compensate for energy loss of the pulses during their round trip travel.

When the two pulses arrive within the resolving time of the coincidence circuit in Fig. 5.64, the combined amplitudes are assumed to be large enough to cause the bow to slip resulting in a pulse traveling through to the bridge. The time between coincidence pulses is determined by the number of round trips made in both loops

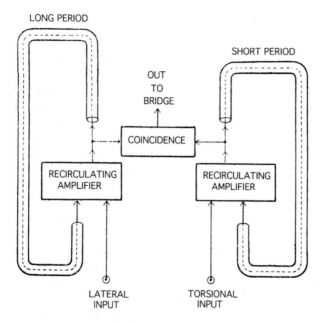

Fig. 5.64 Schematic model of a subtone generator that might produce results similar to those obtained by Kimura. (The schematic arrangement is similar to one devised by Peter Kindlmann in 1966 to measure lifetimes of excited atomic states except that the delay lines in his experiment were closely spaced and coincidences generally occurred in times much less than the "Long Period.")

before the next coincidence occurs. If that time is greater than the circulation time for the slow loop, a subtone is produced. For example, if the long-to-short pulse frequencies are in the ratio of 2:3, the output of the coincidence circuit will be at half the circulation frequency in the long loop. Hence, if the long loop corresponds to the normal lateral vibration resonance for the G string, the output signal will be at precisely half the normal open-G frequency. It is also probable that the harmonics of the subtone in that case would be phase-locked to the harmonics of the normal open G-string frequency. (See Fig. 5.65.)

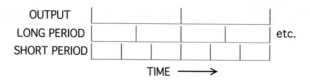

Fig. 5.65 Pulse sequences for the model shown in Fig. 5.61 for the case where the long-to-short pulse frequencies are in the ratio of 2:3. Note that in this case, the output coincidence pulse rate is half that for the long period pulses. Hence if the long period pulse rate corresponded to lateral vibrations excited near the open G-string frequency, the output frequency would be at the sub octave

5.22.3   Other Stick-Slip Models

There are ways in which a subtone frequency might conceivably be produced simply by lateral or torsional waves alone. Consider the diagram illustrated in Fig. 5.65 in the case of lateral waves (although much the same model could be used with torsional waves). Here, we consider the difference that might be produced between "soft" and "hard" reflections. At step 1 (the top of the figure), the bow slips and a pulse is launched in the direction of the nut. A "hard" reflection (steps 2 and 3) must be produced at the nut; i.e., where the amplitude of the reflected wave changes sign so that the sum of the incident and reflected wave at that point is zero. But with strong down bow, the bow is moving in the same direction as the incident pulse in step 4. Hence, a soft reflection may occur at that point (step 5). That wave is reflected back toward the nut where it again undergoes a "hard" reflection (steps 6 and 7.) But now when the reflected wave from the nut reaches the bow (step 8), its polarity is opposite to the motion of the bow, causing the bow to slip. A pulse is then transmitted to the bridge and a positive pulse is sent toward the nut.

There are ways in which a subtone frequency might conceivably be produced simply by lateral or torsional waves alone. Consider the diagram illustrated in Fig. 5.66 in the case of lateral waves (although much the same model could be used with torsional waves). Here, we consider the difference that might be produced between "soft" and "hard" reflections. At step 1 (the top of the figure), the bow slips and a pulse is launched in the direction of the nut. A "hard" reflection (steps 2 and 3) must be produced at the nut; i.e., where the amplitude of the reflected wave changes sign so that the sum of the incident and reflected wave at that point is zero. But with strong down bow, the bow is moving in the same direction as the incident pulse in step 4. Hence, a soft reflection may occur at that point (step 5). That wave is reflected back toward the nut where it again undergoes a "hard" reflection (steps 6 and 7). But now when the reflected wave from the nut reaches the bow (step 8), its polarity is opposite to the motion of the bow, causing the bow to slip. A pulse is then transmitted to the bridge and a positive pulse is reflected toward the nut.

The point in this model is that the pulses travel the long length of the string four times before getting through to the bridge. Hence, the frequency of pulses arriving at the bridge would be about half that for the normal lateral bowing frequency. Much the same argument could be made for torsional wave pulses.

This model is similar to the one proposed by Shigeru Yoshikawa (1997, Fig. 2).

5.22.4   Heterodyne Model

It is well-known that nonlinearities can result in difference and sum frequencies between two applied signals. This phenomenon results from the multiplication of sine waves of different frequencies in the nonlinearity and can be understood by application of simple trigonometric identities. If the nonlinearity involves a power series in the sum of the input signals at frequencies F_A and F_B, one can expect output signals at additional frequencies given by

$$F_{\text{Out}} = nF_A \mp mF_B \quad \text{where} \quad n, m = 1, 2, 3, \ldots. \tag{5.12}$$

(This phenomenon is used to advantage in many radio receivers where a local high frequency oscillator output is mixed with radio wave frequencies to produce signals at lower frequencies.)

As shown in Fig. 5.60, the third and fourth harmonics of the subtone frequency are strongly excited in the acoustic spectrum in most cases. For example, in the case of the open G-string resonance, if the torsional wave velocity were adjusted by bowing pressure to produce a torsional resonance frequency of 294 Hz, or three times the observed subtone frequency (96 Hz), the fourth harmonic of the subtone

Fig. 5.66 Slip-stick model for generating a subtone at half the normal string frequency

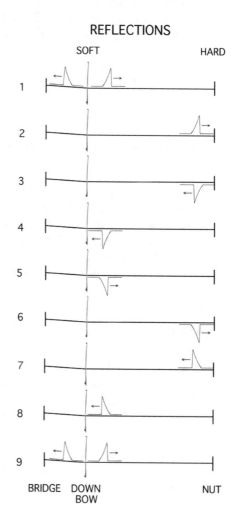

Table 5.7 Relation between observed open-G harmonic frequencies and octave subtone harmonics

Open G harmonic (n)	1	2	3	4,	etc				
Frequency (Hz)	196	392	588	784					
Subtone harmonic (n)	1	2	3	4	5	6	7	8,	etc
Frequency (Hz)	98	196	294	392	490	588	686	784	

would tend to phase lock with the second harmonic of the normal open-G frequency. Similarly, the sixth harmonic of the subtone would lock with the third harmonic of the normal open-G frequency. The difference between the second harmonic of the normal open-G frequency and the third harmonic of the subtone would then provide the sub-octave frequency. (See Table 5.7.)

Such difference frequencies could be provided by the nonlinearity introduced by the heavy bowing process. Similarly, that mechanism would also produce sum frequencies of the type

$$\text{Open G} + \text{third harmonic of subtone} = \text{fifth harmonic of the subtone} \qquad (5.13)$$

and

$$\text{Second Harmonic of Open G} + \text{Third harmonic of subtone} = \text{Seventh harmonic of subtone} \qquad (5.14)$$

Presumably, similar relations would hold for the other subtones reported above. Unfortunately, it is exceedingly difficult to make precise determinations of the torsional resonance frequencies during the actual bowing process and the above interpretation is not conclusive. However, it is the only interpretation advanced so far (other than direct subharmonic production) that would provide equivalent results for symmetrically placed bowing positions about the center of the active string length.

Problems

5.1 If a violin is tuned to A=440 Hz and the other strings are tuned in fifths starting with the A, what are their frequencies? What would these frequencies be on the Well-Tempered Scale? (Hint: when musicians say two strings are tuned in fifths, the second harmonic of one is equal to the third harmonic of the other. Also see Appendix D.)

5.2 The strings on a certain Stradivarius violin are all 33 cm long. If the A string is tuned to 440 Hz, what is the lateral wave velocity on the string? What is it for a G string tuned to 195.55 Hz?

5.3 (3) A silver-wrapped G string 33 cm long requires a tension of 10.4 lbs = 4727 g to tune the string to 195.55 Hz. What is the density of the string? (Hint: the wave velocity is 12,906 cm/s.)

5.4 The combined weight of the two volumes on violin research by Hutchins and Benade (1997) is 10.4 lbs. A curious student drills a small hole through the centroids of both volumes and suspends the two on a silver-wound violin G string from a clamp 33 cm above the top volume. Noting that the density of the string is 2.838×10^{-5} g/cm^3, what is it frequency?

5.5 Eugene Commins, a violinist and physics professor, plays a double third consisting of the first E on the D string and the open G string. What subtone does he hear in his head?

5.6 After playing his bagpipe in "Pervertimento for bagpipes, bicycle and balloons," Maurice Eisenstadt picked up his violin for the next piece. He tuned his D and A strings in a perfect fifth with A=440 Hz. Why didn't he notice a subtone at the difference frequency?

5.7 K.S. Bostwick and R.O. Prum (Science, vol. 309, July 29, 2005, p. 736) reported that the club-winged manakin bird gets a violin-like sound using one ridged feather as a bow to excite other hollow feathers, producing fundamental frequencies of 1.49 kHz lasting 1/3 s with at least four harmonics. Assuming there are eight ridges on the "bow," how many strokes per second does the bow make?

Chapter 6
The Voice

The human voice is almost certainly the oldest and possibly the most beautiful of all the musical instruments. The unique physiological characteristics of the human voice have also led to the development of speech and spoken language. Indeed, most written languages and forms of music have the characteristics of human speech thoroughly imbedded in them from the Lieder of Schubert to more contemporary Sprechstimme.

6.1 The Vocal Tract and Speech

6.1.1 The Vocal Cords

The ancient Greeks recognized the larynx as the source of vocal sound and thought the "glottis"—the space between the vocal cords—was a kind of fluttering tongue. ("Glottis" is Greek for "tongue.") However, it was not until 1744 that Antoine Ferrein showed that the sound produced in singing actually originated through vibration of the vocal cords (cordes vocales)—a term that he himself coined. In 1854, a Spanish singing teacher named Manuel Garcia used sunlight and a dental mirror placed at the back of the throat to study the vocal cords of one of his students directly.[1]

Pioneering studies of the vocal tract were conducted in the early days of the Bell Laboratories. There in 1940, D. W. Farnsworth extended the earlier work of Manuel Garcia by viewing the vocal cords with a high-speed motion picture camera especially designed for that purpose and capable of taking pictures at up

---

[1] The historic references are from Arnold (1973), Kaplan (1971), and Sataloff (1992). For more detailed properties of the larynx in recent accounts, see Sundberg (1987), Davis and Fletcher (1996), and Stevens (1998).

© Springer Nature Switzerland AG 2018
W. R. Bennett, Jr., *The Science of Musical Sound*,
https://doi.org/10.1007/978-3-319-92796-1_6

Fig. 6.1 Method used by Farnsworth for taking motion pictures of the vocal cords (from Farnsworth 1940; reproduced with the permission of Lucent Bell Laboratories)

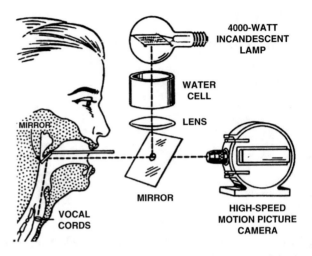

to 4000 frames per second. The film was then played back in slow motion at about 16 frames per second. Thus, if the vocal cords were originally vibrating at 250 Hz, one cycle of vibration was spread out over a time of about 1 s in playback. More recently, Farnsworth's technique has been superceded by stroboscopic video technology using fiber-optic cables through the nasal passages that accomplish the same objective with greater simplicity in the case of periodic motion. There, the stroboscope is adjusted to flash at a slightly different rate than the resonant frequency of the vocal cords so that one can see the vocal cords moving directly in slow motion (Fig. 6.1).

The first thing one notices from Farnsworth's original photographs (Fig. 6.2) is that the vocal cords are not really "cords" at all but consist of a pair of muscular bands that open and close in the air passage from the lungs. Hence, the eighteenth-century term "vocal cord" has been replaced in recent years by the more physiologically accurate term, "vocal fold." The vocal folds have a resonant frequency that can be varied over a sizable range by muscles in the larynx and play a similar role in producing sound in the human vocal tract to that of the lips on the mouthpiece of a brass instrument.

In the normal breathing position, the vocal folds are widely separated and form a triangular opening as viewed from above. When "voiced" sounds are produced, the folds are drawn close together by muscles of the larynx (Fig. 6.2). Steady flow of air from the lungs at low velocities through the narrowed opening produces an oscillation in the vocal fold that results in a nearly sinusoidal variation of sound pressure in the throat. This fundamental resonance (which we will designate F_0) is typically about 110 Hz for adult males, an octave higher for adult women, and in the order of 300 Hz for children. (See Stevens 1998.)

Figure 6.3 shows the opening and closing of the glottis over one complete cycle at a frequency of $F_0 \approx 110$ Hz. The vocal folds move in opposite directions laterally inside the throat. In the fundamental spatial mode of oscillation, the shape of each

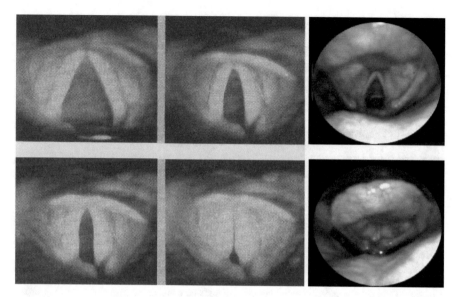

Fig. 6.2 The "vocal cords" as seen from above and back. The pictures illustrate the change in shape of the vocal folds from normal breathing (upper left and extreme upper right) to "voicing" (lower middle and far right). Black and white photos are from Farnsworth 1940, reproduced with the permission of Lucent Bell Laboratories. Color photos (right) of the author's vocal folds were photographed by Lauren Campe using fiber optics

band at maximum amplitude is roughly half of a sine wave and similar to that of the lateral motion of a vibrating string in its first resonant mode. The controlling muscles raise the pitch of the voice by stretching the length of the vocal fold, sometimes by as much as 30%. That increases the tension, hence the resonant frequency. However, because the length of the folds is also increased in this process, the situation is more complex than in a simple vibrating string. The second spatial mode of vibration (the "falsetto" register) has a frequency that is not as high as the second harmonic of the speaking voice. Normally, the vocal folds oscillate in the lowest, "heavy register" mode. The motion of the folds in the different registers of the voice is complicated and will be discussed in more detail later.

An often-quoted model due to Benade (1976) of the vocal fold consists of a mechanical oscillator with a mass moving against a spring as shown in Fig. 6.4. Friction slows down the motion and the mass is driven by the Bernoulli effect.[2] Air coming from the lungs passes through a constricted region where the conservation

---

[2]The Bernoulli principle applies to steady streamline flow, which requires that

$$S\rho V = constant$$

where V is the flow velocity, ρ is the fluid density, and S is the cross-sectional area within a tube bounded by streamlines (here, the walls of the wind pipe and larynx). Conservation of energy requires that

$$P + \rho V^2/2 = constant$$

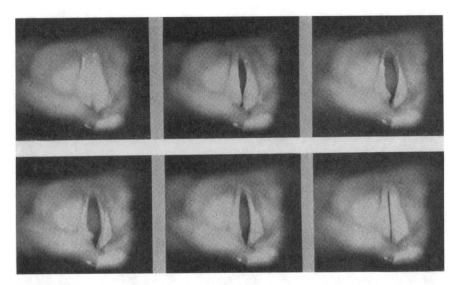

Fig. 6.3 The opening and closing of the glottis over one cycle at the resonant frequency of the voice (Farnsworth 1940; reproduced with the permission of Lucent Bell Laboratories)

of both energy and air molecules in the flow causes the velocity to speed up and the pressure to decrease. The effect is similar to fluid flow in a laboratory "aspirator" or in the "Venturi tube" of an automobile carburetor. The decreased pressure in the constriction pulls the mass to the right in Fig. 6.4 until the restoring force from the spring (the muscles of the larynx) becomes large enough to reverse the motion. In the case of the real larynx (Fig. 6.5), the muscles in the sidewalls play the role of the restoring spring.[3] The sharp edge at the top of the "vocal fold" (Fig. 6.4) in the real larynx can also produce turbulence in the air flow that contributes to the broadband noise source used when one whispers. At very low air flow rates, a nearly sinusoidal motion of the vocal folds occurs at resonance. However, as the air flow increases, the motion of the mass actually cuts off the air flow altogether during more than half of the cycle, producing a periodic waveform that can contain dozens of harmonics. The harmonic content is extremely important both in determining the tonal color of the sound and as a source of frequencies that can be filtered out by the cavities of the mouth and nose to produce recognizable speech.

---

which is known as Bernoulli's equation, where P is the pressure. When the streamlines are closely spaced (as in the larynx), ρV is large, hence P is small. Thus, the pressure is reduced in the larynx.

[3]See Davis and Fletcher (1996) for a discussion of more elaborate mechanical models.

Fig. 6.4 Mechanical model
of the larynx (after Benade
1976)

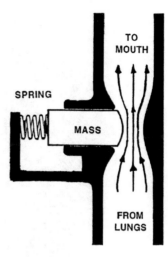

Fig. 6.5 Anatomical drawing
of the larynx from the front
(Gray 1918)

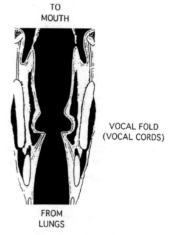

6.1.2   Formants or Resonances in the Vocal Tract

The overall working of the vocal tract is shown schematically in Fig. 6.6. The main
energy source is air flow from the lungs. The air next runs through the constricted
region of the larynx where the Bernoulli effect produces forced oscillation of the
vocal folds. The sound wave created there, now rich in harmonics of the vocal fold
resonance frequency F_0, is next coupled to resonant cavities ("Formants") in the
mouth and nasal passage. Traditionally, these cavity resonances have been regarded
as broad transmission filters, through which the sound passes to the outside world
by way of the mouth and nostrils. When the palate is closed, the sound wave travels
over the tongue and to the mouth, where it may be joined by a hissing sound from
turbulent air flow over the front teeth. With the tongue in the right position, that
turbulence produces sibilants—the "sss" sounds in speech.

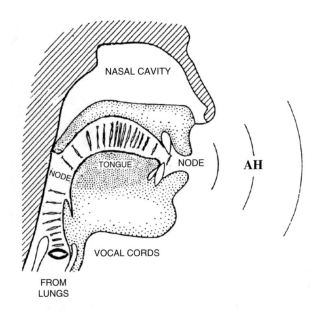

Fig. 6.6 The vocal tract, showing the nodes and pressure maxima for the second closed-pipe resonance, or F_2 formant. (The palate has closed off the nasal resonances in the figure.) The tongue position is that for pronouncing the vowel sound "AH." By arching the tongue in the middle, one shifts the resonant frequency up and obtains the vowel sound "EE." As indicated schematically by the spacing of the curved lines, the sound flux is greater inside the vocal tract than that radiated through the mouth. (Here, close spacing means high standing wave pressure amplitude.) That occurs because some of the sound is reflected back into the vocal tract at the speaker's mouth to create a standing wave resonance, which here has two nodes and two pressure maxima. The sound inside the speaker's head may be much louder than the sound coming out of his mouth (leading to the phenomenon called "mumbling")

These resonances in the vocal tract are similar to the modes in a closed organ pipe. The principal mode is distributed over the region shown in Fig. 6.6, which has a length L (vocal folds-to-mouth) of about 17–18 cm in the adult male. As shown in Chap. 1, the fundamental resonance of such a pipe is given by $c/4L$, where c is the speed of sound and L is the pipe length. Hence, the first resonance (formant F_1), or principal mode of the vocal tract, is at about 500 Hz. The reader will note that this resonance is much higher than the pitch of the voice in an adult male. It is typically at about the fourth harmonic of F_0.

That means the coupling to the air of the fundamental frequency of the vocal folds is very poor, a matter of special importance to bass singers. The next closed-pipe resonances of the vocal tract are at odd harmonics of 500 Hz. Hence, the second resonance (formant F_2) occurs at about 1500 Hz and the third (F_3) at about 2500 Hz. But, these resonances or formants are very broad (typically about 200 Hz wide) in the normal speaking voice because of the large fractional energy loss by the sound waves coming out of the mouth.

As indicated by the pairs of curved wave fronts in Fig. 6.6, standing waves occur in the vocal tract resonances. (As shown in Chap. 1, a standing wave is made up by the sum of two running waves of the same frequency going in opposite directions.) Each resonance has a node (pressure minimum) at the mouth and an antinode (pressure maximum) at the vocal folds. Because the tongue is located at the pressure maximum for the second closed-pipe resonance, that resonance (formant F_2) is strongly tunable by moving the tongue. As we will see later, the second formant plays a dominant role in determining different vowel and diphthong sounds. When the soft palate is open, another set of resonances in the nasal cavities is coupled to the vocal folds and to the outside air through the nose. The nasal resonances are not readily tunable because the cavity dimensions are determined by bones in the skull. You can easily detect the presence of the dominant formants in your own vocal tract by letting air blow into your mouth, thereby creating "unvoiced" sounds, or "whispers." Turbulence is created at the mouth and plays a role analogous to the vocal folds, but in reverse. If you wear ear protectors to shield your ears from outside noises, you can easily detect those resonances while riding in an open vehicle at low speeds. Let some air enter your mouth and then try mouthing various vowels or syllables. The method is especially sensitive because sound travels from the back of the throat through the Eustachian tubes to the middle ear. (It is also something to do to relieve boredom while mowing the lawn on a tractor; in that situation, the ear protectors are a good idea anyway.)[4]

The harmonics of the vocal folds are sharply defined and are uniformly spaced by F_0 throughout much of the lower audio range up to perhaps 7 kHz. In contrast, the turbulent hissing from air moving over the front teeth produces a broad spectrum of noise extending from about 5 to over 10 kHz. Hence, people who develop hearing loss at the upper end of the spectrum often have trouble detecting sibilants. Although the resonant widths of the principal cavities of the mouth, nose, and throat are typically in the order of 200 Hz, the frequencies at resonance can be varied by factors in the order of two (especially in the case of the second formant, F_2) by moving the tongue and the muscles controlling the face and mouth. The vocal folds themselves may, of course, be varied substantially in pitch using the muscles around the larynx.

6.1.3 Talking Birds and Other Animal Sounds

In contrast, the voiced sounds emitted by animals tend to be fairly uniform in relative spectral distribution, although they can vary in pitch, as with the howling of a wolf. There are two basic requirements for producing speech: (1) a sound source that has a

---

[4]Erwin Hahn (of "spin echo" fame in nuclear magnetic resonance) had a novel way to demonstrate these resonances. When an aspiring physicist once asked him for a selection of his work, Prof. Hahn made a fist and used it to play the William Tell Overture on his forehead.

Fig. 6.7 Mozart's musical
bird

waveform rich in harmonic content over the audio spectrum; (2) a group of formant resonances that can be varied at will by the animal in question.

Song birds can usually be ruled out on the basis of harmonic content of the waveform. In most instances, they produce a whistle-like sound consisting of little more than a sine wave, although with a frequency that can be varied over an octave range and with interesting rhythmic quality. However, the European starling, Sturnus vulgaris, is an exception. According to West and King (1990, p. 107), a contemporary one learned the phrase, "Does Hammacher Schlemmer have a toll-free number?"

Mozart's pet starling was also quite musical (Fig. 6.7). When the composer walked into the shop to purchase the bird on 27 May 1784, it greeted him with the opening theme from the last movement of his piano concerto No. 17 in G-major—except that the Gs were all sharped. Some inferred that Mozart had actually stolen the theme from the bird, but the truth is obviously different for the concerto was completed some 6 weeks earlier on 12 April 1784. One explanation proposed was that the pet store owner had slyly taught the theme to the bird ahead of time. The Starling has a two-part voice organ, enabling it to sing two different melodies (in different keys) at once. According to Kroodsma (2005, p. 274), they also have perfect pitch. It has been suggested that Mozart's "A Musical Joke," which he wrote shortly after the pet starling's death in June 1787, was inspired by that ability and the off-key singing of the bird. (See West and King 1990.)

A reed-like tone quality such as the "caw" from a crow offers greater possibility for speech. Some say that one can facilitate a crow's ability to talk by slitting its tongue. But, the best facsimiles of human speech in the bird family are parakeets, Mynah birds, and African gray parrots. Although these birds do not have vocal folds, they have an organ called the syrinx, which consists of membranes that produce sound with tonal quality much like that from a kazoo or a krummhorn; that is, they, too, have lots of harmonics.

Parrots also have unusually thick tongues that must provide the tunable difference in mouth cavity resonances. According to the Guinness Book of Records, a female

African gray parrot named Prudle learned a vocabulary of nearly 800 words. Another parrot named Alex, who studied at Purdue University, seemed able to recognize a variety of objects and colors and answer questions about them in English with 80% accuracy, including correct use of the word, "No." (See Pepperberg 1991; Perrins and Middleton, 1985, p. 225.) However, the characteristics of most talking birds are largely imitative. Probably, the most remarkable case is that of the Australian Lyre Bird reported by Attenborough (1998). As well as mimicking the sounds from 20 other song birds, that bird could imitate sounds ranging from a camera shutter (including one with a motor drive) to a car alarm and a chain saw.[5] Some birds are able to produce sounds at two different pitches at once, apparently enabled by a division in the breathing tube within the chest. Most of these produce a sound like that of a closed-pipe resonance with first and third harmonics. The minor third in the case of the Mourning Dove is especially noteworthy, since that interval cannot be made up from one harmonic series alone.

Most mammals other than Man seem unable to vary the relative position of resonances in the vocal tract with much facility. Surprisingly, relatively unsuccessful results were obtained from those closest to Man. In the 1940s, Catherine and Keith Hayes took a 6-week old chimpanzee named Viki into their home and tried to teach her to talk. After 6 years, Viki could only utter mama, papa, cup, and up. Yet, some other chimpanzees were able to learn over one hundred symbols in American Sign Language (Hayes 1951; Gleason 1997; and Kent 1997).

There are several major physiological differences in vocal tract anatomy between human beings and other land-dwelling animals that probably affect speech acquisition. The vocal folds in human beings are in a fairly direct line to the back of the throat—something which can be life threatening due to accidental aspiration of food. In the case of dogs and other animals, the vocal fold is less directly accessible; hence, they can "wolf" down their food without risk. As indicated in Fig. 6.6, human beings can close off the nasal cavities from the vocal tract with the soft palate. This ability may be indicative of an aquatic background in the early stages of human evolution since the soft palate can prevent water from entering the throat through the nose while swimming under water. It is suspected that early humans learned to emit loud vocal sounds by keeping that flap closed, using the basic closed-pipe resonances, and cutting off absorption from the soft nasal tissues. Other features that help in the speech process are a short snout, not to mention a large brain with lots of synapses. (See the review by Kent 1997.)

---

[5]According to bird neurological expert Dr. Fernando Nottebohm (private communication), these exotic sounds may actually just be part of the Lyre Bird's normal vocabulary and not picked up by imitation of chain saws, and so forth. The sounds provide a kind of auditory Rorschach Test for the human listener. However, there are birds such as the Greater Racket-tailed Drongo that definitely do imitate the calls of others. (See the note in the American Scientist 94, No. 2 March–April 2005, p. 191.)

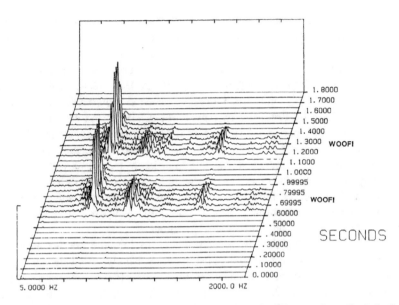

Fig. 6.8 Spectrum of an adult male German Shepherd in B-flat. The sound amplitude is shown versus the frequency and time for two barks

Animals often want to communicate, but their vocal signals generally consist of the same envelope of relative frequencies.[6] As an example, the spectrum from the author's German shepherd is shown in Fig. 6.8. The message, consisting of "woof, woof," always meant that he wanted something (e.g., ice cream, or to go out) or that something annoyed him. He especially hated the sounds of applause and of our door bell. But, the relative frequency structure in his bark never varied significantly and always had a fundamental pitch of B-flat. Ironically, the dog (named Mozart by our children) had perfect pitch. He would bark in response to the sound of a door bell in G (either live or recorded); but, if the pitch were changed by as little as a half-step on the musical scale (by varying the playback speed of a tape recording), he would not bark. (I discovered that fact accidentally while trying to get him to perform for a class on Live Fourier Analysis at Yale, using a tape recording of our doorbell to trigger his bark; but, the machine was running too slowly and the dog would not bark until I sped up the tape to its normal speed.) The particular frequency of our doorbell had evidently been burned into his memory from early puppyhood as a source of displeasure. Perhaps he did not like the minor third relationship to his own voice. He did appear to have discerning musical taste. When our son started

---

[6]Physicist Richard Feynman once spent an entire day in the cellar with Robert Serber's German Shepherd, Nicka, after which Feynman announced that the dog had a vocabulary of 80 English words. Unfortunately, the dog could only communicate by barking ferociously and by grabbing Serber's pants leg to pull him to the scene of the accident when his wife had a bad fall (Serber and Crease 1998, pp. 161, 162).

playing the oboe, Mozart would flee to the farthest corners of the house. But, as our son got better, the dog seemed to like the sound and eventually would curl up near the instrument. Similarly, the dog seemed very fond of the chorale settings by Bach in the Orgel-Büchlein. We had a small pipe organ on the third floor of our house in New Haven and that became a favorite resting spot.

6.2 Speech Synthesis

The earliest voice synthesizer may have been the 70-ft high statue of Memnon at Thebes, built $\approx$1490 BC. At sunrise, the statue was said by the early Greeks to emit a curious high note as a greeting by King Memnon to his mother, Eos, Goddess of the Dawn. The voice was created by air passing through a fissure leading to resonant cavities in the stone but was stilled by masonry repairs of the Romans during the reign of Septimius Severus, circa 200 AD.[7]

Albertus Magnus (1198–1280) was credited with the design of a mechanical speaking head, but the design appears to have been lost. (It was said that St. Thomas Aquinas was so terrified by the device that he broke it to pieces.) The thirteenth-century English monk, Roger Bacon ($\approx$1220–1292) was also known to have had a "talking head." Although these things had mechanically moving parts, it is thought that they were all activated by speaking tubes coupled to the human voice.

6.2.1 von Kempelen and Wheatstone Voice Synthesizers

The earliest detailed description of a real speech synthesizer that has survived was given by the Hungarian, Wolfgang Ritter von Kempelen (1734–1804).[8] Von Kempelen was a very inventive man and did quite a number of interesting things: He designed the fountains for the Schönbrunn Palace outside of Vienna and the Royal Castle at Buda, Hungary. He also designed a mechanical chess player which beat Napoleon at the game. But, he cheated on that one: a midget, who was expert at chess, was kept inside the apparatus. (Edgar Allan Poe later wrote an essay "exposing" that fakery, which by then was well known.) Von Kempelen worked for some 22 years on his speech synthesizer and did much research on the mechanism of pronunciation in the process. His pioneering work was described in a 456-page volume with 25 plates which was published the year Mozart died (von Kempelen 1791).[9]

---

[7]Drower (1974, p. 264).

[8]In Hungarian, his name was Kempelen Farkas Lovag.

[9]A copy of the von Kempelen book is in the Beinecke Library at Yale University under call number 2000 888.

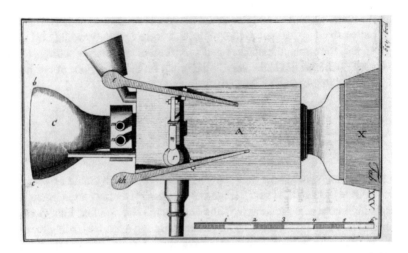

Fig. 6.9 Von Kempelen's speech synthesizer from his book Mechanismus der menschlichen Sprache, 1791, p. 439 (reproduced courtesy of the Yale Collection of German Literature, Beinecke Rare Book, and Manuscript Library)

In the von Kempelen machine (Fig. 6.9), the operator rested his right arm on a bellows (X) which was pumped up and down to provide a source of air that fed a wind box (A), which in turn led to a vibrating reed (the "vocal fold"), which was followed by a narrow throat section containing two "nostrils" (m and n). The output of the throat led to a rubber bell-shaped mouth (C). The two levers (S and Sch) were operated by the right hand and opened and closed odd-shaped passages to produce fricative sounds.

The left hand was placed in front of the "mouth" (C) to manipulate resonances in the manner of someone playing the French horn with his hand in the bell. It is said that von Kempelen could utter short sentences in both German (including a trilled rrr) and French ("je vous aime de tout mon coeur") with the machine and boasted that anyone could learn to use it in 3 weeks. Although von Kempelen's ultimate dream was to develop a machine operated by something like a piano keyboard, the device he actually built must have required a great deal of skill to use. In 1846, a Professor Joseph Faber of Vienna demonstrated a still-more elaborate version using a variable pitch source that permitted it to sing airs such as God Save the Queen. (It was during Victoria's reign.)

The English scientist Sir Charles Wheatstone produced a simplified model of the von Kempelen apparatus in which the vowel sounds were made by squeezing a leather resonator tube at the outlet. Starting in the fall of 1862, young Alexander Graham Bell (who gave his name to the old Bell System) spent a formative year living in London with his grandfather (also named Alexander), who was a close friend of Wheatstone.[10] Wheatstone demonstrated his apparatus to young Alec and his grandfather, both of whom were interested in teaching the deaf to speak, and loaned them a copy of the plans for the von Kempelen machine. The grandfather-and-grandson team decided to build one for themselves. In their version, the wind was supplied by blowing down a tube containing a reed and the human tongue was simulated with six sections that could be raised or lowered by the fingers to produce different vowel sounds. (It was probably a little like playing a krummhorn.) After much effort, they made a device that said "Mama" so realistically that one of the downstairs tenants rushed up to see "what can be the matter with that baby" (Mackay 1977, p. 38; also see Wheatstone 1879; and Dudley and Tarnoczy 1950).

6.2.2 The Jew's Harp

Sound generation in the Jew's harp works in a very similar way to sound production in the vocal tract, except that the vibrating tongue of the harp replaces the larynx. The nomenclature for the instrument has varied from country to country and seems to have nothing to do with the Jewish people.[11].

Although the Jew's harp today is thought of mostly as a folk instrument, it became a true virtuoso instrument during the nineteenth century for which serious composers wrote serious works. Among them, Beethoven's orchestration teacher Johann Georg Albrechtsberger wrote two concertos for Jew's harp and orchestra. Probably, the greatest living virtuoso of the period was Karl Eulenstein, who toured Europe and England, even playing before King George IV on several occasions. Due to the limitations on tonal range of a single instrument, the nineteenth-century virtuosi often utilized several Jew's harps, changing between them with lightning speed. It is said that Eulenstein himself used sixteen instruments during his performances, all tied together with silken string. (See Eulenstein 1892.)

A photograph of an original Jew's harp made by J. R. Smith (regarded by some players as the "Stradivari" of the instrument) is shown in Fig. 6.10. Smith was a

---

[10]There is a touch of Wheatstone in the modern telephone. His electrical bridge is the basis of the side-tone rejection circuit used to prevent deafening yourself with your own voice. Campbell and Foster (1920) enumerated some 446,234 ac versions of the bridge for the Bell System Patent Office.

[11]Some maintain that it is really "The Jaw's Harp"—an instrument made from the jaw bone of an ass (sometimes played on that of another), but the assertion is doubtful. In north-east England, it is called the "Gewgaw," a word possibly derived from the Swedish word "munngiga," meaning "mouth fiddle." In the Saar region of Germany, it is called the "Maulgeige," or "mouth violin." The French call it "La Guimbarde," a name suggestive of its actual sound. The Italians call it "Il Scacciapensieri," literally something that drives away thoughts. Some Russians call it the "BAPAH" (pronounced "vahrgahn") derived from the verb VAPAHITb ("vahrgahneet") meaning "to botch or bungle." Strangely, it is feminine in French and masculine in Italian. The instrument was evidently appreciated by the American Indians for in 1630 Peter Minuit bought Staten Island from the Tappans for a number of Jew's harps and other small items (Shorto 2004, p. 56)

Fig. 6.10 An original J. R. Smith Jew's harp made circa 1900. Note the graceful tapered cross-sectional shape of the harp, which was forged from iron by the master craftsman himself. This instrument had an especially strong spectrum rich in harmonics (see Fig. 6.11)

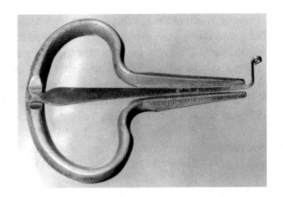

craftsman working in up-state New York circa 1900. In use, the narrow section of the "harp" is placed horizontally against the upper and lower front teeth, with a spacing just adequate to allow the vibrating tongue of the instrument to move back and forth between the teeth. (The narrow spacing provides optimum coupling to the vocal cavities, but the novice tends to produce nerve-shattering clicks caused by the tongue of the Jew's harp colliding with the player's teeth.) The vibration has a low resonant frequency similar to that of the voice and is very rich in overtones. Because of its tonal similarity to the human vocal folds, one can actually talk over a Jew's harp. It was seriously used by some people as a mechanical speech synthesizer. (See Leipp 1963a,b.) According to John Wright (1980), who is himself a virtuoso on the instrument, the Jew's harp serves as a voice for spoken communication between courting couples in south-east Asia. If you have access to one, try producing the common vowel sounds "a," "e," "i," "o," and "u," mouthing them as you would in normal speech, but exciting them by striking the tongue of the harp.

In serious performance on the instrument, the tongue is struck with the index finger of one hand, but not for each separate note. (Unlike the violin, it works equally well with left-handed and right-handed players.) Articulation is also accomplished by drawing air in and out of the mouth from the lungs. The working of the instrument is illustrated by the spectral surfaces shown in Fig. 6.11, where the sound amplitude as a function of frequency is again displayed linearly against time (in seconds), which recedes diagonally off to the right in these three-dimensional plots. At any instant in time, one can see the first 11 harmonics of the tongue spaced at its resonant frequency of 174.6 Hz. (The Jew's harp was tuned to F below middle C.) In addition, there was some real sound produced at half the lowest pitch of the instrument, probably due to a subharmonic mode of oscillation. That effect does not seem to have been reported elsewhere in the literature. The tongue of the Jew's harp is deliberately tapered to produce nonlinearities that enhance harmonic production, and, as discussed in Appendix A, a quadratic nonlinearity can also introduce subharmonics.

However, the main intensity is selectively peaked at different mouth resonances, one centered at about 300 Hz, another at 1100 Hz, and a third close to 1800 Hz. These are probably the F_1, F_2, and F_3 formants mentioned above. The piece played

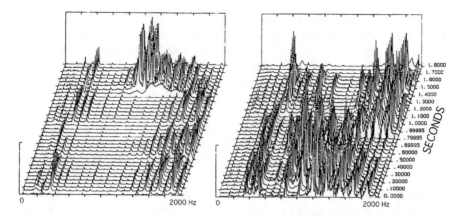

Fig. 6.11 Spectral surfaces illustrating the variation of sound amplitude versus frequency as a function of time for the "Stradivarius" Jew's harp shown in Fig. 6.10 when played by the author's father, William Ralph Bennett. Time recedes toward the upper right. (**a**) Left: "Bonaparte's Retreat." (**b**) Right: "The Irish Washer Woman." The frequency scale in the figures runs from 5 Hz (the limiting resolution of the analyzer) to 2000 Hz and is marked in ten equal steps of 200 Hz

in Fig. 6.11a is a section from the traditional dirge known as Bonaparte's Retreat. The basic rhythm is a quarter note, followed by a triplet, repeated over and over. The Jew's harp tongue has been struck only on the quarter notes (only once in the figure) and the triplets are created by breathing in and out through the tongue of the harp. A sudden change in the mouth resonance—hence, perceived pitch—occurs at 1.4 s in Fig. 6.11a. The sound distribution obtained on a different piece (a folk tune called the Irish Washer Woman) by the same player on the same instrument is shown in Fig. 6.11b. The fundamental pitch of the vibrating tongue of the harp is unaltered throughout both examples. The player's formants have simply been changed by moving his tongue to enhance different overtones at different points in time. The human voice works much the same way in normal speech. Analysis of the Jew's harp dates at least to the time of Wheatstone (1828). For an extended discussion of the acoustics of the Jew's harp as a musical instrument, see Leipp (1963b, 1967).

6.2.3   The Dudley Vocoder and Electronic Speech Synthesis

From its very inception in 1925, one primary goal of the Bell Telephone Laboratories was understanding the nature of speech and speech reproduction. To this end, many electronic devices were developed, ranging from the sound spectrograph (an instrument that would display the intensity and frequency distribution of sound as a function of time) to methods of electronic speech synthesis.

The Vocoder (an acronym standing for "VOice CODER") was conceived by Dudley in 1931. It was first demonstrated privately to a group of engineers at the Bell Labs in 1935 and then publicly at the Harvard Tercentenary in September, 1936. It together with the Voder (another acronym standing for "Voice Operated DEmonstratoR") were the earliest electronic speech synthesizers ever made. Using other developments by Dudley and his colleagues (in particular, the "Sound Spectrograph"), Dudley had observed that the spectrum of the human voice could be broken up into the time variation of a small number of frequency bands, now called "formants." The spectrum analyzer he used consisted of a band-pass filter that could be swept slowly through the audio spectrum from near 0 to 8000 Hz. The original signal was recorded magnetically on the periphery of a cylinder (the equivalent of a magnetic tape loop), which was rotated continuously while a single filter was tuned through the entire spectral range. The intensity coming out of the filter was then recorded by darkening a piece of paper that had been wrapped around the cylinder. The dynamic range was fairly limited (perhaps to 10 dB), but one could still see the dominant frequency components in the audio signal. (There are, of course, much faster and better methods for accomplishing the same objectives available now that are based on the use of Fast Fourier Analysis.)

Displaying the formants from the frequency spectra required some subtlety in filtering technique. If you just looked at the output from a swept narrowband filter (of say ≈45 Hz width), what you would see is a large number of discrete harmonic components of the vocal fold's fundamental frequency. (See the upper part of Fig. 6.12, where some 24 harmonics are resolved.) However, as you increase the bandwidth of the swept filter to 300 Hz, a characteristic structure develops that is the basis of speech recognition. (See Fig. 6.12.) Typically, about five dominant dark bands evolve whose variations in height represent the variations in the center frequencies of the resonant cavities or formants in the vocal tract. One can actually learn to recognize the syllables displayed in such "visible speech" and, thus, "read" the message contained in the original audio recording. The human ear recognizes speech through the relative positions of the formants, which vary with the different vowel sounds and, in the case of diphthongs, change in position relative to each other during pronunciation.

Because the time variation of those formants is quite slow (they typically vary at less than 25 Hz) compared to the sound vibrations in the normal human voice (which extend up to perhaps 8000 Hz), only a very small telephone transmission bandwidth would be needed to transmit the information contained in a spoken message. For example, the first five formants, plus signals indicating the presence of the random "hissing" noise from sibilants and frequency of the vocal fold pitch would only require about a 350 Hz bandwidth. Since the normal telephone bandwidth was about 5000 Hz, one could expect to send at least 14 different messages over one telephone channel! The trouble with that approach is that when you substitute the harmonics from a relaxation oscillator for those of a person's voice, the sound becomes unrecognizable. You could understand what grandma was saying on a transcontinental call, but she would sound like a mechanical robot. Dudley's initial work on the vocoder was probably stimulated by the challenge to send speech over

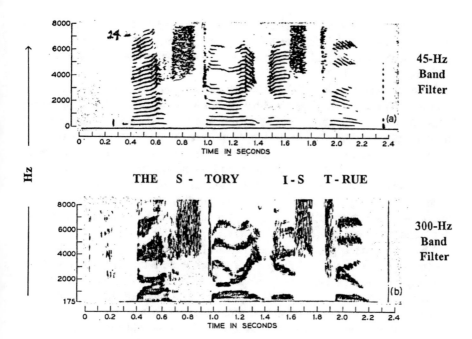

THE S - TORY I - S T - RUE

Fig. 6.12 (a) (Top) Spectrum from the message "THE STORY IS TRUE" as seen through a 45 Hz filter. (b) (Bottom) The same spectrum as seen with a much broader 300 Hz filter. Here, five separate formants are clearly visible, and the individual harmonics of the vocal folds are no longer seen (reproduced from Dudely 1955 with the permission of Lucent Bell Laboratories and Jean and Richard Dudley)

the (1928) transatlantic telegraph cable that had a bandwidth of only about 100 Hz. (He had been asked to work on the problem of voice transmission over transatlantic radio, long telephone lines, and submarine cables during the 1920s.)

The real utility of the discovery came during World War II when the Bell Labs developed a secret telephone system used by Roosevelt and Churchill. In that system, a random key (used only once) was added digitally to the different vocoder channels at the transmitting end and subtracted at the receiving end. The recovered signals then operated a vocoder to recreate the speech sounds. The system was kept highly classified until 1976. By 1982, my father, who had been a close friend of Dudley's and who was the only person still left alive who had worked on the system, was asked by the editors of an IEEE journal to write a description of it. That paper was became his last publication. (See Bennett 1983.)

Dudley had set about to simulate the entire vocal process electronically. One interesting analogy he proposed was "the carrier nature of speech." He noted that the modulation impressed by the vocal formants by the sound from the vocal folds was analogous to the way in which radio waves were modulated. The speaker was pictured "as a sort of radio broadcast transmitter, with the message to be sent out originating in the studio of the talker's brain and manifesting itself in muscular wave motions in the vocal tract." These contractions impressed the formants of speech on the audible sound which then acted as a higher-frequency carrier transmitted to the listener, who then detected the lower-frequency message contained in the formant variation (Dudley 1940).

With the Vodor (a device demonstrated at the 1939–1940 New York and San Francisco World's Fairs), ten band-pass filters were used to process the output amplitude from a relaxation oscillator (see Appendix A), and the attenuation in front of each was controlled by the fingers of an operator (logarithmically with key deflection) in such a way as to mimic the formant frequency variations. (Without this control, one just heard the raw spectrum from the oscillator itself.) In addition, random noise amplitude for sibilants and the oscillator pitch were controlled by the operator. (See Fig. 6.13.) After considerable practice, the operator could produce understandable speech just by hitting keys on the machine; the apparatus was the first "talking typewriter." (See Dudley 1939.)

Visible speech recognition became a field in itself during that period. The formant patterns characteristic of various parts of speech were categorized and a stylized shorthand for those patterns was developed. (See Potter et al. 1947.) Astonishingly, the Bell Laboratories trained stenographers to take dictation from voice prints displayed on a moving screen in real time. (See Fig. 6.14.) The more advanced students could then reproduce the speech on a Vodor. Although a remarkable feat, that approach to stenography certainly did not catch on generally in the rest of the world. However, the subsequent development of computer-controlled speech synthesizers based on the principles initially developed by Dudley and his colleagues has found many applications, especially with handicapped people. The techniques developed are also valuable tools for understanding properties of the singing voice.

With the Vocoder, a compromise technique was used because it was difficult to extract the formant variation electronically. The output from a microphone was passed through an analyzer section which covered the principal formant bands from about 250 to 2950 Hz in channels 300 Hz wide. Pitch and sibilant signals were also extracted which controlled the pitch of a relaxation oscillator and turned on a random noise generator in the synthesizer.

Harmonics from the relaxation oscillator were then transmitted to an identical set of band-pass filters whose outputs were controlled by the intensity in the original bands. The remade speech was then sent out to a loudspeaker (Fig. 6.15). Ten band-pass circuits of the type shown in Fig. 6.16 were used. Later developments in this field at the Bell Laboratories included extensions of the original Dudley concept to vocoders with more channels, formant vocoders that incorporated electronically tunable filters and oscillators, the use of predictive encoding, and methods for electronic speech recognition. (See Flanagan 1972, and Schroeder 1999.)

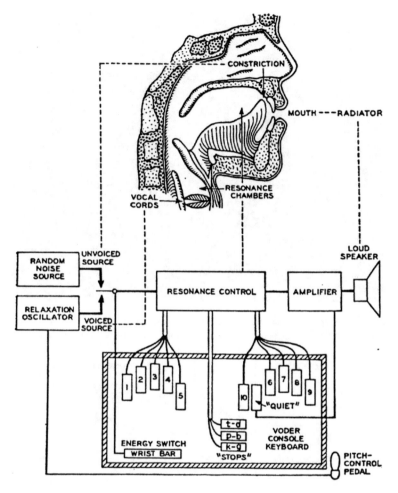

Fig. 6.13 The Vodor and its analogy to the human voice (reproduced from Dudley 1939 with the permission of Lucent Bell Laboratories and Jean and Richard Dudley)

6.2.4   The Vocoder as a Musical Instrument

With the advent of both analog and digital music synthesizers and the easy availability of MIDI ("Musical Instrument Device Interface") keyboards, some novel applications of the vocoder have occurred in the composition of music. For example, one can replace the periodic relaxation oscillator in the original Dudley vocoder by the output of a musical instrument or a synthesizer controlled from a piano-type keyboard. Then, merely by speaking into the microphone while playing a melody on the keyboard, one can simulate singing. One of the most impressive demonstrations of that technique I have heard was done by an undergraduate

Fig. 6.14 A training class in visible speech recognition held at the Bell Laboratories in the 1940s (reproduced from the Bell Laboratories Record with the permission of Lucent Bell Laboratories)

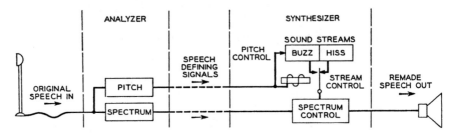

Fig. 6.15 Functional plan of the Vocoder (reproduced from Dudely 1955 with the permission of Lucent Bell Laboratories and Jean and Richard Dudley)

electrical engineering major and amateur composer at Yale University named, David Dana (Class of 1984). During his senior year, he composed a piece in that fashion which he called I am not Singing, that combined the musical treatment of a rhapsodic melody played on a Moog synthesizer with a delightful sense of humor. Surprisingly, the technique has not been used more often in the world of serious musical composition. Early commercial applications of the Vocoder were devoted to such tricks as advertising the headache remedy Bromo Seltzer with the sounds

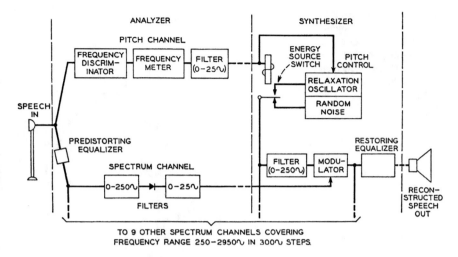

Fig. 6.16 Block diagram of the Vocoder circuit (reproduced from Dudely 1955 with the permission of Lucent Bell Laboratories and Jean and Richard Dudley)

from a steam locomotive.[12] More recently, modifications of the vocoder have been used for special effects by popular entertainers (Figs. 6.17, 6.18, and 6.19).

6.3   The Voice as a Musical Instrument

6.3.1   Operatic Singing Voice

One often wonders if all that girth is really necessary for the successful Wagnerian singer and indeed many modern singers are quite thin (Fig. 6.21). Certainly, they need powerful voices, for the traditional opera house is pretty "dead" acoustically and the singer does not get much help from reverberation. Some have argued that an enormous chest provides additional resonance. However, the lungs are composed of very spongy, absorbent material that would prevent any significant high-Q resonance. Large lung capacity would, of course, produce the wind power necessary to drive the vocal fold at high volume. But, if that were the only explanation, one would expect the Helden Tenor to be built like an Olympic swimmer—mostly chest and upper body. The principal resonances utilized by the voice are in the throat, mouth, and nasal passages. So, if internal acoustic resonances are really the main

---

[12] One early proposal left on the cutting room floor was to use the device in the movie, Mr. Smith goes to Washington. The vocoder was supposed to make the steam engine from Jimmy Stewart's passenger train repeat the movie title over and over. Perhaps, the director feared that the movie would just become another Bromo Seltzer ad.

Fig. 6.17 The Wagnerian
Soprano (after a drawing by
Katie H. Maguire). Overheard
at a performance of the
"Liebestod" from Tristan und
Isolde: "When you saw her
walk out on stage, you knew
she had to be good!"

Fig. 6.18 The acoustic horn
in an adult hippopotamus
produces 115 dB (after a
photo by Jessie Cohen)

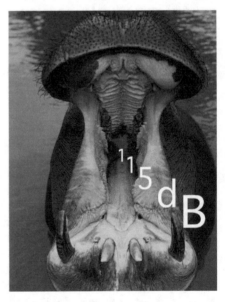

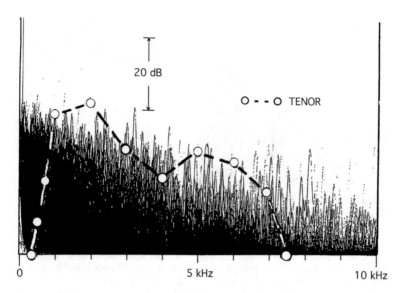

Fig. 6.19 Average energy spectrum of a large symphony orchestra during an entire performance of Le Sacre du Printemps by Igor Stravinsky. Frequency runs from left to right in the plot. The data were taken by playing a high-quality recording of Le Sacre du Printemps repeatedly through a spectrum analyzer. The darkening represents the average intensity, but there are occasional higher peaks at almost any frequency throughout the audio band. The average spectrum of an operatic tenor is indicated by the dashed lines, based on data by Cazden (1993)

need, one would expect the successful Wagnerian singer to have an unusually long neck, large mouth, and enormous head. The mouth structure, if it were big and long enough, might also serve as an efficient horn to couple acoustic energy to the outside world. That seems to be the case with the adult hippopotamus, which typically can produce low-frequency noises in excess of 115 dB.[13] (See Schwartz 1996, and Fig. 6.22.) Similarly, it has been found recently by X-ray tomography that the Parasaurolophus walkeri (a duckbill dinosaur) had two long curved, trombone-like air passages in its head leading from the nostrils to the back of its throat which probably would have enabled it to emit extremely loud sounds (Browne 1996; Diegert and Williamson 1998). But, those characteristic features are not found in the typical opera singer.

I know of several svelte singers who are capable of producing a very beautiful sound in moderate-size halls at levels of at least 115 dB. Hence, it seems unlikely that enormous body mass is really required. The main secret seems to lie in the

---

[13] As discussed in Chap. 2, the "dB" (standing for "decibel") is a logarithmic measure of the sound intensity ratio given by 10 Log10(Intensity Ratio). Here and in the following paragraphs, the ratio sometimes tacitly refers to a reference level in which 0 dB corresponds to 2×10^{-4} dynes/cm^2, roughly the threshold of human hearing. (120 dB is about the threshold of pain.) An increment of 1 dB is about the smallest change in intensity ratio that the average human ear can detect.

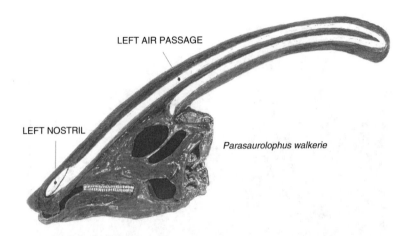

LEFT AIR PASSAGE

LEFT NOSTRIL

Parasaurolophus walkerie

Fig. 6.20 The Parasaurolophus walkeri. Although not really an invention by Richard Wagner, this duck-billed dinosaur would probably have made an excellent Helden tenor. It had two folded internal air tubes inside the crest of its skull (the left one shown here in white) going from the nostrils to the back of the throat having an estimated unfolded length of about 6 ft. They each would have produced a fundamental open-pipe resonance of about 90 Hz (drawn by the author from the fossil on exhibit in the American Museum of Natural History in New York City)

distribution of energy from a powerful set of vocal folds that is also unusually rich in harmonics. There is a technique which singers themselves refer to as "focusing the sound." It is seldom clear from their own descriptions what they really mean by that term. Certainly, sound waves are not being focused. Instead, the terminology has to do with the spectral distribution of energy over the vocal fold harmonics. Professional singers develop something known among laryngologists as the "singer's formant." By lowering the larynx and expanding the laryngeal ventricle, some can tune the normal fourth formant down in frequency to coincide with the third, thereby enhancing vocal energy in the region of above about 3 kHz where the energy level of an accompanying orchestra rapidly falls off. As shown in Fig. 6.23, the average energy spectra of a large symphony orchestra drop with increasing frequency from its peak value at about 100 Hz and are down by about 20 dB at 3 kHz. Another benefit of the "Singer's Formant" is that it shifts energy from lower frequencies having nearly omnidirectional radiation patterns to higher frequencies that are more directional. Hence, the voice can be aimed more efficiently at the audience. Combining these techniques, a moderately powerful singer can manage to be heard above a very loud orchestra. (See Sundberg 1987, Chapter 5.)

Of course, another possibility would be to produce a new breed of singers by genetic control. For example, imagine the possibilities that might be attained through gene splicing in the DNA molecule. (See Fig. 6.20.)

Normal singing relies heavily on the use of vowel sounds where most of the energy is contained in the first four formants, the first two of which are usually the strongest. (The vocal tract, of course, also provides articulation and consonant

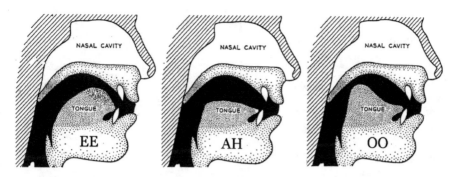

Fig. 6.21 Extreme positions of the tongue for three different vowel sounds. Limitations on the ability to sing or speak different sounds rapidly arise from the physiology of the tongue. For example, try to stick out your tongue and say, "OO" (as in "f<u>oo</u>d")

sounds.) It is important to remember that these formants are physical resonances in the vocal tract determined by the geometry of the mouth and nasal passages, the opening of the mouth and jaw, and the position of the tongue. (See Fig. 6.21.)

Representative frequencies and formant patterns for six of the common vowel sounds used in a normal speaking voice are shown in Fig. 6.22. The lower part of the figure shows voice prints for the vowel sounds [ee] as in "seat," [eh] as in "set," [aa] as in "sat," [ah] as in "calm," [aw] as in "ought," and [oo] as in "food." The top part of Fig. 6.22 shows typical frequencies in Hz for the first three formants for women's (W) and men's (M) voices. Frequency increases from bottom to top in each of these figures. The fact that the relative positions of these formant frequencies can be duplicated and recognized by a wide variety of people of both sexes is what makes oral communication possible. The second formant, which is sometimes called the "Hub," is the resonance that changes the most in going from one vowel sound to another. Its frequency is highest for [ee] and lowest for [oo] and can change by a factor of almost three. The effect, of course, is very striking in the case of diphthongs which consist of a transition between two vowel sounds. (See Fig. 6.23.)

The trick in producing a powerful singing voice in which the text can also be understood is to adjust (or "tune") these resonances, especially the first three formants, to coincide with harmonics of the vocal folds. This tuning process can increase the sound level of the voice by as much as 30 dB (a factor of 1000 in. intensity!)[14] Ideally, one would like to obtain a match with at least four, but that is hard to accomplish in practice for low voices and nearly impossible with the soprano voice. As noted above, the fourth is often tuned down to match the third. But, these resonances in the vocal tract are physically interrelated and cannot all be tuned independently. The problem is least severe with bass voices where the

---

[14]Tenor Franco Corelli, after singing a duet with Eileen Farrell, said: "Who is this woman? She has made me deaf!" [See the obituary of Eileen Farrell by A. Tommasini, The New York Times, 3/25/2002, p. B7.]

Frequency (Hz)

Formant

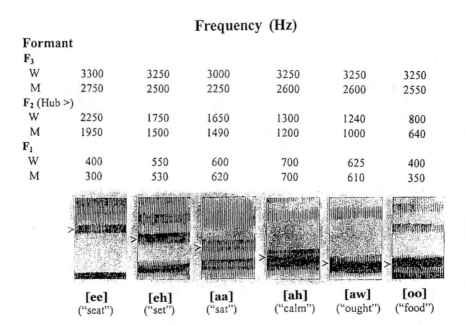

F_3		[ee]	[eh]	[aa]	[ah]	[aw]	[oo]
	W	3300	3250	3000	3250	3250	3250
	M	2750	2500	2250	2600	2600	2550
F_2 (Hub >)							
	W	2250	1750	1650	1300	1240	800
	M	1950	1500	1490	1200	1000	640
F_1							
	W	400	550	600	700	625	400
	M	300	530	620	700	610	350

[ee]	**[eh]**	**[aa]**	**[ah]**	**[aw]**	**[oo]**
("seat")	("set")	("sat")	("calm")	("ought")	("food")

Fig. 6.22 Frequency distributions and formant patterns for common vowel sounds. Frequencies increase from bottom to top. The second formant F_2 is called the Hub (indicated by >) and plays a principal role in identifying individual vowels. Note that the Hub (>) decreases from left to right in the above examples. Typical values of frequency are shown for women (W) and men (M) (sources: Appelman 1967 and Potter et al. 1947)

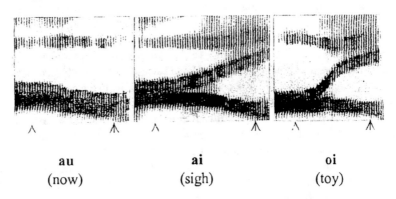

au	**ai**	**oi**
(now)	(sigh)	(toy)

Fig. 6.23 Formant patterns for several diphthongs. Note that each diphthong has two Hubs (ʌ) corresponding to the two vowel sounds involved (source: Potter et al. 1947)

vocal fold harmonics are closely spaced and cover a large range. But, the problem becomes more difficult with the soprano voice where a lot of the singing involves a pitch which is higher than the first formant of the speaking voice. As a result, the soprano singer is often forced to match the second formant to the pitch of her voice and then try to tune the higher formants to match the third or fourth harmonic of the vocal fold—for example, by moving the tongue and opening the jaw. One can sometimes hear this transition occurring in a singer after the note has been initially sounded, thus producing a kind of "blooming" effect loosely analogous to a note being sounded on an oboe or violin with delayed introduction of vibrato. In this tuning process, the formant pattern is often changed from that for the original vowel sound with the result that identification of a particular vowel is blurred. The sound can be very beautiful, but it often becomes hard to understand the text.

6.3.2 Registers of the Voice

As briefly mentioned in connection with Fig. 6.2, there are different possible lateral modes of oscillation of the vocal folds, and, because of the elastic nature of the tissue, these different modes may have nonharmonically related fundamental frequencies with little overlap of their fundamental resonances. But, the harmonics produced in the sound of the voice arise from the time-dependent shape of the bursts of air transmitted through the vocal fold and have no simple direct relationship (other than fundamental frequency) to the spatial modes of vibration. The sound spectrum, of course, is modified by the resonant formants in the vocal tract.

Singers refer to these spatial modes of vibration as different "registers" of the voice, but the terms are not the same as those used to describe registers in wind instruments. The names given the vocal registers are largely subjective and are often derived from sensations felt by musicians while singing.

These sensations naturally arise from different parts of the anatomy related to the production of sound: Hence, they involve the diaphragm, the muscles controlling the lungs, the vibration in the throat, and in the resonant formants in the neck, mouth, and head. One finds descriptions from singers such as that given by Birgit Nilsson (1984): "You must always be open, here and there [touching her forehead and lower abdomen]... I always want a very, very deep support for the breath. The whole body has to work... The key to producing the sound lies in support from downstairs."

The vocal folds are much more than a two-dimensional pair of elastic bands. They are three-dimensional in character, covered with mucosa, and can take on complex variations in shape and thickness during "phonation" (the production of sound). In the "Heavy Register" (see below), glottal closure is complete and starts in the lower part of the fold, where the tissue is drawn together initially by the action of muscles around the larynx and the closure is then completed by the Bernoulli effect. While the glottis is closed, the two halves of the fold effectively roll upward against each other to the top part of the fold before the two halves separate. (See Fig. 6.24.)

In the "Falsetto Register," the upper part of the fold stretches out tightly on each side of the glottis, making each half of the fold very thin and preventing the glottis from closing completely. When the vocal fold is viewed from above in stroboscopic illumination (using a frequency slightly different than the vocal resonance so that the folds appear to open and close slowly), it is easy to distinguish visually between those two registers of the voice. With the "falsetto" (high-pitched) register, the vocal

Fig. 6.24 Vibration of the Vocal Folds shown in a vertical cross-section through the middle of the vocal folds during the production of a single sound. The perspective is from the front of the larynx. Before the vibration starts (1), the folds are together. They separate as the air is forced upward through the trachea (2–7) and then come together again as the sound ceases (8) (figure reproduced from Robert T. Sataloff (1992), with the permission of the author)

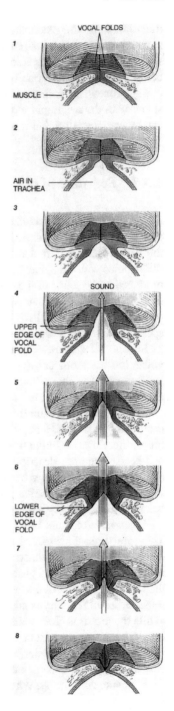

fold is stretched and appears at maximum closure as a long narrow slit that is slightly open. In contrast, with the "heavy" register, the length of the fold is noticeably shorter and the closure of the glottis is complete along most of the fold, although there is a slight opening at the front.

Laryngologists prefer to talk in terms of "laryngeal registers," of which only three have definitely been identified according to Baken (1998):

1. The Heavy Register or Modal Register is most commonly used by untrained singers and runs from about 75 to 450 Hz in men and 130 to 520 Hz in women. This register may include the "chest," "head," "low," "mid," and "high" registers described by singers.
2. The Pulse Register, occurs in the low end of the frequency range, from 20 to 45 Hz in men and 25 to 80 Hz in women. The term is used synonymously with "vocal fry," "glottal fry," and "strohbass" (German, for "straw bass"). Here, the output from the larynx tends noticeably to have a double-pulse character. (See later discussion of the Tuvan Throat Singer.)
3. The Falsetto or Loft Register (sometimes called the Light Register) occurs at the high end of the spectrum (275 to 620 Hz in men and 490 to 1130 Hz in women).

The Pulse Register seems associated with massive vocal folds that are completely relaxed—so much so that the ventricular folds may actually lie on top of the vocal folds, thereby increasing the effective mass and lowering the resonant frequency (see Allen and Holien 1973.)

The Falsetto or Loft Register is just the opposite. The tension is extremely high and causes the vocal folds to be thinned out so much that often no vocal fold contact occurs during phonation. That is, the glottis does not quite close during the fundamental singing cycle. However, the glottal opening is spread out fairly uniformly over the length of the fold. Both the intensity and harmonic content in this register are less pronounced than in the Heavy Register and the intensity is determined almost entirely by the rate of air flow through the vocal fold, rather than the muscles of the larynx (Hirano 1970).

The Heavy or Modal Register is characterized by a larger opening at the front of the vocal fold, producing a slightly "-shaped" aperture in the glottis as viewed from the top at near-maximum closure. Here, because the glottis closes almost completely during part of the cycle, both the intensity and the harmonic content tend to be much greater than in the "falsetto" mode. Switching from the "Heavy" register to the "Light" (Falsetto) register is primarily controlled by the vocalis muscle.[15] From stroboscopic images of the glottal shape, one can easily recognize changes between these two registers. The vocalist's art consists in part of striving to develop smooth transitions between the Heavy and Light registers. (See Fig. 6.25.)

---

[15]See Gray's Anatomy (Williams and Warwick, 1980, pp. 1234–1238) for illustrations and a discussion of the musculature of the larynx. There are three principal, but complex muscles controlling the vocal fold: the vocalis, the cricothyroid, and the crico-arytenoid. Hirano (1970) has reported substantial research on the role of these three muscles in controlling the singing voice. Also, see Sundberg (1987) and Sataloff (1992).

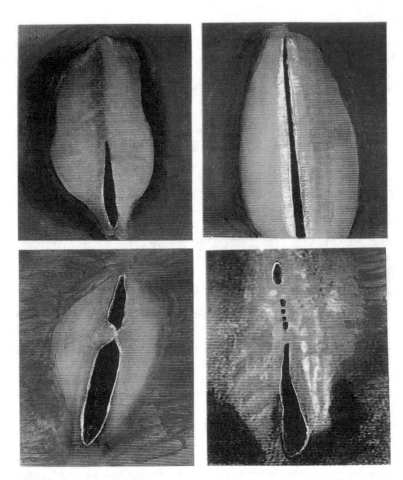

Fig. 6.25 Upper Left to Upper Right: Transition from the "Heavy" register (left) to the "Falsetto" or "Light" register (right) during phonation in the vocal folds of a professional soprano at the point of maximum closure of the glottis. The lower figures show the results of vocal abuse. Lower Left: Nodule on a male singer's vocal folds that arose from improper vocal practice. (Note the posterior glottal chink.) Right: Vocal fold from a male rock singer. Note the nodules, the scar tissue, and gnarled appearance (drawn by the author from the Morrison 1983 stroboscopic tapes)

Surprisingly, there is little significant difference between the maximum vocal intensities of trained and untrained singers. But, there is a major difference in the damage done to the vocal fold from singing in the two cases. Due to bad vocal attack and "vocal fry," the vocal folds of country western singers and rock singers who have never had formal voice training are often gnarled, scarred, swollen, and covered with nodules and lesions that produce posterior glottal chinks and prevent complete closure of the glottis (Morrison 1983). Surprisingly, some of those problems are curable over a period of time just by expert vocal instruction. However, in severe cases such as that of Elton John, surgery has been required to restore the singing voice (Sataloff 1992).

In addition to the harmonic content, vibrato is generally added to the singing voice which impresses a frequency (and sometimes amplitude) modulation on the sound. That can occur in different ways: one by actually varying the pitch (and harmonic frequencies) of the voice by muscles controlling the vocal fold as generally done by operatic singers; the other by varying the formant resonances (waggling the tongue or opening and closing the jaw) as done by some popular singers. (Examples come to mind ranging from the vocal sounds portrayed by Hollywood "Indians" to festive, ululating Arab women.)

6.4 Spectra of the Operatic Voice

In this section, spectra of several leading operatic soloists are presented with the thought that they should "speak [i.e., sing] for themselves." The following examples were all taken from the Verdi Requiem, where isolated solo notes are few and far between.[16] One has to catch those moments "on the fly," as done here by the use of a real-time Fast Fourier Transform analyzer (often abbreviated FFT).[17] I chose that work as an example because Verdi was one of the great masters at writing vocal music with orchestral accompaniment. It also happens to be one of my favorite pieces and the particular performance one of the finest I have heard.[18] Some purists may argue that the Requiem is not really an opera, but in the hands of Verdi it certainly presents the characteristic features of operatic singing. One main point of the present examples is to illustrate how singers can manage to be heard over an orchestra, not to mention an accompanying chorus. In each of the four following cases, the main components of the spectrum of the singer are indicated in music notation above the orchestral accompaniment that follows one measure after the solo vocal note. The harmonic amplitudes displayed in the following examples are all on a linear, rather than a logarithmic scale. (They are not in dB.) Similarly, the frequency is on a linear scale in order to make it easy to identify the harmonics (which are always equally spaced in frequency).

Because of the transient nature of the selections chosen, the waveforms are not precisely periodic over the observation intervals. However, they are at least quasiperiodic and provide a realistic indication of the spectra found in performance. Another problem in interpreting the data arises from the fact that the singers

---

[16]London recordings CD 1 (411945-2) and Cd 2 (411946-2) recorded in Vienna in October, 1967 with soloists Joan Sutherland, Marilyn Horne, Luciano Pavarotti, and Martti Talvela. Sir Georg Solti conducted the Vienna State Opera Chorus and the Vienna Philharmonic; however, all but the soloists have been cut out of the spectra presented here.

[17]See the discussion of Fourier analysis in Chap. 1 and Appendix C.

[18]But, the reader should be warned that the text deals with some pretty grim stuff.

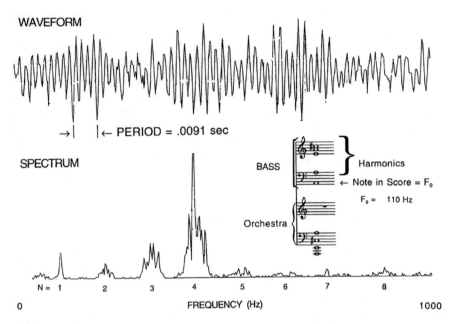

Fig. 6.26 Bass: Martti Talvela. The isolated low A occurs in the shattered silence at the end of the Tuba mirum from the Dies irae. The music has just changed key in the Molto meno mosso section. The note is the last one in the section and becomes the bottom of an A-major chord when the orchestra comes in on the last measure. It is the "pe" in the phrase "Mors ... stu-pe-bit" (meaning "Death ... benumbing"). Since the part is marked "pianissimo," one might not expect much harmonic content. However, most of the energy is actually in the treble clef in the fourth harmonic of the voice at $A = 440$ Hz. The presence of the second, third, fourth, and fifth harmonics gives the impression that the pitch is 110 Hz, even though the first few harmonics are much weaker than the one two octaves above the fundamental. The spectrum illustrates how a low-voiced singer can "focus" his voice so as to provide harmonics above the pitch of the orchestra when it enters on the final chord

generally use vibrato, hence the harmonics (or overtones) are smeared out somewhat in the following spectra. The harmonic content of the voice, of course, changes as the singer moves from one register to another. As is the case with most instruments, as the pitch goes up, the harmonic content generally goes down. Relatively low notes have been selected here to illustrate the large harmonic content that can occur in the singing voice. With the lower-pitched voices (including the mezzo-soprano), much of the energy is contained in the first few even harmonics (especially, the fourth and sixth).

The spectra of operatic singers shown in Figs. 6.26, 6.27, 6.28, 6.29, and 6.30 have several interesting features. First, there is a predominance of even harmonics (especially, the fourth) of the vocal fold resonance, one minor exception being the presence of a strong fifth harmonic component in the sample of the tenor voice (Fig. 6.27). In the case of both the bass (Fig. 6.26) and the tenor (Fig. 6.27), the strongest harmonic is not at the frequency of the note sung. With the bass, the

WAVEFORM

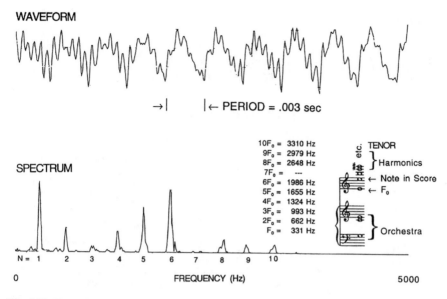

→| |← PERIOD = .003 sec

Fig. 6.27 Tenor: Luciano Pavarotti. There are almost no isolated tenor notes in the entire Requiem. The spectrum is from the Offertory where a note on E natural is accompanied by a "pianississimo" (ppp) violin tremolo. The text at that point was "fac-eas, Domine" ("make them to pass, Oh Lord...[from Hell unto life everlasting]"). The spectrum has more higher harmonics than any other voice studied. There is almost no seventh harmonic. The fundamental frequency (F_0) is an octave below the note in the score. The illusion of the treble E is created by the second, fourth, sixth, eighth, and tenth harmonics of F_0, which create a harmonic series based on the treble E in the score

strongest component was the fourth harmonic—two octaves above the note in the score. The impression of a low A at $F_0 = 110$ Hz in that instance comes about from the numerous harmonics of F_0 in the spectrum. At the same time, the presence of the strong fourth harmonic (at 440 Hz) makes it easier to hear the bass above the orchestra when it enters in the following measure. With the tenor (Fig. 6.27), the strong harmonics also extend far above the notes in the orchestral score that enter after the solo voice. Interestingly, the strongest component is one octave below the note in the score. The illusion that the tenor is singing a treble E at 662 Hz is created by the second, third, fourth, and fifth harmonics of 662 Hz in the overtone series of $F_0 = 331$ Hz.

With the mezzo soprano (Fig. 6.28), the fundamental ($F_0 = 440$ Hz) and its fourth harmonic are of about the same intensity, with much additional energy in the sixth. The voice easily stands out above the following orchestral chord (inverted dominant seventh in A major) based on C♯. The soprano spectrum in Fig. 6.29 consists mostly of first and second harmonic of $F_0 = 623$ Hz and is easily heard over the other solo voices. The unusual appearance of the waveform is due to varying amounts of second harmonic during the sample interval.

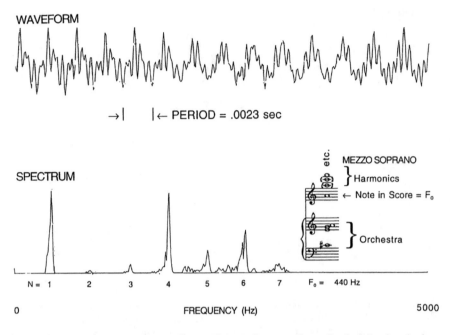

Fig. 6.28 Mezzo Soprano: Marilyn Horne. The opening note (immediately following the bass note in Fig. 8.27) starts the phrase "Liber scriptus proferetur" (meaning "The written record shall be cited"). As with the Bass spectrum in Fig. 6.28, the fourth harmonic is strong

6.4.1   Castrati

Starting in the early 1600s at about the time of Jacobi Peri's Euridice (1600) and Monteverdi's Orfeo (1607) until the middle of the nineteenth century, the castrato played a major role in opera. To put it medically, these people had been subjected to a bilateral orchiectomy (surgical removal of both testes) before their voices changed at puberty. They became so prevalent in the musical world that the very word musico (Italian for "musician") became synonymous with castrato. It has been argued that at least two of them, Farinelli (née Carlo Broschi, 1705–1782) and Gasparo Pacchierotti, were the greatest singers ever known. The stage name Farinelli was probably derived from the Italian word pharinge meaning "pharynx" or "throat;" thus, "Farinelli" would mean "Little Throat." Mozart used castrati in at least two of his operas (Idomeneo and La Clemenza di Tito) and one might argue that the "Queen of the Night" aria from the Magic Flute could have been written with a castrato in mind.[19] (The aria goes up to a high F, which would have been much easier for a castrato to hit than for an adult female soprano and the opera was

---

[19] According to operatic soprano (Bartoli 2002), arias written for castrati are "very demanding" and require an "incredible range."

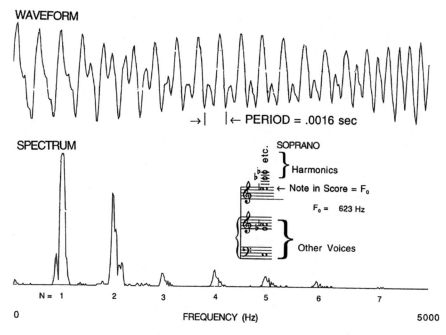

Fig. 6.29 Soprano: Joan Sutherland. The E-flat is near the end of Offertory ("ad vietam," meaning [to pass unto life] "everlasting"). Principally, first and second harmonic with a trace of third through sixth. The unusual waveform arises from varying relative amounts of second harmonic

written just after La Clemenza di Tito, which featured the castrato role of Sextus. However, die Zauberflöte was written in German rather than Italian—the language of the great castrati, and the role was sung by Mozart's first love and sister-in-law, Josepha Weber Hofer, on opening night.[20]) Mozart clearly was intrigued by male sopranos, having met the Florentine-born castrato Giovanni Manzuoli (1720–1782) during his early visit to London circa 1764. During his later trip to Italy in 1770, he renewed his acquaintance with Manzuoli and met the legendary Farinelli as well. One of Mozart's encounters with young castrati singers was described in a letter from Milan dated 3rd February, 1770 from his father to his mother. At that time, the young Mozart was "composing two Latin motets for two castrati, one fifteen and the other sixteen years old. . . whom he could refuse nothing because they. . . sing beautifully." Mozart was just fourteen himself, and, though he abhorred the operation, he wished he could have taken the two fellows back with him to Salzburg. In 1771, Manzuoli came out of retirement at Mozart's request ("like a good castrato") to sing at Milan in the first performance of Mozart's early opera

---

[20] According to his former student, Ignaz von Seyfried, Mozart's last words while delirious in his final illness were, ". . . hush! now Hofer is taking the high F . . ." He thought he was attending a performance of the Magic Flute and listening to the "Queen of the Night" aria. (See Neumayr 1994, p. 166.)

Ascanio in Alba (K.111), written for the wedding of Archiduke Ferdinand of Austria and Maria Ricciarda of Medina. (See Mersmann 1972; Heriot 1975; Hansell 1980; Sadie 1980; Walker 1980; Pleasants 1981; Schonberg 1985.)

Castration not only produced male sopranos, but an array of other physical and psychological deformities.[21] There were other reasons for the practice of castration than the role of castrati in Italian opera. Those ranged from the creation of eunuchs to serve as guards in the harems of Asia, through the penalty for adultery in ancient Egypt, to religious reasons (see Mathew 19:12). Only in Italy, however, was it done for the sake of music. Because the practice was illegal in Italy, descriptions of the procedure are unavailable. It is thought that the victims were given opium and placed sitting in a hot bath to soften the tissues of the scrotum before the surgeon's knife was applied. At the onset of puberty, the pliable membranous part of the vocal folds of both boys and girls is of about the same length. During puberty, the androgen hormones released from the testes stimulate the rapid growth and lengthening of the vocal folds, with the result that in the adult male they have nearly doubled in length. As noted earlier, the frequency of vibrating strings varies inversely with their length and the same is true with the vocal folds. Consequently, the adult male voice typically drops by nearly an octave from its childhood value. The female vocal folds also increase somewhat in length after puberty but by much smaller amounts. However, if castration is performed well-before the onset of puberty, the male vocal fold does not increase significantly in length at all. Consequently, the adult castrato could sing with a pitch that was actually higher than that of an adult female soprano. This feature was much sought after by a number of the famous operatic composers—especially, Monteverdi, Scarlatti, HÃd'ndel, Glück, Mozart, and Bellini. The results were so popular that by the eighteenth century 70% of all male opera singers were castrati. It is thought that in the eighteenth century, as many as 4000 boys were castrated in Italy for this purpose, with the usual age for the operation being at 6–8 years. But, the operation was against the law in Italy and was performed in a clandestine way. Most of the children had been sold by poor parents to singing schools. One of the strong influences behind this practice was the Roman Catholic Church: Women were not allowed to perform in musical services because St. Augustine had forbidden their active participation. Castrati provided an alternative to conventional soprano singers and were employed in the Church starting in the early fifteenth century. References to their presence in the Sistine Choir date to about 1565 and their use was reported in churches in Spain and Munich from about the same period (Walker 1980; Peschel and Peschel 1987).

In spite of popular rumors that flourished in the eighteenth century, it seems improbable that the castrati could have been active in either heterosexual or homosexual affairs. Instead, they were wracked with medical and psychological disabilities. For example, the castrato Filippo Balatri actually wrote about the torment that asexuality caused him. Although in love with a beautiful girl at one

---

[21]Farinelli's body was exhumed recently to study the physical characteristics of the singer (NY Times 7/12/2006, p. AR2).

point, he felt obliged "to give up women" because he had been emasculated. The other cause of psychological torment among these people arose from abnormal physiological development. The same androgen hormone that causes the vocal folds to double in length after puberty and the penis to develop in normal males, also stops bone growth in the legs and arms, a process called "osseous maturation." As a result, the castrati acquired freakish proportions with limbs much too long for their body lengths, something defined medically as "eunuchoid appearance." In addition, they tended to develop large fatty deposits localized in the hips, breasts, and buttocks. Consequently, the more famous castrato singers were caricatured and otherwise lampooned (Peschel and Peschel 1987).

The use of castrati singers in opera pretty well died out in the early nineteenth century. Neither Verdi nor Wagner were interested in writing for them. Giovanni Battista Vellutti, who died in 1861, was probably the last great male operatic soprano. The use of castrati was outlawed in the Roman Catholic Church by edict of Pope Pius X in 1903. The last known castrato was Alessandro Moreschi (1858–1922), who was a member of the Sistine Choir from 1883 to 1912 and was said to have had a voice "as pure and bright as the sound of silver." Although he was not one of the great singers, he lived long enough to have made cylinder phonograph recordings between 1902 and 1904. His recorded selections included the Bach-Gounod Ave Maria, Mozart's Ave verum, and the Crucifixus from Rossini's Petite Messe Solenelle. Although the recording technology was "early Edison" and the singer himself must have deteriorated from his prime, his recordings do provide some sense of the unusual upper range of the castrato voice. He could still hit a high C in 1902, but it was on a scale where $A \approx 412\,Hz$, and the then high C was approximately the present B natural.

Although there are no completely isolated spots in the recording where Moreschi was singing alone, there was one place in the Ave Maria where he hit a high B (written C, but actually 988 Hz) under conditions where I was able to filter out the piano, violin, and clarinet accompaniment by inserting a high-pass filter that fell off at 20 dB per octave below 1000 Hz. Comparison with and without the filter showed that the recorded spectrum of Moreschi's voice was unaffected by this filtering process.[22] Although there was some variation between different samples taken from the same section of the record, it was clear that the strong fundamental frequency at close to 1 kHz was definitely accompanied by at least weak second and third harmonics. The results are shown in Fig. 6.30, where I have added the spectrum obtained from a recent digital recording of a boy soprano (Max Emanuel Cencié) singing the same note. Although one would expect the vocal fold frequencies to be roughly the same for a boy soprano and a castrato, the head resonances in the castrato would probably tend to be lower in frequency. (As discussed above, the

---

[22]The filtering also removed an annoying low-frequency "rumble" in the original cylinder recording. The early Edison recordings used the "hill and dale" method of modulation in which the stylus moved up and down, rather than horizontally and were much more susceptible to mechanical rumbling noise than later lateral recordings. The filter also removed a similar rumbling noise in the Valente (1960) recording which was traced to singing by the pianist Rudolph Serkin.

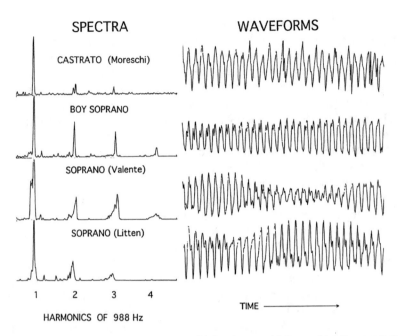

Fig. 6.30 Comparison of four different singers hitting the same high note (high B natural). From top to bottom: Moreschi 1902; Cencié 1990; Valente 1960; and Litten 1963 (see text)

bones of the castrato tended to keep on growing long after puberty.) The difference between those two spectra is probably due largely to high-frequency fall-off in the early (1902) Edison recording of Moreschi. The actual spectrum of the castrato was probably intermediate between the limits shown in the upper two traces of the figure.[23] Representative spectra obtained from recordings of the same piece and note sung by two different female sopranos are also shown in the bottom of the figure. The three lower traces were all taken on the high B natural that occurs toward the end of the slow section of Schubert's Der Hirt auf dem Felsen ("The Shepherd on the Crag") using the same low-frequency rejection filter (sources: Moreschi 1902; Schonberg 1984; Cencié 1990; Valente 1960; and Litten 1963).

It is of interest to see what a modern soprano does with even higher notes. Towards the end of the Queen of the Night aria in Mozart's opera The Magic Flute, there is a long melismatic sequence ending on an arpeggio going up to a high F. Understandably, relatively few recordings of that aria are available. However, data from one sung by the Russian soprano Edita Gruberova are shown in Fig. 6.31. She not only hit the high F, but exceeded it slightly (the note is actually closer to $F\sharp$)

---

[23]One should note that the spectral distribution varies throughout the vocal range. For example, spectra taken of the boy soprano on the note D natural just before the high B showed nearly equal amounts of first and third harmonic, but almost no second harmonic, with some fourth and very little fifth, sixth, and seventh.

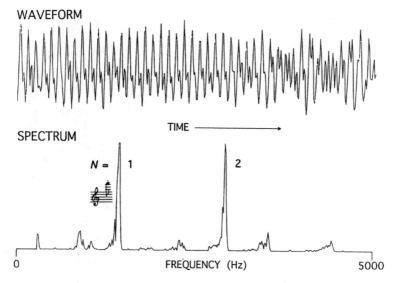

Fig. 6.31 The first and second harmonic of the high *F* near the end of the Queen of the Night aria as sung by Edita Gruberova (1988). The same low-frequency rejection filter was used here as in Fig. 6.30

and produced a second harmonic of equal strength. Hence, it is clear that there are occasional sopranos who are well-enough equipped to sing parts that could have been originally written for a castrato. However, the quality of the voice could differ significantly between the male castrato of previous centuries and the present-day female soprano: The "Heavy" register in the castrato would have contained more harmonic content than the "Falsetto" register of the female soprano. That would come about because the glottis closes more completely in the lower register. (See the previous discussion of the laryngeal registers.) Because the adult female soprano would have a voice significantly lower in pitch than a castrato (or a boy soprano, too, for that matter), she would be obliged to use the falsetto register for some parts that might have been sung in the lower register by the castrato of Mozart's time.

A movie entitled Farinelli was made in 1994 of the romanticized life of Carlo Broschi (1705–1782), one of the great Italian castrati (Fig. 6.32), using more current recording technology to simulate his voice. There was, of course, no castrato soloist available to record the sound track, and the singing consisted of a skillful editing job combining the range of a male counter tenor (Derek Lee Ragin) and of a female soprano (Ewa Maijas Gadilewska) going up to high D (near two octaves above middle C). But, it did provide some sense of the type of sound that would have been produced by the famous castrato. (See Farinelli 1994.) With the disappearance of the castrato from the operatic stage, the problem remains of how to perform the parts originally written for them. In most cases, sopranos now take over the role. As shown in Fig. 6.31, some are fully capable of doing so. But, those listeners who have a penchant for performance on original "period instruments" must find that

Fig. 6.32 Caricature of the castrato Farinelli, née Carlo Broschi (from the Bettman Archive)

solution frustrating. Once in a while, a courageous counter tenor who has not had the operation will attempt the parts, sometimes with disastrous results.

6.4.2 Die Kunst der Jodel

Yodeling (from the German verb, Jodeln) is a type of singing often characterized by rapid transitions through wide intervals on the musical scale using jumps between different registers of the voice. The technique is very popular in the Swiss and Austrian Alps and is also a basic feature of African Pygmy music, which was generally admired by the ancient Egyptians. There, the voice often leaps by intervals of fourths, fifths, and sixths. (See Cooke 1980.) In recent years, it has also been adopted as a form of folk music in the southern USA, not to mention the performance of classical music in Australia. In a sense, yodeling involves the opposite goal to that of the operatic singer, who strives for seamless transitions between the Heavy and Falsetto registers. In yodeling, one exaggerates the difference (Fig. 6.33).

In looking for recorded examples of yodeling, I was struck by the fact that some "country western" singers who practice the art seem to have burned out their normal

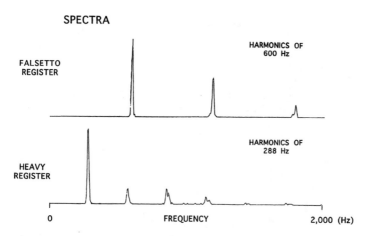

SPECTRA

FALSETTO
REGISTER

HARMONICS OF
600 Hz

HEAVY
REGISTER

HARMONICS OF
288 Hz

0   FREQUENCY   2,000 (Hz)

Fig. 6.33 Spectra from the Falsetto and Heavy registers of a popular country-western yodeler

voices. Although they may have a very pleasant and smooth falsetto range, it is often
without much harmonic content. Similarly, their Heavy registers sometimes suggest
the symptoms of inadequate vocal training discussed previously. As an example,
the spectra (on a linear amplitude scale) from a CD recording of one anonymous
popular country-western yodeler are shown in Fig. 6.35.

In contrast, equivalent spectra from a digital recording of the champion Australian classical yodeler Mary Schroeder (1999) are shown in Fig. 6.34. Purists
probably will not care much for her settings of Rossini's William Tell Overture
or Mozart's Symphony No. 40 in G Minor, but her vocal pyrotechniques are truly
amazing. The remarkable thing is that she can jump back and forth rapidly between
a falsetto spectrum with eight strong harmonics to a Heavy register showing at least
three formants and perhaps 20 harmonics.

(Note that the frequency scale in Fig. 6.34 extends up to 5 kHz, in contrast to the
2 kHz range in Fig. 6.33.) Nevertheless, one senses that the normal Heavy register
must suffer from the strenuous and rapid register changes in yodeling.

6.5 Vocal "Chords"

6.5.1 The Gyütö Tantric Choir

Buddhist Monks from the Gyütö Tantric Choir (originally from Tibet, but now in
forced exile at a monastery in Bomdi-La, India) use a style of singing that at least
dates to the fifteenth century in which each individual voice appears to produce three
widely spaced tones simultaneously. It is said that it takes the novice 3 years to learn
the technique and that the practice may have been brought to Tibet from India during
the eleventh century. Digital recordings of a number of chants by the Tantric Choir

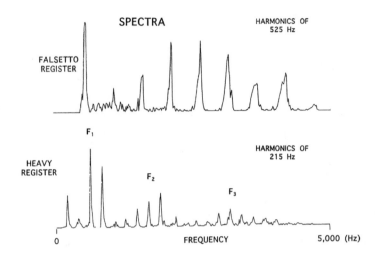

Fig. 6.34 Spectra from the Australian champion yodeler, Mary Schroeder (1999)

are available on a Compact Disc. (See Gyütö Monks 1986.) Unfortunately, with the annexation of Tibet by the Chinese mainland government, the Gyütö Monks have already become an endangered species. A basic problem in studying the sound from these recordings is that the Monks generally sing in groups. One really needs an isolated voice to study the vocal mechanism. Fortunately, brief passages for solo voice sometimes mark the beginning of a chant. Such a section occurs at the start of the Windham Hill CD recording where the lead monk begins alone. That excerpt was used to obtain the data shown in Fig. 6.37.

Their singing is reminiscent of the sound of bag pipes due to the constant presence of a bass drone accompanying the chords that are two octaves higher. The apparent drone pitch ranges chromatically throughout the recording from Bb to C, corresponding to frequencies ranging from about 58 to 65 Hz. The "drone" stands out very clearly because its pitch is so widely separated from the higher tones.

Analysis of the spectrum suggests that it simply consists of a vocal fold pitch of $F_0 = 60$ Hz with widely spaced harmonics—the strongest components being at $N = 2, 4, 5$, and 7. From Fig. 6.35, the most important frequencies are 60, 120, 240, 300, and 420 Hz, which approximate notes on the well-tempered scale indicated by the music notation shown in the figure. The notes form a diminished seventh chord on B, with a missing F.

I was prepared to believe that the 60-Hz drone frequency actually was present until I looked at the waveform and spectrum of the sound. As can be seen from the upper part of Fig. 6.35, the waveform is closely periodic at a frequency of 60 Hz, but there is almost no energy in the fundamental. (See the lower part of the figure.) The result is not too surprising since it would take a closed pipe about four feet long to resonate at that frequency. (The vocal fold could, of course, merrily buzz away at 60 Hz without any significant acoustic coupling to the outside world.)

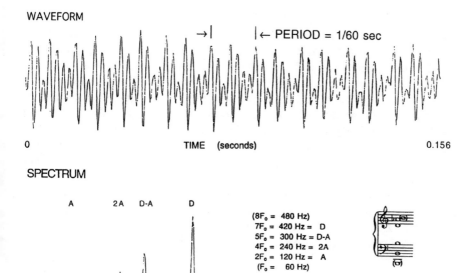

Fig. 6.35 Spectral data analyzed for a single monk, using one stereo channel of a CD recording by the Tibetan Tantric Choir. (See Gyütö Monks 1986.) The waveform (upper part of the figure) is periodic at 60 Hz, but the spectral amplitudes (lower part of the figure) contain almost no energy at 60 Hz

Unless these monks suffered from an extreme form of Marfan syndrome (a condition in which bones are abnormally elongated and heart valves are often defective),[24] significant acoustic coupling to the air at that frequency would not be expected. Yet, the sensation of the low pitch is clearly present and 60-Hz components permeate the spectrum. The presence of the second, fourth, fifth, and seventh harmonic of F_0 in Fig. 6.35 not only provides the impression of a strong 60-Hz pitch (frequency F_0) but also results in widely spaced peaks in the spectrum suggesting the sound of a chord. As discussed in Chap. 5, psychoacoustic properties of the ear can create the impression of a low-frequency tone when only a few higher harmonics of that frequency are present. The psychological effect is similar to that used to create the illusion of a very low-pitched organ pipe through use of shorter pipes tuned to the harmonics of the missing fundamental; if enough harmonics are present, the ear thinks that the fundamental is actually there.

As is well known in the electronics world, a nonlinear element (that is, one whose output amplitude is not simply proportional to the input, and hence is not a linear function of the input) can effectively multiply two oscillating waves at

---

[24] See Braunwald (1984, p. 1666).

two different frequencies together. That process produces harmonics of the two separate frequencies and additional terms at the sum and difference frequencies of the input signal components involved.[25] There is little doubt that the vocal tract contains nonlinearities. For example, the sound produced by the vocal folds contains numerous harmonics which must arise through some nonlinear process. Some simple physical models favor even harmonic production in the human voice, the most straightforward being one that involves a quadratic nonlinearity.[26] The peak labeled "A" in Fig. 6.35 is the second harmonic of the vocal fold resonance (F_0) and "2A" is the second harmonic of that. Those two frequencies would normally be expected and are also characteristic of the bass, tenor, and mezzo-soprano operatic voices discussed before.

The strong peak labeled "D" in the figure is probably the result of carefully tuning a narrowed F_1 formant to the seventh harmonic of F_0. The monks seem to be able to reduce the width of the cavity resonances in the vocal tract enough so as to isolate individual harmonics of the vocal fold (here, the seventh). That might be done by narrowing the mouth opening (as in sounding the vowel OO with the tongue arched as in Fig. 6.22) to cut down radiation loss. It is possible to show that the full width at half-maximum intensity response of the first resonance in the closed-pipe model of the vocal tract is $\Delta F_{cav} \approx F_1 L / \pi$, where L is the fractional energy loss per round trip of the running wave. (See Appendix A.) For ΔF to be less than $F_0 \approx 60\,\mathrm{Hz}$ (the vocal fold harmonic spacing), the round trip energy loss only has to be less than about 45% and would provide a cavity Quality Factor of Q ≈ 7. (See Appendix A.) Hence, by partially closing the mouth, isolating one vocal fold harmonic in an F_1 resonance (at $\approx 420\,\mathrm{Hz}$) seems possible. However, isolating a single harmonic in an F_2 resonance ($\approx 3F_1 \approx 1260\,\mathrm{Hz}$) would probably be out of the question.

The remaining important spectral component at D–A could then result as a difference frequency produced by "nonlinear mixing." Nonlinear mixing of two

---

[25]This result follows by application of simple trigonometric identities. Let the two signals be $\sin A$ and $\cos B$, where $A = 2\pi F_1 t$ and $B = 2\pi F_2 t$ and t is the time. Consider the general identities,

$\sin(A + B) = \sin A \cos B + \cos A \sin B$

$\sin(A - -B) = \sin A \cos B - \cos A \sin B$

Adding the two relations yields

$2 \sin A \cos B = \sin(A + B) + \sin(A - B)$

Hence, multiplying two separate oscillating waves together yields waves at frequencies $F_1 F_2$. To see how the process generates the second harmonic, just let $A = B$ in the above expressions.

[26]The response of any continuously varying physical system can be expanded in a power series in the stimulus. For example, suppose the lateral opening x of the vocal cords in respect to its equilibrium position x_0 is a function of the time-varying component of the pressure amplitude p above the larynx. The corresponding power series (or Taylor series) could be written

$x = x_0 + ap + bp^2 + cp^3 + dp^3 + \dots$, where $a, b, c, d, \dots$ are numerical coefficients which might be determined from the derivatives dx/dp, d^2x/dp^2, and so on. Then, ap represents the linear term, and bp^2 represents the first nonlinear term. The requirement for making this expansion is that the various derivatives are finite and continuous. The pressure-amplitude transmission and reflection coefficients would be simply related to x (i.e., to the area of the opening). Similarly, the effective spring constant in a mechanical model of the vocal fold could be expanded in a power series in the deflection, as discussed in Appendix A.

modes in the vocal tract can come about as follows: The constriction at the vocal fold increases the reflection of sound waves coming down the throat from the mouth. (See Fig. 6.6.) Hence, the reflected wave going back up to the mouth from the vocal fold is "amplitude modulated" (multiplied) by the wave motion in the vocal fold and therefore by the pressure above the fold. (The reflected waves moving upward contribute to the formation of pressure standing waves in the vocal tract, whereas the downward moving running waves that are not reflected by the vocal folds in Fig. 6.6 are simply transmitted through the larynx and absorbed in the lungs.)

There is a striking similarity between the present case and that discussed in a later chapter regarding multiphonic tones produced on the English horn and oboe. Both cases are characterized by a highly periodic waveform with missing fundamental, strong second harmonics and difference-frequency terms. Because there is no vibrato in either case and the harmonics are locked in phase, the resultant sound also has a strong "mechanical" tone quality.

6.5.2   The Tuva Throat Singer: A Siberian VASER?

A related style of singing is practiced by the so-called throat singers of Tuva (pronounced "TuVAH"), who inhabit a region of southern Siberia. Most singers use their throats, so the terminology may seem redundant. However, these singers use the throat in a remarkable way. They not only produce two apparent tones at once, but the low pitch can be even more startling than in the Tantric Choir. As with the Gyütö Monks, a bag pipe-like drone is present which gives the impression of a fundamental pitch of about 65 Hz, or a C two octaves below middle C on the piano, and one higher tone which varies over a one-octave range starting at about the G above middle C. That component carries the melody. Beyond that, the waveforms they generate sometimes have a component at half the vocal fold resonance!

Fortunately, a high-quality digital recording is available of one isolated Tuvan singer. (See TUVA 1990.) The waveforms and spectra shown in this section were made from one channel of the first band of that recording. There are several interesting features of the spectra.

First, the harmonics of the "drone" are very constant throughout the first band of the recording and uniformly separated by about 65 Hz. As with the Gyütö Monks, there is negligible energy radiated at that fundamental frequency. The lowest frequency directly observable on the linear scale used in the present figures from the recording was at 130 Hz, or the second harmonic of the perceived drone frequency. As with the Gyütö Monks, weak harmonics of the vocal fold resonance are spread throughout the spectrum. If you look at the spectrum with enough amplification, you can see that there actually is a little energy radiated at 65-Hz, but that component is down by about 15 dB from the adjacent signal at 130 Hz, the weakest component observable in the present figures. Examples of two higher-pitched tones a fifth apart are shown in Fig. 6.36 ($N = 8$) and Fig. 6.37 ($N = 12$).

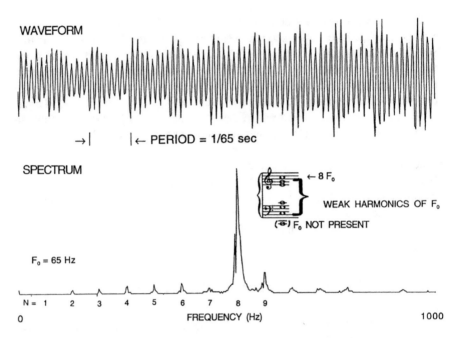

Fig. 6.36 Waveform and spectrum of a tone produced at 520 Hz = $8F_0$ (treble C) by one Tuva singer. Weak harmonics spaced at 65 Hz create the illusion of a bass tone at $F_0 = 65$ Hz which is not actually present in the audio spectrum

As shown in Fig. 6.38, there are two strong formants (F_1 and F_2) involved and a third one (F_3) that is weakly excited. The singer is able to reduce the resonance width of the lowest (F_1) formant to about 65 Hz and tune that formant to individual harmonics of the 65-Hz fundamental frequency. The second formant is about an octave higher than the first, but has a much broader resonance in the order of 450 Hz wide. From the results in Figs. 6.36 and 6.37, the first formant clearly can be tuned over the range of at least 520–780 Hz, or four harmonics of F_0 spaced at 65 Hz. According to Levin and Edgerton (1999), the singer can tune the notes from about 392 to 784 Hz. The notes do not change continuously in frequency but jump from one harmonic of F_0 to another. During this process, the apparent 65-Hz pitch is kept remarkably constant.

There is still another unusual property of the data in Fig. 6.38. As can be seen by visual inspection of the waveform, it is actually periodic at a frequency of 32.3 Hz or about half the value of the vocal fold resonance at $F_0 = 65$ Hz. This singer may actually be producing the first subharmonic of his normal vocal fold resonance. An alternative explanation is that he has learned how to relax his vocal fold sufficiently to produce the Pulse Register ("Strohbass") in which the effective mass of the vocal fold is increased (lowering the resonance frequency) by collapsing the ventricular folds from above. One clue in support of that interpretation is the appearance of a double pulse in the waveform. (See Fig. 6.38.) Again there is no significant energy

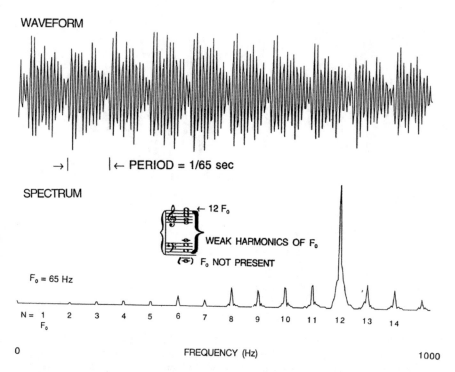

Fig. 6.37 Waveform and spectrum of a tone produced at 780 Hz = $12F_0$ (treble G) by one Tuva singer. Again, weak harmonics spaced at 65 Hz create the illusion of a bass tone at $F_0 = 65$ Hz which is not significantly present in the audio spectrum. The note at $N = 12$ is a fifth higher than the note in Fig. 6.38, but the other spectral components are about the same

radiated at that extremely low frequency and there are certainly no resonances that low in the vocal tract. It would take a closed pipe about 8-ft long to resonate at 32.3 Hz. This apparent subharmonic was not present on most notes, suggesting that it was hard to produce. By applying enough magnification, one can actually see the 32.3 Hz peaks in the spectrum, but they were down by about 35 dB from the sixth harmonic of F_0 at 390 Hz in Fig. 6.38. The latter in turn is about the smallest component that shows up on the linear scale used in the spectral plot.

The tones sound very much like those from a Jew's harp. Indeed, since the Tuva singers are fond of the Jew's harp as a musical instrument, it seems probable that they are actually imitating the tone quality of the Jew's harp in their singing. The harmonics of F_0 are analogous to those of the tongue of the Jew's harp and changing between different high-Q resonances of the vocal tract is similar to the method used to produce tunes on that instrument. One can also get subharmonics on the Jew's harp. (See Fig. 6.12 and associated discussion.) In each case, cavity losses can be reduced greatly by keeping the mouth nearly closed. That, in turn, can provide the narrow resonances observed in the lowest formant. In the present case, the Q of the F_1 resonance is given by $Q \approx F_1/\Delta F \approx 10$, where ΔF is the full resonance width

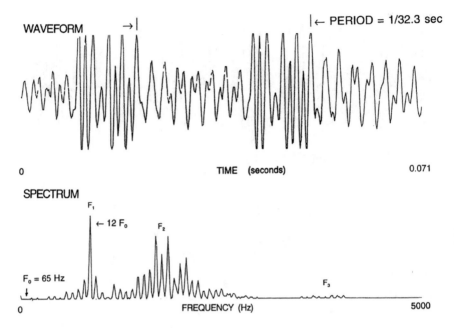

Fig. 6.38 Waveform and spectrum of a tone produced at 780 Hz = $12F_0$ (treble G) by one Tuva singer, but shown on a much broader frequency scale. As before, weak harmonics spaced at 65 Hz create the illusion of a bass drone at $F_0 = 65$ Hz which is not significantly present in the audio spectrum. The example here is remarkable in that the waveform is periodic at half the drone frequency. Note the double-pulse character which may be indicative of the "Strohbass" register. The spectrum indicates the presence of three formants: The very narrow, tunable one labeled F_1, together with a very broad formant (F_2) about an octave higher, and a weakly excited formant (F_3) in between the fourth and fifth harmonics of F_1

at half maximum response on an intensity scale. (For a discussion of cavity Q, see Appendix A.) The narrow resonances achieved here are similar to those encountered with the Gyütö Monks and probably produced in the same manner.

With such narrow resonances, the vocal tract would tend to interact nonlinearly with the vocal folds in a way analogous to that in which the resonances in a brass instrument couple with the vibrating lips to produce isolated notes. As discussed in a later chapter, one usually needs at least two high-Q modes resonant on harmonically related frequencies to produce stable oscillation in brass or reed instruments. However, since there are no other high-Q modes present in the spectral data of the Tuva singer shown here, the phenomenon of "gain narrowing" encountered in laser physics may be involved. The vibrating vocal fold may provide enough gain in the lowest mode to reduce the normal radiation loss through the mouth. The presence of a round trip fractional energy gain G in the vocal tract would reduce the cavity width to $\Delta F \approx F_1(L - G)\pi$.

In the limit that $G \to L$, a condition required for steady-state oscillation in a laser, the spectral width would be limited only by the time duration of the coherent oscillation, or by random noise present in the mode. (See Bennett 1962, p. 54; 1977, p. 89.) It would indeed be ironic if these singers had been practicing a form of vocal laser oscillation (or VASER—standing for "Voice Amplification by Stimulated Emission of Radiation") in southern Siberia for the last thousand years!

The main puzzle left is how the Tuva Throat singer can produce such low sub-bass frequencies. As mentioned above, the sub-bass frequency may just be an example of the "Strohbass" register in which the effective mass of the vocal fold is increased substantially. However, the result might also come about from subharmonic generation due to a quadratic nonlinearity in the vocal fold mechanism. (See the discussion of the subharmonic oscillator in Appendix A.) One really needs stroboscopic pictures of the vocal fold vibration from the throat of a Tuva singer to determine what is actually happening.

Problems

6.1 An elephant has a trunk 6 ft long. If it is used as a closed pipe, what is the fundamental resonance?

6.2 The Parasaurolophus walkeri had two long curved, trombone-like air passages in its head leading from the nostrils to the back of its throat. (See Fig. 6.26.) If the unfolded length of each air passage was about 6 ft, what would the fundamental resonances and the first several modes have been? The two nostrils would not have precisely the same length, so the sound from the two nostrils would probably produce beats. A musically talented Parasaurolophus might be able to produce harmonic intervals simultaneously like double stops on a violin (thirds, fourths, fifths, etc.), in which case beats produced between the two nostrils could have a mellowing effect like vibrato.

6.3 A resourceful SCUBA diver goes to a Halloween party disguised as Henry Kissinger, whose speaking voice is at about low D♯ $= 77.8$ Hz. Unfortunately, the pitch of his voice is about 120 Hz. To compensate for that failing, he fills his SCUBA tank with a mixture of oxygen (molecular weight $= 32$) and krypton (molecular weight $= 84$). (a) What is the average molecular weight of the gas in his SCUBA tank? (b) What must mixture of oxygen and krypton be? Take the molecular weight of air to be about 29. (Warning: Do not try this! It is very hard to get the krypton out of your lungs.)

6.4 A counter tenor who normally can only reach high G at 1568 Hz wants to sing the "Queen of the Night" aria from the Magic Flute, which goes up to a high F (2793 Hz), almost an octave higher. If he fills a tank with a mixture of 20% oxygen and 80% helium, will he be able to hit the note? (The molecular weights of air $= 29$, helium $= 4$, and oxygen $= 32$.)

Chapter 7
Pipe Organs

7.1 Historical Background[1]

7.1.1 Introduction

Called "The King of the Instruments" by Mozart, the pipe organ is the oldest of the keyboard instruments. Historians tell us they date to at least 246 BC. One of the earliest (from the Hellenistic period) was in the form of a tree containing whistling birds. (See Sonnaillon 1985.)

Organs operated by keyboards similar to those on harpsichords were prevalent in Europe by the mid-1400s. Typically, there was one wind chest containing several different kinds of pipes for each keyboard. Each such unit is thought of as an "organ" by itself and, indeed, as more keyboards and pedals were added, more and more "organs" were contained in one instrument. Thus in French, the word for a large organ (grandes orgues) is nearly always used in the plural form (Fig. 7.1).

The sixteenth century "positiv" was a single keyboard instrument with one wind chest and an air supply provided by manually operated bellows (Fig. 7.2). Several different stops were often contained on such an organ, each containing a separate "rank" or set of pipes with pitches covering the compass of the keyboard. As the complexity of organs grew, more keyboards and wind chests were added, often with couplers between the keyboards similar to those used later on two-manual harpsichords. The "positiv" organ was frequently retained on large organs with the exception that it was often placed behind the organist—hence the term "ruckpositiv." Especially in Germany in the sixteen hundreds, the simple positiv was enhanced with more elaborate structures ("brustwerks" and "oberwerks" or "hauptwerks,"

The original version of this chapter was revised: Equation on page 319 was corrected. The correction to this chapter is available at https://doi.org/10.1007/978-3-319-92796-1_8

[1]For histories of the organ, see (1766), Vol. 1, Audsley (1905), Vol. 1 and Sonnaillon (1985). The author is indebted to Bernard Sonnailion for permission to reproduce illustrations from his book.

© Springer Nature Switzerland AG 2018
W. R. Bennett, Jr., *The Science of Musical Sound*,
https://doi.org/10.1007/978-3-319-92796-1_7

Fig. 7.1 Bellows-pumped
bird whistles thought to be
the earliest form of the pipe
organ

Fig. 7.2 A sixteenth-century
positiv by an unknown
builder (courtesy of the
Historiches Museum, Basle)

for example) placed in front of and above the organist, some with wind chests
containing pipes of 16- to 32-ft length. During Bach's time, the tallest pipes were
confined to the pedal section of the organ and often placed in so-called pedal towers
on either side of the console. It is thought that Bach himself did not have pipes
of more than 8-ft length accessible from the keyboards. Producing the powerful
low frequency notes was delegated to the "pedal" section and operated by his feet.
Indeed, for most of his life as an organist in Leipzig, Bach only had an organ with
two manuals and pedals. Of course, he had choir boys to change the stops and pump
the bellows. Some very large baroque pipe organs employed teams of grown men
working foot-operated bellows (See Fig. 7.3). Contemporary organs usually employ

Fig. 7.3 Possible "Winding" for a very large baroque pipe organ

large motor-driven blowers, often located in the basement of the church or concert hall in order to reduce extraneous sounds from the machinery.[2]

Organists have the curious convention of labeling the different stops according to the length of an open pipe required to produce the same pitch as the lowest C on the keyboard (or pedal board). Because closed pipes produce a fundamental pitch that is half that of open pipes of the same length, one finds, for example, that the active length of a 16-ft "gedekt" or "bourdon" (closed pipe) is actually only about 8-ft long. (The "foot" of the pipe, which tapers from a small opening on the pipe board to the much wider diameter of the pipe itself, often adds another few feet to the overall length.) On the other hand, a 16-ft open principal or "diapason" pipe will generally have an active length that really is 16-ft long.

---

[2]One of the dangers of putting the blower in the basement was illustrated at Cornell University in the late 1960s. Unknown to the organist, the basement had been flooded by spring rains. He went directly to the console and flipped the switch to turn the motor on. Within a few minutes the entire organ was filled with water! Another problem with such a location for the blower is that the basement air is apt to have rather different temperature and humidity content than that in the room containing the wind chests with the result that the pipes may be thrown out of tune.

7.2 Structure of an Organ

A common design for baroque organs is outlined in Fig. 7.4. The organist sits on a bench in front of the "ruckpositiv" at a console below the "oberwerk" or "hauptwerk" and in front of the "brustwerk." In early organs, actions involving mechanical linkages called "trackers" were used to couple the depression of a key on the console or pedal to the opening of a "pallet valve" under the wind chest that let air into a channel for a particular note on the keyboard to be sounded. Slider valves placed perpendicularly to the note channels and operated by draw knobs at the side of the console would allow air to enter any pipe in a specified rank for which the pallet valve had been opened. The slider-stops consisted of strips of wood with holes cut in them at the pipe spacing for each rank.

The basic mechanism[3] was fairly simple and some of these tracker-action organs are still working some 300 years after they were made.

Figure 7.5 illustrates the working of a tracker action wind chest. Vertical rods (trackers) pull down pallet valves when the corresponding keys are depressed on a manual. That lets air into the appropriate note channel. If the slider valve for a

Fig. 7.4 Schematic drawing of the position of the organist with respect to the ruckpositiv (left), the brustwerk (middle right), the hauptwerk (upper middle), the pedals (middle bottom), and the wind supply (right) on a baroque organ. As some organists have sadly learned, the blast of sound from the brustwerk, which hits the organist in the face, can lead to deafness (The drawing is partly based on an illustration in 1768, volume II)

---

[3]For a more complete discussion of these mechanisms, see Audsley (1905, Vol. 2) and Sonnaillon (1985).

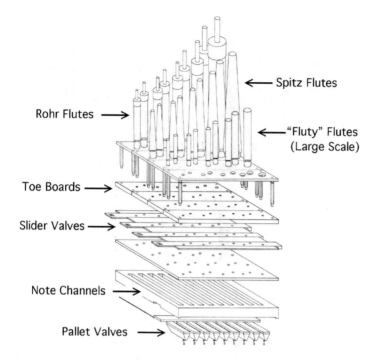

Spitz Flutes

Rohr Flutes →

"Fluty" Flutes
(Large Scale)

Toe Boards →

Slider Valves →

Note Channels →

Pallet Valves →

Fig. 7.5 "Exploded" view of a tracker-action wind chest

particular rank of pipes is open (pulled to the left in the drawing), air is let into the pipe corresponding to the key depressed on the manual.

The operation of a pipe organ involves rather complex switching problems which were solved with great ingenuity by baroque builders such as Arp Schnitger (1648–1719) and Andreas Silberman (1678–1734), who made pipe organs at the time of Bach. One wants to insure that only the notes depressed on the keyboard and the ranks of pipes selected are sounded. The tracker action wind chest in Fig. 7.5 clearly accomplishes this objective. However, the connecting linkages were amazingly complex in design. As an example, the arrangement of trackers for one such German organ is shown in Fig. 7.6.

As evident from Fig. 7.6, the vertical motion of the tracker rods is translated laterally from the note spacing on the keyboard to the pipe spacing on the wind chest. That is accomplished through a structure of roller bars on the large triangular frame in Fig. 7.6 of the type illustrated below (Fig. 7.7).

Figure 7.8 is a schematic drawing for the air supply of an organ. The main difference between this one and its older predecessors is the incorporation of a blower powered by electricity. Air comes in from the left passing through a curtain valve regulator and from there into a large reservoir with an expansion bellows on top. Pressure regulation arises from the connection by a rod between the top of the bellows and the roller curtain in the regulator valve. As the top of the bellows rises,

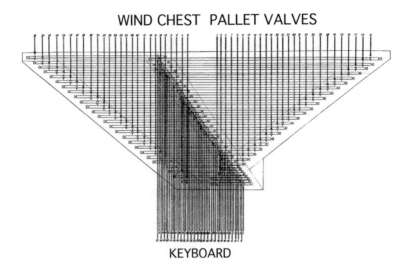

Fig. 7.6 The connection between vertical trackers and the wind chest pallet valves on one German organ containing a 52-note keyboard (after Audsley 1905, vol. 2)

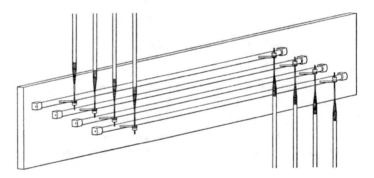

Fig. 7.7 Method of displacing vertical pull-down motion horizontally on a tracker-action organ (after Audsley 1905)

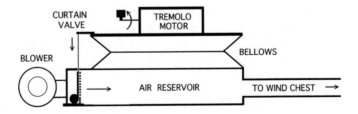

Fig. 7.8 Air supply for a modern organ

the curtain is pulled up gradually, closing off the air path from the blower. (It is shown nearly closed in Fig. 7.8.) As the pressure in the reservoir is consumed by playing the organ, the top of the bellows slowly drops down allowing the curtain to be retracted on a spring-loaded roller. The assembly works somewhat like an upside down window shade and the restoring spring on the roller keeps the curtain under moderately constant tension. The mechanism affords a particularly simple method of controlling the air pressure. A weight placed on top of the bellows determines the air pressure.

In addition to maintaining constant wind pressure in the organ (an important function since the pitch of the pipes varies with the pressure), the mechanism provides a simple way to introduce vibrato (or tremolo) in the organ. One way of achieving that goal is to place a motor on top of the air reservoir that turns a shaft with an off-center weight mounted on it as shown in the figure. As that weight moves up and down a periodic variation in the air pressure is introduced at the motor rotation frequency. Generally the speed of the tremolo is fast compared to the normal rate at which pressure is consumed by the organ and the curtain valve doesn't work fast enough to remove the vibrato. The tremolo rate varies a good deal among organs. Some makers prefer a rather slow tremolo, whereas others shoot for something more approaching that of a good violin or oboe vibrato (perhaps 6 Hz).

It should be acknowledged that there are many variants in use for producing and controlling the wind pressure on a pipe organ. (See, for example, Audsley 1905, Vol. 2.) In some cases cones are raised and lowered in a conical regulator valve, and in others various complex pneumatic oscillators have been designed as tremolo engines. However, the ones discussed above are among the simplest (Fig. 7.9).

7.2.1 Baroque Organs

It may help to understand the switching operation by thinking of the situation schematically in terms of the matrix outlined in Fig. 7.10. This figure could actually be part of the wiring diagram for an organ using direct-electric pipe valves.

Suppose we regard each of the solid circles as an electromagnetically operated (solenoid) valve for each pipe. (We, of course, assume these are all mounted on the inside of a wind chest containing air at the pressure needed to operate all the pipes.) For example, to turn on the 8-ft rank, a switch is closed that allows current to flow from a power supply along the lowest horizontal wire from the left at the "8-ft" designation. We also assume that when any key is depressed, it closes a switch that can return current to the power supply from the wire running vertically above the key. Then, when the 8-ft rank is turned on and the lowest C on the keyboard is depressed, current flows through the first junction at the lower left in the diagram, down through the vertical wire through the switch on the C key, and back to the power supply. That current causes the solenoid valve below the first pipe in the 8-ft rank to open, letting air flow into the pipe. The same process clearly works in principle for any number of ranks and for any note on the keyboard. This approach

a

Fig. 7.9 Examples of Baroque organs. (**a**) Sixteenth century organ rebuilt by Andreas Silbermann in 1711 and located in the Peter Church of Basel, Switzerland. The organ has since been renovated by several other builders. (**b**) Examples of Baroque organs. Large tracker-action organ in the

b

Fig. 7.9 (continued) Nikolaikirche at Altenbruch, Germany, originally built in 1497–1498 by Johannes Coci. According to Sonnaillon (1985, p. 100), substantial modifications were made by several builders, including Johann Werner Kapmeyer, a pupil of Art Schnitger in 1730. The organ was restored by Rudolf von Beckerath in 1967 (photograph courtesy of Bernard Sonnaillon)

Fig. 7.10 Diagram to illustrate the switching process used to open stops and sound individual notes on a wind chest (see text)

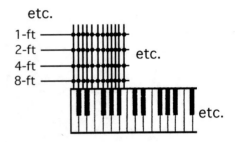

actually works quite well for low wind pressures, although the solenoid valves usually require transistorized switches to handle the necessary currents.[4] In practice, there are usually only 61 notes on any one contemporary organ manual (as opposed to the 88 on the piano keyboard) and only a half-dozen or so different ranks on a given wind chest. Some additional switching methods are required to couple the pipes on different wind chests together.

7.2.2 Modern Organs

For higher-pressure organs (say more than 5-in. of water[5]), various mechanical and electromechanical methods have been devised for accomplishing the same objectives. Although mechanical linkages ("trackers") were used for hundreds of years, other methods briefly mentioned above for handling the switching problem were incorporated in the nineteenth century. As the wind pressure and the number of stops in different wind chests were increased, it became harder and harder to push the keys down.[6] This difficulty became more noticeable in French organs such as those initiated by Aristide Cavaillé-Coll (1811–1899) in Paris, who designed the instruments in Notre Dame, the Basilica of Ste. Clotilde and the Church of St. Sulpice. Toward the end of the nineteenth century larger and larger air pressure was required to increase the loudness of the pipes in huge cathedrals and went from 2 or 3 in. of water characteristic of the pressure produced by blowing a pipe

---

[4]The Peterson Company in Alsip, Illinois, for example, makes such valves and control circuits in conveniently applicable form for low-pressure organs.

[5]This curious unit for pressure is based on the method used to measure air pressure in organ wind chests. A U-tube manometer is partially filled with water and one end connected inside the wind chest. The pressure above local atmospheric pressure then corresponds to the difference in water level between the two halves of the manometer. Typically, a person blowing a flute or recorder only generates a pressure of a few inches of water and early organ pipes were adjusted for that range. However, that is not the same for modern reed instruments. Oboist Robert Bloom told me that Arthur Benade once hooked up a water-filled manometer to a rubber tube in order to measure the pressure in the oboist's mouth. He then asked Bloom to play a note on the oboe. Bloom obliged and shot all the water out of the manometer up to the ceiling!

[6]Leonard Brain once told me that his brother, Dennis—the famous British horn player—had practiced regularly on a stiff tracker action organ to develop strong finger muscles.

with the human mouth, to pressures of more than 20 in. of water needed to operate some modern reed pipes. By 1900, electropneumatic actions were incorporated to reduce the mechanical effort required to play such a large organ. However, with many large organs using such electromechanical amplification, long time delays are encountered after the keys are depressed before the pipes sound. Professional organists somehow adapt to this situation, but the secret seems to be to avoid listening to the music. Some have argued that large delays are inevitable in big cathedrals anyway due to the finite velocity of sound. The problem becomes still worse when different organs controlled by the same console are positioned all over a large building.[7]

One way of characterizing an organ is simply by the total number of pipes. Some have a horrendous number. The new Disney Hall organ in Los Angeles has 6325 pipes, and the new organ in Verizon Hall in Philadelphia has 6932 pipes. You might think that those are really large numbers until you examine some of the organs made in the early twentieth century. For example, the Newberry Organ in Woolsey Hall at Yale has 12,617 pipes and the Wanamaker organ in Philadelphia has 30,067 pipes. Such large organs have pipe rooms that seem like small cities. As an example, the layout in Woolsey Hall at Yale University for the Newberry organ is shown in Fig. 7.11—a space 80-ft wide, 40-ft high and 10-ft deep. This organ was originally built in 1901 by Hutchings, was nearly doubled in size in 1915 by Steere, and greatly enlarged in 1928 by the Skinner Organ Company.[8]

Pipe organs of this size are especially appropriate for compositions of the late nineteenth century such as those by Charles-Marie Widor in which huge masses of sound are desired. I once measured sound pressure levels of close to 100 dB from the Newberry organ over the entire first balcony at the rear of Woolsey Hall while University organist Charles Krigbaum was playing a Toccata by Widor. (The 656,000 cubic-ft hall was filled with some 2700 people—graduating students and their friends and relatives.) This organ has two 20 HP blowers (one for back-up purposes) in the basement that can each produce enormous amounts of air at pressures of over 15 in. of water. Indeed, the fluctuating air pressure coming out of the top of the bass reeds provides a hazard for low-flying wild life. On at least one occasion, a 32-ft bombarde was put out of commission when a bat was sucked into the end of the pipe! The list of stops for the Newberry organ is contained in Appendix E.[9]

The console for the Newberry organ with its four keyboards and huge array of stops is pretty awe-inspiring, but it nevertheless pales in comparison with that for the Wanamaker organ (Fig. 7.12). We won't even attempt to provide the stop list for

---

[7] One Italian organ virtuoso said very proudly that he had mastered a technique for playing on two different manuals controlling pipes at the front and back of his church so that the sound arrived at his location in phase. (Pity the poor people in between!)

[8] The author is indebted to the Yale University Curators of Organs, Nicolas Thompson-Allen and Joseph Dzeda, for helpful discussions regarding the Newberry organ.

[9] A study of loudness contours for the Newberry organ was given by Harrison and Thompson-Allen (1996).

Fig. 7.11 The Yale University Newberry Organ Bottom: View from inside Woolsey Hall (Photograph courtesy of Joseph Dzeda) Top: View of the organ behind the front pipes (Drawing by Kurt Bocco of West Haven, CT)

that organ, although it is available from the "Friends of the Wanamaker Organ" in Pennsylvania.[10, 11] The original design of the Wanamaker organ was by the famous British architect George Ashdown Audsley. It was built by the Murray Harris

---

[10]The "Friends of the Wanamaker Organ" (a non-profit organization) is located at 105 Charles Drive G-3, Bryn Mawr, PA 19010–2313.

[11]According to the Guinness Book of Records, the largest and loudest organ in the world is the one in the Atlantic City Convention Hall; that one is said to have 33,114 pipes and a console with seven keyboards.

Fig. 7.12 Console of the Wanamaker organ designed by William Boone Fleming (added in 1928) (photograph courtesy of the "Friends of the Wanamaker Organ")

Company in Los Angeles, and was featured in the 1904 St. Louis World's Fair. (It took 13 railroad freight cars to bring the organ from its factory to Philadelphia early in the twentieth century!) The organ occupies three floors in the old Wanamaker department store building and has a console reminiscent of the cockpit on a large jet plane.

However, "biggest" is not necessarily "best"—especially for playing the organ works of Johann Sebastian Bach. There, a relatively small number of carefully chosen, low-pressure pipes can be more effective. For example, such pipework can provide a much clearer development of a multi-voiced fugue or of a four-part chorale than the muddier sound obtained from high-pressure organs with multiple "expression" pedals contained in large cathedrals.[12]

---

[12]"Expression pedals" control the opening and closing of Venetian-blind-like "swell louvers" surrounding some organs and provide crescendo and diminuendo changes while the organist is playing. Bach did not have (or possibly want) such things and depended instead on having stops changed while playing.

A movement back to lower-pressure organs began early in the twentieth century, led by people such as Albert Schweitzer and E. Power Biggs. Schweizer said, "With the pneumatic action one communicates with his instrument by telegraph."[13] Biggs in particular played the role of a kind of "Johnny Appleseed" and went about persuading numerous churches in the New England area to turn in their huge electro-pneumatic organs for low-pressure tracker-action instruments reminiscent of those used at the time of Bach.

Companies such as the Andover Organ Company in Methuen and that of Charles Fiske in Gloucester (both near Boston, Massachusetts) sprang into existence and began installing such instruments all around the country. Of course, builders such as Pflentrop in Holland and Rudolph von Beckerath and the Hillebrand Brothers in Germany had been making tracker-action instruments all along.

During the 1960s, New Haven went from a city largely dominated by electro-pneumatic organs made by the local Hall Company to a place containing at least four large tracker-action organs of both domestic and European origin.

One wonders what the future will offer. Digital electronics has already been felt and companies such as Allen Organs in Macungie, PA offer so-called "computer organs" that feature digitally encoded waveforms captured from real pipes. These can sound at least as good as recordings played over a "Hi-Fi" system. However, the best way to simulate the sound from an organ pipe is still to get a pipe and blow air through it.

Organ control circuitry has already benefited largely from the MIDI industry.[14] Instead of requiring huge bundles of wire characteristic of the early electropneumatic organs, one can now use standard MIDI interfaces with just a few wires to couple wind chests to large, moveable consoles.

Miniaturized transistor circuits have also made the associated electronic hardware much more compact. At the moment, the most effective use of digital techniques seems to be merely through the control of more conventional organ pipes and wind chests. Small churches on tight budgets might use MIDI techniques to play back recordings through their pipe organs of the pulse sequences generated by occasional outstanding musicians. Of course, if enough purely electronic organs permeate the landscape, the public may forget what a real organ sounds like. The listener might want to see whether the output from a MIDI keyboard and synthesizer is preferable to the sound from a real piano. (As Arthur Schnabel once implied, the most important thing about playing music anyway is the space between the notes.)

---

[13]Carl Weinrich, "Albert Schweitzer's Contribution to Organ Building" in Albert Schweitzer Jubilee Book, edited by A. Roback (Greenwood Press, 1970), p. 222.

[14]MIDI stands for "Musical Instrument Digital Interface," a technology now used in the electronic keyboard and synthesizer world.

7.3 Organ Pipes

A rudimentary discussion of open and closed pipes was given in Chap. 1. Here, we will consider properties of different organ pipes in more detail.

7.3.1 Edge Tones and Air Reeds

Flue pipes are excited by the vortex motion produced when air impinges on a wedge, as illustrated in Fig. 7.13. Here a jet of air (actually a small sheet of air perpendicular to the paper) travels from the left at velocity V, initially hitting the lower side of the wedge a distance L from the orifice. (That distance is often called the "cut-up" by organ builders.) It then creates a vortex that circles backward and deflects the air stream upward as indicated by the curved arrow in the diagram. The sheet of air then hits the top side of the wedge (at the right in Fig. 7.13) where another vortex is created, circling in a reverse path and deflecting the air stream back downward in the diagram. In the meantime, the original lower vortex has moved further to the right. This process results in the stream of air being switched back and forth in the space between the original orifice and the tip of the wedge, producing a kind of vibrating "air reed." (Similar action is used to excite the resonances in an orchestral flute or to produce the sound you hear when you blow over the top of a soda bottle.) As time continues, a periodic sequence of vortices travels to the right along both the top and bottom of the wedge. (Such a sequence of vortices is known by fluid dynamicists as a "von Kármán street"; for example, see Van Dyke (1982).)

The frequency with which the air stream vibrates is at least roughly proportional to the ratio V/L and for greatest efficiency that frequency should be tuned to the fundamental resonance of the pipe. For example, to make high-pressure pipes work well on a low-pressure organ, one needs to "lower the cut up" (i.e., shorten the length L) so that the time taken for air to travel that distance is about the same in

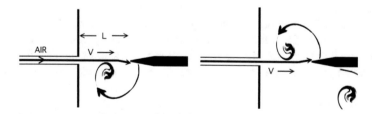

Fig. 7.13 Generation of an edge tone (see text)

both pipes. That is just one of several adjustments that may be required in "voicing" the pipes on an organ.[15]

When the wedge occurs at the entrance to an organ pipe, periodic running waves from the vortices inside the pipe result in standing waves that reinforce the oscillation of the air reed. In practice it takes many cycle periods at the fundamental pipe resonance frequency for the oscillation to build up to a steady-state value. (See Fig. 2.8 of Chap. 2.) That time is easily estimated from the Q or width of the pipe resonance, as discussed in Appendix A, and is determined by the energy loss in the resonator. With low-pressure pipes, this start-up transient has a characteristic sound similar to that produced when air rushes over the teeth in the production of sibilants. It is often described by the onomatopoetic term, "chiff," where the syllable "chi" is pronounced as in the word "children." Some people like this sound and some do not. In high-pressure organs the turbulence that produces "chiff" may also result in unstable pipe oscillation. That effect is often reduced by a process known as "nicking." One takes a set of needles or a narrow file to produce several parallel deep diagonal scratches across the "languid".[16] The "languid" is the insert—often made of lead—used in pipes to produce reflections at the entrance to the resonator.

7.3.2   Pipe Construction

The time-honored way of making metal pipe walls is illustrated in Fig. 7.14. Mixtures of tin and lead (typically 1 part tin to 4 parts lead) are melted together in a furnace to create something called "spotted metal" and moved in a bucket like the one shown hanging from a chain at the left of Fig. 7.14. The liquid metal is then poured into the wooden trough marked A that has a slotted opening at the bottom. The man at the right then moves the trough (now labeled B in the lower figure) to the right at a uniform rate while the metal flows out the bottom onto a long wooden table. After the metal cools, the resulting spotted metal sheet is rolled about the lengthwise axis and soldered along the seam to form a cylindrical metal pipe. The design of the toe attached to the cylinder depends on whether it is to be a flue pipe or a reed pipe.

Lead is used in many parts of pipe organs—in the material for the walls of metal pipes, in the languids of flue pipes and the base of reed pipes. Lead is also very effective in damping wall vibrations in organ pipes. Recently, the European Union has declared pipe organs to be health hazards due to their lead content.[17] But this

---

[15]In a pinch, lowering the cut-up (reducing L) can be accomplished for small pipes by gluing a piece of heavy lead tape across the top of the "edge."

[16]One can reduce the effects of nicking by taking out the languid and covering the scratches with the same sort of heavy lead tape used to lower the cut-up. Alternatively, the lead surface of the languid may be scraped flat with a sharp knife, or for wood pipes, the nicks may be filled with glue or plastic wood.

[17]Alan Cowell in the New York Times, 3/22/2006, p. A8

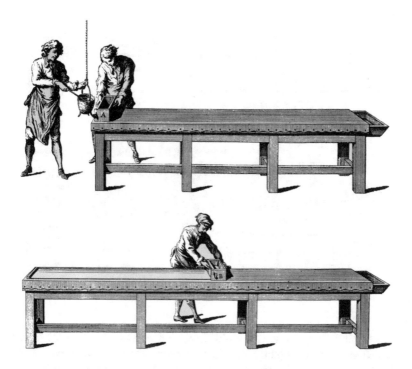

Fig. 7.14 Method of making the wall material for a metal pipe (from Bédos de Celles, 1766-1778, reproduced by courtesy of the Beinecke Music Library at Yale University)

amazing ruling only applies to organs in which electrically pumped air is forced through the pipes. Organs using hand-operated bellows pumps are still allowed! The basic conjecture is nonsense. The vapor-pressure of lead at room temperature is almost immeasurably small (probably about 10^{-12} Torr). Of course, if you heat it, the vapor pressure will go up. For example, it hits about 3×10^{-9} Torr at 600 K. (Standard atmospheric pressure is 760 Torr.)

Narrow scale (small diameter compared to the active length) open pipes are the basic building blocks of a pipe organ and give the instrument its characteristic "organ" sound. The smaller the scale, the larger the harmonic content of the pipe for a given wind pressure. These higher modes of the pipe are excited by the vortex motion illustrated in Fig. 7.13 and are phase-locked by the nonlinear characteristics of the air flow about the edge. As the scale becomes wider (larger ratio of diameter to length), the higher modes are less excited and a "flutier" sound is produced.

Of course, pipes differ in pitch as well as harmonic content or timbre. The typical large pipe organ has a set of open-principal (or "diapason") foundation stops at 8, 4, 2, 1-ft pitch. These successive pitches, of course, can be used to create a sound emphasizing even harmonics of the basic 8-ft pitch on the keyboard. In addition to those, many organs contain so-called "mutation stops" at pitches such as 22/3, 13/5, 11/7, and so on. These correspond to the odd harmonics (3rd, 5th, 7th, etc.)

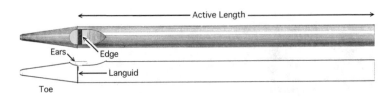

Fig. 7.15 Open flue pipe

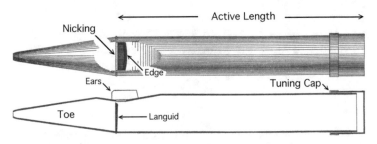

Fig. 7.16 Closed metal pipe

of the 8-ft pitch. Thus, one can build up a kind of Fourier series of harmonics to produce various tone colors. Very interesting contrasts can then be effected by the choice of stops on different keyboards. Although it is probably misleading to think of synthesizing the sound of orchestral instruments too literally in this way, a choice of 8, 22/3, 13/5 -ft stops will produce a clarinet-like sound, whereas the combination of 8, 4, and 1-ft stops will sound more like an oboe. Some organs have "mixture" stops in which depressing a single key automatically sounds several ranks that are carefully tuned to be harmonically related. Some of these sound almost indistinguishable from reed stops (Fig. 7.15).

As discussed in Chap. 1 and Appendix F, closed pipes only support odd-harmonic modes of the fundamental pitch. Hence, if one wanted to simulate a clarinet sound, it would pay to start with a closed pipe before adding mutations. As with open pipes, closed pipes are made with both metal (as in Fig. 7.16) and wood. Really large pipes are generally easier to make out of wood and for larger lengths can be fairly stable. Short wooden pipes tend to fluctuate a great deal in pitch due to humidity changes. Hence, there is a tendency to make pipes of metal when 4-ft or less pitch is required. The pipe "ears" are put in to minimize the spreading of the sheet of air and to avoid interference from neighboring pipes. In some cases, bending the ears can provide fine tuning.

Wooden pipes of either closed or open variety are usually made with rectangular cross sections for ease in construction. Some people believe that wooden pipes have a mellower tone quality than metal ones. There is apt to be more damping of the higher-frequency modes by the wooden wall vibrations. However, this difference may depend on the thickness and density of the material. Wooden pipes usually have thicker walls than metal ones, and with some metal pipes there is considerable wall vibration. For example, wrapping layers of lead tape around the middle of a metal pipe often changes the radiated tone quality. But there is no doubt that metal pipes tend to remain in tune longer than wooden ones.

A variant on the closed metal pipe is the "Rohr flute" or "Chimney Flute." ("Rohr" means "tube" in German, whereas "chimney" is more self-evident.) Examples of the two pipes are shown in Figs. 7.17 and 7.18. There is relatively little difference in tone quality between these two ways of dealing with the cavity. Having the tube project inward does make the pipe less susceptible to mechanical damage. In both forms of the resonator, the narrower diameter tube tends to enhance the fifth harmonic from the cavity, producing a very pleasing bell-like, clarinet sound.

Helmholtz found experimentally that the fifth harmonic is much stronger in the chimney flute than in a closed pipe of the same fundamental length, but according to Audsley (1905, Vol. 2, p. 538) no one has yet provided a simple theoretical explanation of the phenomenon.[18] Although it is not clear that that situation still holds, consider the following approximate explanation.

In the lower portion of Fig. 7.18, the pipe consists of two coupled cavities in series, the larger-diameter partially closed pipe of length L and the shorter open pipe of length ℓ. (In the drawing, $l/L \approx 0.3$.) It is a property of the wave equation that governs the solutions for the normal pipe modes that the pressure standing wave must be continuous in its value and slope at the boundary between the two media

Fig. 7.17 Enlargement of the Rohr flute drawing showing where the nicking of the languid would occur

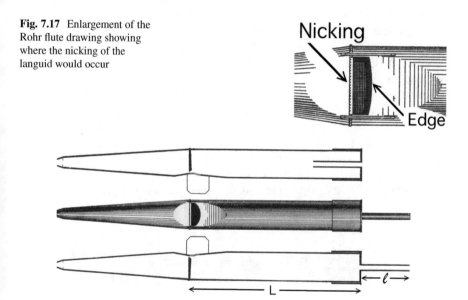

Fig. 7.18 Examples of a rohr flute and a chimney flute (after Audsley 1905)

---

[18]Perhaps in response to this challenge, Bouasse and Fouché (1929) devoted an entire chapter (number IX) to the analysis of this question, but their explanation is not really simple.

at any instant in time. The fifth harmonic of the basic cavity of length L has a wavelength of $4L/5$. (As shown in Chap. 1 the lowest mode for the closed pipe has a wavelength of $4L$.) Because the two cavities are open to each other, the wavelength for the fifth harmonic must be the same in the short section and match in value and slope. The solution in the short section must go to zero at the opening of the short tube to the air at the right and be a maximum at the junction of the two pipes where the air pressure for the larger pipe is a maximum. For the waveform to be continuous at that junction, the short section must have a quarter of the wavelength for the normal lowest mode, which for an open pipe of that length is 2ℓ. For the wavelengths in the two sections to be equal at the fifth harmonic of the large partially closed pipe, $4L/5 = 2\ell$, which means that $l/L \approx 0.4$. (The discrepancy from the ratio obtained from the drawing may arise from the difference in tube diameters.) Audsley notes that the sound is about the same when the short tube is inverted, as in the top of the drawing. In that case, the short tube protrudes into the larger one by a distance l equal to half the large tube wavelength for the fifth harmonic (or $2L/5$) and its opening is again at a pressure maximum within the large closed pipe. Hence, the same enhancement of the fifth harmonic should occur in both configurations shown in Fig. 7.17. In either case, the small tube provides a resonant transmission filter for the fifth harmonic. In practice the small tube length is adjusted for best sound by the pipe maker.

Wooden pipes are usually made of rectangular cross section, as shown in Fig. 7.19. The closed pipe (upper figure) has a movable tuning plug at one end. These produce primarily first and third harmonics, have a somewhat clarinet-like tone and go under the name "gedeckt" (or "gedakt" in old German) or "pedal bourdons." They are often used in places where vertical space is at a premium because the fundamental pitch is half that for an open pipe of the same length. They are often found in mechanical organs such as those programmed for use in "merry-go-rounds." Some open pipes are carefully sawed off to the correct length for a desired pitch. Others (especially, "melodia") have a soft metal flap fastened to one side at the open end that can be bent at a slight angle across the opening to change the pitch. There are slight changes required in the position of the air sheet and edge between open and closed pipes. The edge must protrude inward somewhat on closed pipes and is usually flush with the outer wall on open pipes.

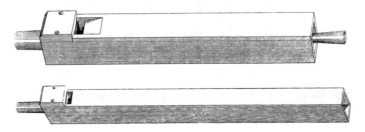

Fig. 7.19 Closed (top) and Open (bottom) wooden pipes (after Audsley 1905)

Fig. 7.20 Pressure variation
along the z axis of a conical
pipe of length L for the first
three modes

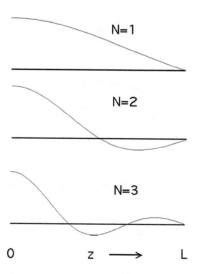

7.3.3   Conical Pipes

Tapered pipes of several varieties are found in different pipe organs, among them
the spitzflute ("pointed flute") and gemshorn ("goat's horn"). If the pipe is made of
a perfect cone of narrow vertex angle, it supports exactly the same set of frequencies
as an open pipe of the same length. (See Appendix F.) This comes about because the
pressure has a maximum at the closed pointed end and a minimum at the large open
end. To excite such a pipe with an "air reed" or flue, the edge tone must be placed
at the wide, low-pressure end. (In contrast a tapered reed pipe would have the reed
at the vertex.) The pressure variation along the pipe axis is more complex than with
an open cylindrical pipe. There, the variation is sinusoidal. Here, the pressure along
the pipe axis varies as

$$P(z) \propto \frac{\sin k_z z}{k_z z} \tag{7.1}$$

where z is the distance from the vertex and k_z is the propagation vector
along the length of the pipe from which the resonant frequencies are determined.
The boundary condition that determines the allowed wavelengths (hence resonant
frequencies) is simply that the pressure must go to zero at the large end while being
nonzero at the vertex. Thus

$$k_z L = n\pi \, , \text{ and } f_n = n\frac{c}{2L} \text{ where } n = 1, 2, 3, \ldots \tag{7.2}$$

The pressure variation along the pipe axis is shown in Fig. 7.20 for the first three
modes of the conical pipe.

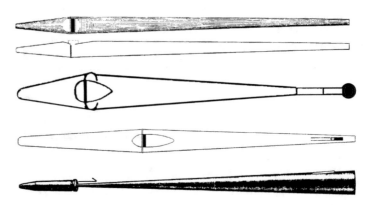

Fig. 7.21 Representative conical organ pipes. From top to bottom: spitzflöte (two views), spitzgedackte, gemshorn, and a typical reed pipe

The air flow is greatest where the acoustic pressure is near zero and least where the acoustic pressure is a maximum. Thus, flue pipes have the edge tone at the large open end of the cone, whereas reed pipes (trumpets and tubas) have the reed at the high-pressure pointed end. High-pressure is needed to force a mechanical reed to vibrate, whereas low pressure (large air flow) is required for the air reed in a flue pipe such as the gemshorn or spitzflöte. For practical reasons, conical pipes are generally made from metal. (See Fig. 7.21.)

7.3.4 Reed Pipes

The reed-pipe assembly shown at the left of the bottom example in Fig. 7.21 consists of the number of components illustrated in Fig. 7.22. The assembled "boot" has a metal outer casing that protects the more delicate inner parts. The protruding tuning wire is attached to a mechanical "embouchure" for the reed. The shallot (left in Fig. 7.22) is made of brass, is roughly similar to the mouthpiece on a clarinet to which a flat brass or bronze reed (center in Fig. 7.22) is attached, and is mounted in a lead block (at the right in Fig. 7.22) containing the tuning wire. However, in contrast to clarinet reeds, the brass reed has to be curved away from the shallot in order to work. (With the clarinet mouthpiece, the "lay" is curved away from the reed.)

One of the more interesting reed pipes from a scientific point of view is the "rohr schalmei." (See Fig. 7.23.) This pipe is of the opposite configuration to the chimney flute or rohr flute in that the extra resonator is larger in diameter than the tube connecting it to the boot. It is normally used as an open pipe and the extra resonator is tuned to the second harmonic of the full active length of the pipe. It has a single reed of the type illustrated in Fig. 7.22 and produces a sound reminiscent of small crows. In contrast, the Krummhorn uses a similar reed but has a resonator

Fig. 7.22 Components contained in the "boot" of a German-style reed pipe (photograph by the author)

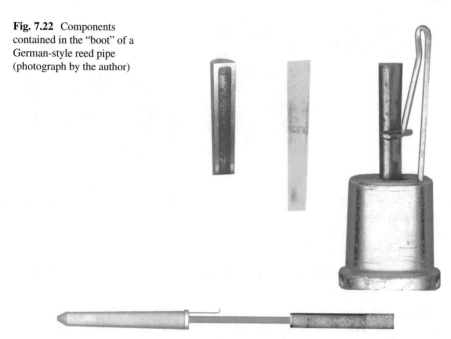

Fig. 7.23 A rohr schalmei reed pipe (as in "rohr flute," the word "rohr" is simply the German word for "tube")

consisting entirely of a long copper tube nearly closed by a cap at the end. In contrast to the rohr schalmei, the krummhorn has a more muffled sound.

7.3.5 Toe Board Channeling

One often-ignored problem has to do with turbulent eddy currents of air produced when a pipe valve is suddenly opened. If the valve is located directly below a hole leading into the toe of the pipe, vortices are formed similar to those produced when a sink or bathtub is drained of water. (See left side of Fig. 7.24.) These vortices have a very audible modulation effect on the sound of the pipe and careful builders take pains to prevent their occurrence. One simple approach is illustrated at the right in Fig. 7.24. Here, a section of a toe board is shown that is made up of three layers of wood that are glued together after appropriate milling. The basic idea is to interrupt the formation of a vortex by breaking up the air path as shown by the arrows at the right in the figure. Many wooden pipes have an offset air channel in their base that can also reduce the effects of vortex formation. In practice, the flexible tubes sometimes used to provide air for offset pedal pipes probably serve the same purpose. However, metal pipes are particularly subject to this problem

Fig. 7.24 Method (right) to reduce vortex formation (left) when the pipe valve is opened

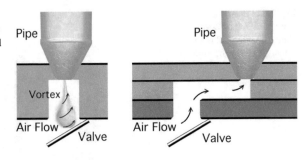

because the conical boot is generally open inside. Some builders fill the boot with loosely spaced steel wool to reduce vortex formation.

7.3.6   Pipe Configurations and Coupling

Three common organ pipe configurations are shown in Fig. 7.25. One might think that the relative desirability of the three is simply a matter of visual appeal. However, there are practical acoustic differences in the three geometries. Closely spaced low-frequency pipes tend to interact with each other by radiation through the air when they are of comparable frequency. The situation is analogous to that of the coupled oscillators discussed in Appendix A and of the interaction between pairs of strings tuned to the same note on the piano, not to mention the interaction between body and string resonances in the violin that result in "Wolf Tones." The closer the two resonators are in frequency, the stronger the interaction—especially with open pipes. As in the previous examples discussed, this coupling effect tends to push the fundamental pipe resonances away from each other. Although configuration (a) in Fig. 7.25 might seem to be the easiest arrangement for tuning purposes, it also has the most closely spaced pipes for adjacent notes on the scale. Configuration (b) has greater separation, except for the two lowest pipes in the middle. But grouping (c) has the largest separation between all the low frequency pipes in closely ascending frequency and should provide the best decoupling between the different resonators. (As a practical matter, configuration (c) also permits putting a catwalk above the shortest pipes to give the tuner better access to individual ranks in a large wind chest.)

Finally, one should acknowledge that the ultimate arrangement for decoupling low frequency pipes is that used by Frank Gehry in the pedal section of the organ in the new Disney Concert Hall in Los Angeles. Here, the pipes are mounted at random angles with respect to the vertical and coupling between them should be entirely negligible. The sound from the different pipes would also be dispersed in

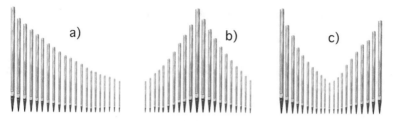

Fig. 7.25 Three common organ pipe configurations

Fig. 7.26 Pedal pipes in the
Disney Concert Hall organ

very different directions, hence there would also be minimum coupling of energy
from different pipes into the same acoustic modes of the hall. (See Fig. 7.26.)[19]

There is another type of interaction that occurs between pipes that are fed by
the same note channel in a wind chest such as that illustrated in Fig. 7.5. When
the different pipes are nearly harmonically related (as they are by definition with
mutation stops), the different ones powered by the same air channel will tend to
lock together in precisely harmonic intervals. This frequency locking effect occurs
through the nonlinear action of the air reeds in the different pipes and makes it very
difficult to tune the individual resonators. However, once the individual pipes have
been tuned, the locking effect is beneficial in that it provides precise tuning of the
harmonic intervals.

7.3.7   Resultant Low Frequency Pipes and Mixtures

When the human ear is presented with a number of harmonics of a low frequency
tone, it hears the fundamental pitch even when the fundamental is missing alto-

---

[19]One humorist commented that the photograph should really be captioned, "After the Earthquake."

Table 7.1 Frequencies and harmonics involved in a 32-ft resultant stop

Phantom 32-ft pipe	17.2,	34.4,	51.6,	68.7,	85.9,	103.1,	120.3,	137.5	154.8 Hz
Open 16-ft pipe	–,	34.4,	–,	68.8,	–,	103.1,	–,	137.5,	– Hz
Closed 5.33-ft pipe	–,	–,	51.6,	–,	–,	–,	–,	–,	154.8 Hz
Harmonic	1	2	3	4	5	6	7	8	9

gether. That fact has been used in the design of many pipe organs to extend the apparent stop list to lower and lower pitch. The technique is sometimes described by the term "resultant." Both the nonlinearity in the ear and the fact that its response falls off rapidly at very low frequencies help in the deception.

To illustrate, suppose you wanted to give the impression of a 32-ft open pipe and only had room for a16-ft open pipe above the wind chest. The first nine harmonics of a 32-ft open pipe are shown in Table 7.1. The ear cuts off completely by about 16 Hz, so the first harmonic of a 16-ft open pipe would scarcely be heard at all, even if it were actually there. We could simulate the apparent sound of a 32-ft open pipe by using two pipes on each note. For example, by starting with an open 16-ft pipe to produce the sound at 34.4 Hz, we could get strong overtones at 68.8, 103.1, and 137.5 Hz. A 5.33-ft closed pipe would provide sound at 51.6 and 154.8 Hz. Hence, these two pipes combined would produce the 2nd, 3rd, 4th, 6th, 8th, and 9th harmonics of our non-existent 32-ft pipe and fool the ear into thinking that the 17.2 Hz tone was actually present. If these two pipes were placed on a common note channel, their fundamental frequencies would also tend to lock in phase on all of the harmonics produced. However, there are some things that are not simulated in this process. A real 32-ft pipe could make the building shake—a sensation some people enjoy. (Earthquake frequencies tend to peak at frequencies of about 16-Hz—a region where you can feel the vibrations even though you cannot hear them effectively.)

The exact mechanism by which the ear senses the low frequency sounds is somewhat in dispute. Helmholz provided an explanation based on the nonlinearities in the ear itself. He argued that those nonlinearities would generate difference tones and among them the fundamental frequency of the missing pipe would dominate. Others think that the ear has the capability of recognizing periodicity in the resultant waveform. If all the real tones are phase-locked (or approximately so), the overall acoustic waveform will indeed be periodic at the missing fundamental frequency. Some digital tuning meters are actually based on the principle of recognizing the fundamental period in a waveform rather than its pitch and the human ear and brain may well do the same thing.

"Mixture stops" utilize similar combinations of pipes and are very effective for simulating the sound of reeds. A reed pipe generally has much a richer harmonic content than a single flue pipe. By choosing three or more different flue pipes for each note in a mixture, it is possible to provide remarkable simulation of the sound of a reed. One might ask, "Why not just use the reed pipe alone?" One reason is that flue pipes tend to be far more stable in both pitch and tone quality than reeds and are much easier to voice. However, the wind chests do take up more space.

7.3.8 Scaling

As is intuitively obvious, the diameters of flue pipes in a given rank must decrease with increasing pitch. Further, the decrease should be monotonic and follow some systematic method to determine the change on successive notes. This problem occupied many of the early pioneers in organ building and a number of different solutions were proposed. Some have the diameter halving at the 16th step and others at the 17th or 18th and even at the 20th, 22nd, or 24th step. The problem was discussed in detail by Audsley (1905, Vol. 2, Chapter 37). According to Audsley, many agreed that the most satisfactory choice consists of halving the diameter at the 16th step, a scaling recommended by J.G. Töpfer of Weimar in 1833. This scaling appears to have been adopted by most German and French builders. Audsley provided many pages of numerical tables of the diameter change for different versions of scaling. However, anyone with a modern pocket calculator can work out the intervals more quickly by calculating the diameter D from an equation of the form

$$D = D_0 e^{-kN} \tag{7.3}$$

where D_0 is the initial pipe diameter, N is the number of pipes above the initial one, and k is determined by the halving point. For example, if the scale halves at the 16th step,

$$k = \frac{\log_e 2}{16} \approx 0.0433217\ldots \tag{7.4}$$

In addition to the point at which the halving occurs, the scale of a rank of pipes is also defined in terms of the flue diameter of the lowest pipe.

One of the nice things about this kind of scaling is that the tone quality changes continuously as you go from one note to the next. That is particularly advantageous in playing fugues on an organ because the tone quality within the same rank, although basically similar, will be noticeably different in voices that enter a fifth apart. For that reason, organists usually play the opening voices of a fugue using the same stop or stop combination.

7.3.9 Unusual Pipes

There are a few unusual pipes worth brief mention:

Strings are generally open flue pipes of narrow scale in which a horizontal wooden rod is placed in the air stream at the middle of the air reed at the flue end of the pipe, often supported by the "ears" of the pipe using small nails driven into the rod through holes in the ears. The location of the rod requires very critical adjustment. The rod produces extra turbulent vortex motion at about twice the normal frequency, thereby enhancing the second harmonic and creating a slight buzzing sound reminiscent of the scraping of a rosined bow on a stringed instrument. The higher pitched strings often have a small brass knife-edge inserted in the air stream instead of a rod. These rods are used on pipes ranging from the 16-ft (open) *violone* to higher-pitched 8-ft *violas* and other pipes. They are also used on closed wooden pipes under the name *Lieblich Gedekt*. But strings are very hard to voice at low pressure levels.[20]

The *Harmonic Flute* is used in many organs and consists of a narrow scale (because the fundamental is suppressed), open flue pipe with a small hole drilled in the sidewall at the mid-point of the active length. The hole, located at the maximum pressure point for the fundamental resonance, reduces the intensity of that mode and the tone quality consists heavily of second and higher harmonics.

The labial reeds are flue pipes designed to enhance the main overtone differences characterizing certain reeds without encumbering the tuner with the necessity of more frequent tunings. The *Labial Clarinet* is a closed flue pipe with resonant cavities designed to enhance the third and fifth harmonics. The *Labial Oboe* is a narrow scale open flue pipe with a slight taper that is adjusted to simulate the sound of an oboe without using a reed.

The *Bärpfeife* is a reed pipe with double- or triple-cone open resonators tuned to 32-, 16-, or 8-ft pitch. The *Rankett* is a folded, closed reed pipe also used at 32-, 16- and 8-ft pitch on Baroque style organs. These pipes occupy far less height than flue pipes of the same pitch and tend to be used primarily in small organs.

The *Cymbalstern* is another stop used on organs at the time of Bach. It consisted of a rotating star placed at the top of the organ on which several high-pitched bells were attached. The stop was often turned on during the final measures of a large fugue to add sparkle to the tone quality.

The *Nachtigall* a stop consisting of two open pipes blown by a common air source with the open end immersed in water to imitate the sound of a nightingale.

The *Weinpfeife* is a stop introduced by Alf Laukhuff of Franfkfurt, a pipe with wine tap where the flue normally goes.

Problems

7.1 Problem. What it the total weight (including the top plate) needed on a 2 3 ft air pressure regulator of the type shown in Fig. 7.5 to provide an air pressure of 2 in. of water? (Note: the density of water is 1 g/cm^3, there are 2.54 cm per inch and 1 kg = 2.20 pounds.)

---

[20]I once purchased a book on Pipe Voicing with the object of learning how to voice string pipes myself. I eagerly read the tome from cover to cover, but was dismayed by the comment on the last page: "String pipes are much too hard to voice and will not be discussed here." However, I learned enough about the problem to see what the author meant!

7.2 Problem. A physicist buys a rank of open flue pipes that have been voiced for 5-in wind pressure. By what fraction must he reduce the cut-up to operate them on his wind chest at 2-in. pressure?

7.3 Problem. The same physicist decides that, rather than changing the cut-up on each pipe, he would increase the toe-hole diameter by about $\sqrt{(5/2)}$ to let more air flow into the pipe. Explain why that might work.

7.4 Problem. Suppose you want to construct a rohr schalmei pipe tuned to middle C. What would the lengths of the copper tube and the large resonant cavity be? Take the velocity of sound to be 1100 ft/s.

7.5 Problem. A certain labial oboe (flue pipe without a reed) is made from a narrow-scale tapered pipe, closed at the vertex. Why would you expect that geometry?

7.6 Problem. Suppose you only have 9 ft of vertical space above the wind chest and want to simulate the sound of a 16-ft open pipe. How could you do it in the available space?

7.7 Problem. Suppose you want to design a chimney flute tuned to A=440 Hz. What cavity lengths would you use? Take the velocity of sound to be 1100 ft/s.

7.8 Problem. A certain organ has a rank of open principals that starts on a C with a diameter of 8.20-in and the C one octave above has a diameter of 5.80-in. How many steps are required for the diameter to be reduced by one-half?

7.9 The Newberry organ at Yale University has a 64-ft "Gravissima Resultant" stop (see Appendix E) composed of two pipes for each note for the first twelve. The first note consists of a 32-ft diapason (open principal) and a 32-ft Bordon at "quint" pitch (i.e., a closed pipe of ten and two-thirds feet actual length, or twenty-one and one-third feet effective pitch). Explain how this combination would simulate a 64-ft pipe.

Correction to: The Science of Musical Sound

Correction to:
W. R. Bennett, Jr., *The Science of Musical Sound*,
https://doi.org/10.1007/978-3-319-92796-1

The inadvertently published equations have been corrected as mentioned below.

Chapter 1
Page 16

The beta symbol in the below equation in this page has been removed and corrected to appear as below:

$$\approx 1449 + 4.6T + (1.34 - 0.01T)/(S - 35) + 0.0216z \text{ m/s}$$

Equation 1.26: (Page 25)

The letter "M" next to the lambda symbol has been made as a subscript, the letter "S" next to "V" has been made as a subscript in two instances and one set of parentheses has been removed in $(c - V_S)$

$$\lambda_M = (c - V_S)T/(f_S T) = (c - V_S)/f_S.$$

The updated version of the chapters could be found at
https://doi.org/10.1007/978-3-319-92796-1_1
https://doi.org/10.1007/978-3-319-92796-1_2
https://doi.org/10.1007/978-3-319-92796-1_4
https://doi.org/10.1007/978-3-319-92796-1_7
https://doi.org/10.1007/978-3-319-92796-1

Chapter 2
Equation 2.8: (Page 40)
A space has been added between "sin nx" and "[n odd]"

$y = \sin x + \frac{1}{3}\sin 3x + \frac{1}{5}\sin 5x + \cdots + \frac{1}{n}\sin nx \ [n \text{ odd}].$

Equation 2.9: (Page 41)
The n exponent of -1 has been made as a whole to the power of $(-1)^n$

$y = \sin x - \frac{1}{2}\sin 2x + \frac{1}{3}\sin 3x - \frac{1}{4}\sin 4x + \cdots + \frac{(-1)^n}{n}\sin nx$

Chapter 4: (Page 108)
In Eq. 4.2, the second sin has been non italicised.

Chapter 7: (Page 319)
In Eq. 7.1, first z in denominator has been made as a subscript.

Appendix A
The equal to symbol between M and $\frac{d^2x}{dt^2}$ has been deleted in Eq. A.1.

The numeral 0 has been added as a subscript to omega squared in the line before equation A.2, as given here: $\omega_0^2 = K/M$ and in Eq A.2: The ω_0 has been squared to appear as given here: ω_0^2

Eq A.3: The indent before the word "where" has been removed. (Page 330)

In the line before eq A.4, the numeral 0 has been added to omega as a subscript and in the line before eq A.6, the numeral 0 has been added to x as a subscript in two instances, as shown below:

$x = x_0 \sin \omega t$, then $v = dx/dt = \omega x_0 \cos \omega t$

Eq A.6: In the second line, a plus sign has been added between $\sin^2 \omega_0 t$ and $\cos^2 \omega_0 t$:

$\frac{1}{2}Kx_0^2(\sin^2 \omega_0 t + \cos^2 \omega_0 t)$

In the line after eq A.6, the contents in parentheses has been changed as shown below:

$\omega_0 = K/M$ and $(\sin^2 \omega_0 t + \cos^2 \omega_0 t) = 1$

Eq A.7: The subscript "1" in "$K_1 x$" has been removed and made to appear as "Kx". (Page 330)

In the last line in code in page 346, the first instance of **NEXT** T has been deleted and made to appear as below

REM Plot **or** Print results here **NEXT** T

The figure referred to has been modified as Fig. A.3 in the last line of page 335 and in page 347, the first equation in the caption for Fig. A.8 has been modified as "$\gamma = 0.5\omega_0$"

Appendix B

In Eq. B.9, the last equal to sign following "$-k^2$", has been deleted. (Page 359) and in Eq. B.11, "cos" has been non italicised. (Page 359)

In page 366 and in Eq. B.27, the numeral "2" in "$T2$" has been made as a subscript the indent near the word "where" in page 367 & below Eq. B.38 (Page 360), has been removed.

In Eq. B.31, the irrelevant characters "$nx2dxdIt2dxor$" has been deleted. (Page 367)

Appendix A
The Harmonic Oscillator

Properties of the harmonic oscillator arise so often throughout this book that it seemed best to treat the mathematics involved in a separate Appendix.

A.1 Simple Harmonic Oscillator

The harmonic oscillator equation dates to the time of Newton and Hooke. It follows by combining Newton's Law of motion ($F = Ma$, where F is the force on a mass M and a is its acceleration) and Hooke's Law (which states that the restoring force from a compressed or extended spring is proportional to the displacement from equilibrium and in the opposite direction: thus, $F_{\text{Spring}} = -Kx$, where K is the spring constant) (Fig. A.1). Taking $x = 0$ as the equilibrium position and letting the force from the spring act on the mass:

$$M\frac{d^2x}{dt^2} + Kx = 0. \tag{A.1}$$

Dividing by the mass and defining $\omega_0^2 = K/M$, the equation becomes

$$\frac{d^2x}{dt^2} + \omega_0^2 x = 0. \tag{A.2}$$

As may be seen by direct substitution, this equation has simple solutions of the form

$$x = x_0 \sin \omega_0 t \text{ or } x_0 = \cos \omega_0 t, \tag{A.3}$$

The original version of this chapter was revised: Pages 329, 330, 335, and 347 were corrected. The correction to this chapter is available at https://doi.org/10.1007/978-3-319-92796-1_8

© Springer Nature Switzerland AG 2018
W. R. Bennett, Jr., *The Science of Musical Sound*,
https://doi.org/10.1007/978-3-319-92796-1

Fig. A.1 Frictionless
harmonic oscillator showing
the spring in compressed and
extended positions

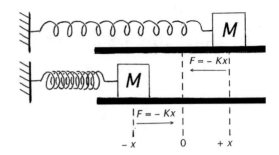

where t is the time and x_0 is the maximum amplitude of the oscillation. The angular
resonance frequency ω_0 is related to the cyclical resonance frequency F_0 and period
$T (= 1/F_0)$ of the oscillator by $\omega_0 = 2\pi F_0 = 2\pi/T$ where

$$\omega_0 = \sqrt{K/M}. \tag{A.4}$$

A.2   Energy

Without dissipation, the total energy in the oscillator at any instant in time is constant
and given by

$$\text{Energy} = \frac{1}{2}(Mv^2 + Kx^2), \tag{A.5}$$

where the first term is the kinetic energy of the moving mass ($v = dx/dt$) and
the second term is the potential energy stored in the spring. The energy oscillates
back and forth between those two forms during alternate half periods. At the turning
points, where the velocity of the mass is zero, the energy is stored in the spring. At
$x = 0$, where the velocity is maximum, the energy is all kinetic. If $x = x_0 \sin \omega t$,
then $v = dx/dt = \omega x_0 \cos \omega t$ and

$$\begin{aligned}
\text{Energy} &= \frac{1}{2}[M(\omega_0 x_0 \cos \omega_0 t)^2 + K(x_0 \sin \omega_0 t)^2] \\
&= \frac{1}{2}Kx_0^2(\sin^2 \omega_0 t + \cos^2 \omega_0 t) = \frac{Kx_0^2}{2},
\end{aligned} \tag{A.6}$$

where again $\omega_0 = K/M$ and $(\sin^2 \omega_0 t + \cos^2 \omega_0 t) = 1$. Regardless of its kinetic
energy the particle is trapped in a potential well of the type shown in Fig. A.2. Since
the restoring force from the spring is

$$F \equiv -\frac{dU}{dx} = -Kx, \text{ the potential is } U(x) = \frac{M\omega_0^2 x^2}{2}, \tag{A.7}$$

where x is defined with respect to the equilibrium position of the oscillator and
$U = 0$ at $x = 0$.

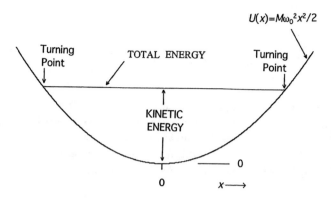

Fig. A.2 Potential well for the simple harmonic oscillator. In this model the particle is bound, regardless of its kinetic energy

A.3   The Damped Simple Harmonic Oscillator

In any real system there will be a "damping" force due to friction that at low velocities takes the form

$$F_{\text{Damping}} = -\Gamma \frac{dx}{dt}, \tag{A.8}$$

where Γ is a damping constant characteristic of the system. The total force acting on the mass M now becomes $F = Ma = -\Gamma dx/dt - Kx$. Hence, the equation of motion becomes

$$M \frac{d^2x}{dt^2} \Gamma \frac{dx}{dt} + Kx = 0. \tag{A.9}$$

Dividing through by the mass and moving everything to the left side puts the equation for the damped simple harmonic oscillator in the more convenient form,

$$\frac{d^x}{dt^2} + \gamma \frac{dx}{dt} + \omega_0^2 x = 0 \tag{A.10}$$

where we have defined $\gamma = \Gamma/M$ as the damping constant per unit mass. Exact solutions to this equation may be obtained in the following straightforward manner. Substituting a trial solution, $x = e^{\mu t}$, gives a quadratic equation

$$\mu^2 + \gamma \mu + \omega_0^2 = 0, \tag{A.11}$$

that has $\mu_\pm$ roots of the form

$$\mu_\pm = -\frac{\gamma}{2} \pm i\omega_0 \sqrt{1 - \left(\frac{\gamma}{2\omega_0}\right)^2} \tag{A.12}$$

where $i^2 \equiv -1$. The solutions without a driving term are of the form $x = Ae^{\mu_+t} + Be^{\mu_-t}$, where A and B are complex constants determined by the initial conditions (i.e., the displacement from equilibrium and the velocity at $t = 0$).

For $(\gamma/2\omega_0)^2 < 1$, which is typical of resonant systems, the roots have an imaginary component that leads to free oscillation at a frequency

$$\omega \approx \omega_0 - \frac{\gamma^2}{8\omega_0} + \text{Order}\left\{\omega_0\left(\frac{\gamma}{4\omega_0}\right)^2\right\} \tag{A.13}$$

that is damped at the amplitude decay rate γ. For $\gamma_{\text{Crit}} = 2\omega0$, no oscillation occurs, the damping is fastest and is said to be "Critical." For $(\gamma/2\omega_0)^2 \ll 1$, one solution is

$$x(t) \approx x_0 e^{-\gamma t/2} \sin \omega_0 t, \tag{A.14}$$

which corresponds to giving the mass an initial velocity kick at $x = 0$ and $t = 0$. In this case the total energy in the oscillator decays with time as

$$\text{Total Energy} \propto |x(t)^2| \propto e^{\gamma t}. \tag{A.15}$$

Note that the energy decay rate is a factor of 2 faster than the amplitude decay rate simply because the energy in this classical motion problem is proportional to the square of the amplitude.

A.4   Resonance or Cavity Width and Quality Factor (Q)

Because the amplitude of the damped oscillator decays with time, the spectrum of its motion is spread out in frequency about $\omega0$. Regardless of the initial conditions chosen, we know that the total energy decays as shown in Eq. (A.15). Since $x(t)$ is of the form $\exp(-\gamma t/2 + i\omega0t)$, the Fourier transform of the amplitude is of the general form,

$$|x| \propto \int_0^\infty x(t)e^{-i\omega t}\,dt \propto \frac{1}{|\gamma/2 + i(\omega - \omega_0)|}. \tag{A.16}$$

Hence, the spectral distribution of energy in the resonant mode will be of the form

$$|x_\omega|^2 \propto \frac{1}{\left|1 + \left(\frac{\omega - \omega_0}{\gamma/2}\right)^2\right|}, \tag{A.17}$$

which is sometimes called a "Lorentzian" function of the frequency and, provided $(\omega - \omega_0)2 >> (\gamma/2)2$, is symmetric about the angular resonant frequency ω_0 with full cavity width at half-maximum energy response given by $\Delta\omega = \gamma 2\pi \Delta F$. The quantity

$$\Delta F_{cav} = \frac{\gamma}{2\pi} = \frac{f}{\pi} \frac{c}{4L} \tag{A.18}$$

is sometimes called the cavity resonance width. In general, for standing-wave resonances in a long cavity such as an organ pipe, $\gamma = f(c/2L)$ where c is the running wave velocity, L is the cavity length, and f is the round trip fractional energy loss. For example, for a closed-pipe model of the first formant (F_1) of the vocal tract,

$$\Delta F_{cav} = F_1 \frac{f}{\pi}. \tag{A.19}$$

The *Quality Factor*, or Q of a resonance, is defined traditionally by the relation

$$Q = 2\pi \frac{\text{Energy Stored}}{\text{Energy Lost per Cycle}} = \omega_0 \frac{\text{Energy Stored}}{\text{Rate of Energy Loss}}, \tag{A.20}$$

where the rate of energy loss is $\gamma \times$ (Energy Stored). Noting that $\gamma = \Delta\omega$, an alternative expression for the Quality Factor is

$$Q = \frac{\omega_0}{\Delta\omega} = \frac{F_0}{\Delta F}, \tag{A.21}$$

where F_0 is the cyclical resonance frequency and ΔF is the full cyclical width at half-maximum energy response, both measured in Hz.

A.5   Driven Damped Oscillation

When a sinusoidal or cosinusoidal driving term of the type

$$E(t) = E \cos \omega t \tag{A.22}$$

is added to the damped oscillator, the equation becomes

$$\frac{d^2x}{dt^2} + \gamma\frac{dx}{dt} + \omega_0^2 x = E(t) \tag{A.23}$$

where $E(t)$ on the right-hand side represents the driving force per unit mass. This equation has the closed-form solution

$$x = \frac{E}{D}\sin(\omega t + \Phi), \tag{A.24}$$

where the denominator D is given by

$$D = \sqrt{(\omega^2 - \omega_0^2)^2 + \left(\frac{\gamma\omega}{M}\right)^2} \tag{A.25}$$

and the phase angle is

$$\Phi = \cos^{-1}\left(\frac{\gamma}{MD}\right) \text{ radians.} \tag{A.26}$$

The introduction of damping causes the oscillator to lag in phase from the driving term. Without damping, $|x| \to \infty$ as $\omega \to \omega_0$.

A.6   Electric Circuit Equivalent of the Damped Driven Oscillator

The equation for the damped, driven oscillator has an exact equivalent in the series LCR (inductance-capacitance-resistance) circuit shown in Fig. A.3. Written in terms of the charge q on the condenser C, the equation for the voltage drop around the loop becomes

$$L\frac{d^2q}{dt^2} + R\frac{dq}{dt} + \frac{q}{C} = E(t), \tag{A.27}$$

where the current $i = dq/dt$, and we have made use of the basic relations for ac voltage drops across the circuit elements: $V_L = Ldi/dt = Ld^2q/dt^2$ (Henry's Law), $V_R = Ri = Rdq/dt$ (Ohm's Law), and $V_C = q/C$ (Faraday's Law). By comparison with the original equation for the oscillator, we find the equivalent relationships in Table A.1.

One may determine the steady-state behavior of a linear resonant circuit using complex ac circuit analysis. In this analysis, one usually assumes that the transient solution has died down to a steady complex current of constant amplitude and phase running through the various resistive and reactive elements of the circuit and that

Fig. A.3 Series resonant
LCR circuit driven by a
sinusoidal EMF
(electro-motive force)

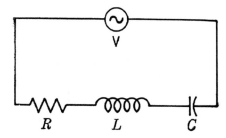

$$R \qquad L \qquad C$$

Table A.1 Electrical and
mechanical equivalent
parameters

AC circuit	Mechanical oscillator
Charge, $q \Rightarrow x$	Amplitude
Current, $i \Rightarrow v$	Velocity
Inductance, $L \Rightarrow M$	Mass
Resistance, $R \Rightarrow \gamma$	Damping constant
Capacitance, $C \Rightarrow 1/K$	1/(Spring constant)
EMF, $E(t) \Rightarrow F(t)/M = E(t)$	Driving force

AC circuit mechanical oscillator

the real part of the resultant voltage drops correspond to values one would measure. Thus, if the current running through a circuit loop is of the form[1]

$$i = i_0 e^{j\omega t} = i_0(\cos \omega t + j \sin \omega t), \tag{A.28}$$

the voltage drops across the elements L (inductance), R (resistance), and C (capacitance) may be written as

$$V_L = L\frac{di}{dt} = i(j\omega L), \quad V_R = iR, \text{ and } V_C = \frac{q}{C} = \frac{1}{C}\int^t i_0 e^{j\omega t} dt = i\left(\frac{-j}{\omega C}\right). \tag{A.29}$$

Note that in the expression for VC we have made use of the fact that the charge on the capacitance is the integral of the current over time; i.e., the current feeding the capacitance is $i = dq/dt$, where q is the time-dependent charge on the capacitance.

In applying this convention, one makes use of Kirchoff's Laws: (1) The sum of the voltage drops around a closed loop equals the sum of the Emfs ("electromotive forces"), which is a statement of the conservation of energy; (2) The sum of the currents at any junction must be zero, which is a statement of the conservation of charge. Using the methods and notation of complex steady-state ac circuit analysis just discussed, the loop equation for the circuit in Fig. A.3 may be written

---

[1]The relation $e^{i\theta} = \cos\theta + i\sin\theta$ where $i^2 = -1$ is known as Euler's formula and may be derived from the infinite series for e^x and those for $\cos\theta$ and $\sin\theta$ by letting $x = i\theta$. This type of ac circuit analysis was invented by Campbell, a research engineer at the A.T. & T Company in 1911, who referred to it as "Cisoidal Oscillations"—an abbreviation for "cos i sin" Campbell (1911).

$$E = i \left[R + j \left(\omega L - \frac{1}{\omega C} \right) \right]. \tag{A.30}$$

The magnitude of the current is

$$|i| = \frac{E}{\sqrt{R^2 + \left(\omega L - \frac{1}{\omega C} \right)^2}} \tag{A.31}$$

and it lags the driving voltage in phase by ϕ where

$$\tan^- 1\phi = -\frac{\omega L - \frac{1}{\omega C}}{R} \approx \frac{(\omega^2 - \omega_0^2)}{\omega L / R} \approx \frac{(\omega_0 - \omega)}{\Delta \omega / 2}. \tag{A.32}$$

Here, the final approximation holds near resonance where ω is close to ω_0.

The power loss in the resistor is

$$P_R = |i|^2 R = \frac{E^2/R}{1 + \frac{(\omega^2 - \omega_0^2)}{(\omega L/R)^2}} \tag{A.33}$$

and is of resonant Lorentzian shape with full width at half maximum given by

$$\Delta \omega \approx R/L \approx 2\pi \, \Delta F. \tag{A.34}$$

The Q or "Quality Factor" of the circuit is then

$$Q = \omega_0 / \Delta \omega \approx \omega_0 L / R. \tag{A.35}$$

These properties are summarized in Fig. A.4.

A.7   A Different Approach

Normally, one uses ac circuit analysis to determine steady-state currents in the presence of applied Emfs. However, useful results are obtained even when $E = 0$ in Fig. A.3. In that case, we can use the loop equation to determine the (complex) frequency, an approach that gives the transient behavior of the circuit. Hence, if the Emf is zero in Fig. A.3, we get

$$i \left[R + j \left(\omega L - \frac{1}{\omega C} \right) \right] = 0. \tag{A.36}$$

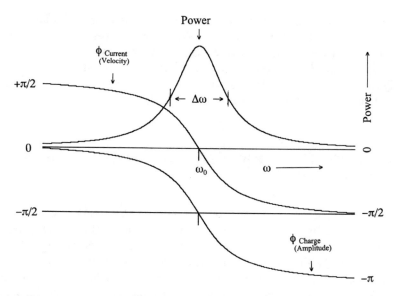

Fig. A.4 Power resonance and phase shifts in the series LCR circuit, related to the properties of the driven, damped mechanical oscillator

Assuming that $i \neq 0$ (i.e., that there is some initial charge or current flowing in the circuit), the large square bracket must equal zero and we obtain a quadratic equation in ω:

$$\omega^2 - j\omega\frac{R}{L} - \frac{1}{LC} = 0, \tag{A.37}$$

whose roots are

$$\omega = j\frac{R}{2L} \pm \omega_0\sqrt{1 - \left(\frac{R}{2\omega_0 L}\right)^2}. \tag{A.38}$$

If we substitute this expression back into Euler's formula, we get

$$e^{j\omega t} = e^{\frac{R}{2L}t}e^{\pm j\omega_0\sqrt{1-\left(\frac{R}{2\omega_0 L}\right)^2}t}, \tag{A.39}$$

where the first exponential describes damping of the current and the second exponential gives the oscillatory solution through use of Euler's formula with a real frequency

$$\omega = \pm\omega_0\sqrt{1 - \left(\frac{R}{2\omega_0 L}\right)^2}. \tag{A.40}$$

Hence, the "steady-state "solution can actually give us the transient behavior of the circuit. (Note that the sign on the real part of the frequency merely alters the phase of the oscillation.) The damped oscillatory solution is, of course, the same one we got earlier for the behavior of the damped mechanical oscillator using a trial solution of the type $e^{\mu}t$. The present approach provides a less tedious way to analyze coupled circuits, if you are comfortable with ac circuit analysis.

A.8   Coupled Oscillations, Wolf Tones and the Una Corda Piano Mode

The double-strung pianos dating from the early work of Bartolomeo Cristofori provide an interesting example of mode coupling with unusual potential for musical expression. The physical processes involved were discussed qualitatively in Chap. 4. We would like to show here that it is easy to derive the important quantitative properties of the coupled system from an electric circuit model. As implied by the term "double-strung," each note on the piano has two strings tuned closely to the same pitch, each of which can be represented as a damped harmonic oscillator. We will only consider the fundamental resonances of the strings here, although the present method could easily be extended to include harmonics of each fundamental frequency. In the piano, the motion of the two strings is coupled through their attachment to the bridge, which in turn transfers energy to the soundboard. We will represent this situation by the pair of coupled circuits shown in Fig. A.5. For simplicity, we will assume that the corresponding circuit elements are identical and that each oscillator is initially tuned approximately to the same frequency given by $\omega_0^2 = 1/LC$. (One might represent a string resonance and the other an air resonance in a violin.)

The coupled circuits in Fig. 4.5 result in two loop equations:

$$i_1 \left[R + j \left(\omega L - \frac{1}{\omega C} \right) \right] - i_2 R = 0 \qquad (A.41)$$

and

Fig. A.5 Coupled circuit model for one note on a double-strung piano. The two current magnitudes may be different

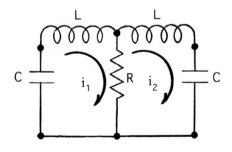

$$- i_1 R + i_2 \left[R j \left(\omega L - \frac{1}{\omega C} \right) \right] = 0. \tag{A.42}$$

Solving both of these equations for i_1 / i_2, we obtain

$$\frac{i_1}{i_2} = \frac{R}{R + j \left(\omega L - \frac{1}{\omega C} \right)} = \frac{R + j \left(\omega L - \frac{1}{\omega C} \right)}{R} \tag{A.43}$$

Hence, we get an equation involving the complex frequency,

$$\left(\omega L - \frac{1}{\omega C} \right) \left(\omega L - \frac{1}{\omega C} - 2 j R \right) = 0, \tag{A.44}$$

which has two distinct solutions for ω^2 yielding roots for ω,

$$\omega = \pm \omega_0 \text{ and } \omega = j \frac{R}{L} \pm \omega_0 \sqrt{1 - \left(\frac{R}{\omega_0 L} \right)^2}. \tag{A.45}$$

The first mode has no loss and oscillates at the resonant frequency, ω_0. The second, lossy mode has the real component of its frequency shifted downward in magnitude to

$$\omega = \omega_0 \sqrt{1 - \left(\frac{R}{\omega_0 L} \right)^2} \tag{A.46}$$

and an amplitude decay rate of R/L, which is twice as large as that for either LCR loop by itself. Note that the Q of each resonant circuit is $\omega_0 L / R \gg 1$. Hence, the second term in the square root provides a small, real correction to the resonant frequency, although the shift is larger than in either single isolated LCR loop alone.

The two coupled modes of oscillation may be interpreted in terms of the coupled circuits shown in Fig. A.6. Here, the same current magnitude i flows in both loops, but in different directions for the two different normal modes. The even-symmetric mode on the left is the high-loss mode (with decay rate of R/L) since the two currents flowing through the resistance are in phase. It corresponds to the normal position of the hammer on the piano in which both strings are struck simultaneously. Striking both strings in the normal hammer position is equivalent to depositing the same charge on each capacitor at $t = 0$. The odd-symmetry mode at the right is the low-loss mode and would correspond to a case in the piano where both undamped strings were hit simultaneously with hammers moving in opposite directions. In the circuit, opposite charges would have to be placed suddenly on the two capacitors at $t = 0$. For the odd-symmetric case (the right side of Fig. A.6), the mode has no loss at all because the two equal currents flowing through the resistor are 180°out of phase. In the piano, that mode corresponds to periodic pulses from the two strings

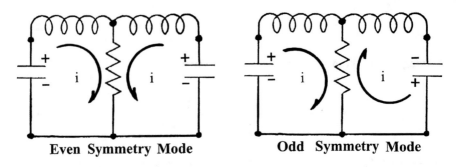

Even Symmetry Mode   Odd   Symmetry Mode

Fig. A.6 The even-symmetric (left) and odd-symmetric (right) modes of the coupled circuit (the current magnitudes are all equal)

arriving at the bridge out of phase. A similar situation arises in the case of "Wolf Tones" on bowed stringed instruments. (See Chap. 5.)

However, pure excitation of the odd-symmetry mode would not occur in a normal piano because only the right-hand string is struck (in the double-strung piano) by the hammer with the una corda pedal depressed. (The hammer is shifted to the right so as to hit only one string, with both strings undamped.) Hence, to simulate the actual una corda case, we need to take the difference between equal amounts of the two normal coupled-circuit modes. Thus, initially,

$$i_{\text{UnaCorda}} = i_{\text{EvenSymmetric}} - i_{\text{OddSymmetric}}. \qquad (A.47)$$

This combination localizes the current on the right side of the coupled circuit when the hammer hits, after which the even-symmetric component rapidly dies out leaving only the long-lived odd-symmetric mode still oscillating. Because the odd-symmetric mode has the least loss (no loss at all for the idealized circuit in Fig. A.6), the excitation would always tend to settle down in the odd-symmetry mode with frequency ω_0. The apparent pitch goes up slightly as the high-loss mode dies out. In reality there is some loss in the vibrating strings themselves due to air resistance. That could easily be included in the electric circuit model by adding a small resistance in series with each coil in the separate current loops. A reactive component could also be added in series with the load resistance R representing loss to the soundboard. Because equal currents flow in the two loops in the odd-symmetric mode, the crescendo effect described in Chap. 4 after the string is struck (produced by damping one string in the una corda mode after the sound decays) would be limited to about $20 Log_{10} 2 \approx 6$ dB, or about the amount observed experimentally.

A.9 Harmonics and Subharmonics

As discussed throughout this book, most musical instruments produce waveforms that are at least quasi-periodic and contain harmonics. These harmonics often result from some nonlinearity in the oscillatory system, for example, the vibrating reed in a woodwind, or the vibrating lips of the brass instrumentalist. In general, by "nonlinearity" we mean that the output amplitude from the system does not rise in simple direct proportion to the amplitude of the input signal as it would if the system were perfectly linear (i.e., if the output simply varied as the first power of the input). Instead, the output amplitude may depend on various powers of the input amplitude. As a general rule, if we feed a sinusoidal signal into such a system, signals at different frequencies that are harmonically related to the fundamental (i.e., are given by integer multiples of the original frequency) will appear in the output. Further, these harmonics are generally phase-locked to the fundamental frequency. This can be understood in terms of basic trigonometry relations. For example, if you square a sinewave, you will get a signal varying as the second harmonic of the input signal which is in phase with the fundamental; if you cube the signal, the third harmonic is produced; and so on. More strikingly, if you feed two signals with different frequencies into the system, the output may contain extraneous signals at the sum and difference frequencies of the two original frequencies.[2] All of these effects can be simply explained (albeit, tediously in some instances) through application of the basic trigonometry identities such as those treated in high school.

Under some conditions, the output may even contain subharmonics of the input frequency; i.e., frequencies that are lower than the input driving frequency. Understanding how to produce subharmonics used to be a mysterious art. To quote one of the pioneers who first demonstrated their existence in electromechanical systems, "Most authors pass over this question in silence, as they take it for granted that the occurrence of subharmonics is an impossibility." (Pedersen 1934)[3]

It is useful to treat the simple mechanical oscillator model of Fig. A.1 with the inclusion of nonlinear terms in the spring constant. As we will show, this model can produce both harmonics and subharmonics, not to mention rectification.

---

[2]To see how these extraneous frequencies come about, just consider the trigonometric identity, $2\sin(A)\cos(B) = \sin(A+B)+\sin(A-B)$. Let $A = 2\pi F_0 t$ and $B = 2\pi F t$ be the higher frequency resonance. Multiplying the two waves together yields "sidebands" at $F = F_0$. If $F = N F_0$, where $N = 1, 2, 3, 4$, etc., "sidebands" spaced by F_0 will occur throughout the entire spectrum, even if there is no strong acoustic wave at F_0.

[3]"Hi-Fi" enthusiasts will want to note that Pedersen got interested in this subject because his Jensen high-fidelity loudspeaker was producing both first and second subharmonics of sinusoidal tones.

A.10   The Nonlinear Oscillator

In general, one expects the oscillator equation will become nonlinear at large enough amplitudes. For example, when x is sufficiently negative the spring will compress, the coils will touch and a large repulsive force will occur. At large positive values of x, the spring constant will begin to decrease and the restoring force will saturate. Assuming that the restoring force from the spring is a continuous function of x with continuous derivatives, it can be expanded in a power series of the form

$$F = -[K_1 x + K_2 x^2 + K_3 x^3 + \ldots] \tag{A.48}$$

where x is the departure from equilibrium position at $x = 0$. K_1 is just the normal spring constant and may be expressed in terms of ω_0. Then, the driven oscillator equation may be rewritten

$$\frac{d^2 x}{dt^2} + \gamma \frac{dx}{dt} + \omega_0^2 \left[x + \frac{x^2}{X_q} + \frac{x^3}{X_c} \right] = E \sin \omega t \tag{A.49}$$

where γ is the damping term from friction per unit mass and the coefficients X_q and X_c have been defined to have dimensions of length and length-squared and describe the amounts of quadratic and cubic nonlinearity. We will ignore terms higher than cubic in x. Here, E is the driving acceleration per unit mass at frequency ω that would be supplied by the Bernoulli effect in the real case. The quadratic term will yield both even harmonics and subharmonics, as well as rectification; the cubic term provides odd harmonics; and with both terms present, additional sum and difference terms arise. The term on the right-hand side of the equation represents the driving force per unit mass. It could be sinusoidal at the normal resonant frequency ω_0. To investigate subharmonic production, it is useful to replace the driving frequency by $2\omega_0$. Then the subharmonic appears at ω_0. Stable solutions to nonlinear equations have very specific phase relationships that are determined by the parameters of the system. Closed-form solutions to this type of nonlinear equation have been investigated in the past, but are extraordinarily tedious to derive and usually only hold for very small driving terms.[4] However, it is relatively easy to solve the equation numerically with a computer.

---

[4]Lord Rayleigh (1877) investigated low-amplitude solutions to nonlinear differential equations of this type in closed form. Also see, Hartley (1939), Hussey and Wrathall (1936) and Pedersen (1934) for approaches to the closed-form solution of similar nonlinear equations.

A.11   Potential Wells for the Nonlinear Oscillator

As shown in Fig. A.2, the potential well for the ideal linear oscillator has even symmetry about the equilibrium point ($x = 0$) and can trap the mass regardless of its kinetic energy. The situation changes drastically when we add a quadratic nonlinearity.

The potential curve shown at the left in Fig. A.7 is for an oscillator containing both linear and quadratic terms. Here, the restoring force is of the form,

$$F \equiv -\frac{dU}{dx} = -K_1 x - K_2 x^2, \text{ hence } U(x) = \omega_0^2 \left[\frac{x^2}{2} + \frac{x^3}{3X_q} \right], \qquad (A.50)$$

where again $U \equiv 0$ at $x = 0$, but we have written U as the potential per unit mass. As can be seen from the figure, the quadratic force term reduces the potential barrier on one side of equilibrium, while increasing it on the other. The result is that the mass can now escape from the well at the left if it acquires enough kinetic energy from the driving force to reach the top of the potential bump.[5] (The top of the bump occurs at $x = -Xq$ and is of magnitude $\Delta U = \omega_0^2 X_q^2/6$.) Consequently, the oscillator with only the addition of a quadratic force term tends to be unstable, except at very low driving amplitudes where the kinetic energy per unit mass is kept less than $\omega_0^2 X_q^2/6$. With a large enough damping constant, the loss of energy inhibits the escape of the mass over the potential bump. However, if γ is made very large, the harmonic (and subharmonic motion) is damped out as well.

Finally, the potential well on the right in Fig. A.7 includes the effect of both quadratic and cubic force terms. Here,

$$F \equiv -\frac{dU}{dx} = -K_1 x - K_2 x^2 - K_3 x^3 \text{ and } U(x) = \omega_0^2 \left[\frac{x^2}{2} + \frac{x^3}{3X_q} + \frac{x^4}{4X_c^2} \right].$$
$$(A.51)$$

The cubic term in the spring constant introduces a stabilizing term to the potential. Hence, with the cubic force term present, stable solutions are obtained at very much larger driving amplitudes than with the quadratic term alone. Surprisingly, with the cubic term present, finite solutions exist at resonance for large driving amplitudes even without the presence of damping. However, in that case, the oscillator never settles down to a steady-state solution.

---

[5]This same phenomenon occurs in the dissociation of diatomic molecules.

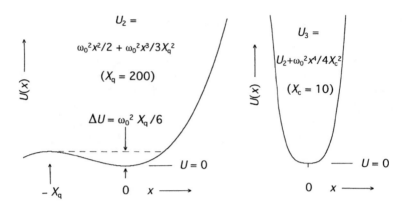

Fig. A.7 Potential wells for nonlinear oscillators containing quadratic (left) and both quadratic and cubic forces (right)

A.12   Numerical Solutions to the Non-linear Oscillator Equation

The driven oscillator equation with nonlinear terms is exceedingly tedious to solve in closed form. However, numerical solutions with a computer are straightforward using a computational method previously developed by the author (Bennett 1976, p. 200). Here, the main trick is to minimize numerical errors that tend to build up in successive integration intervals. That objective may be accomplished by expanding the acceleration in a Taylor series in the time and integrating that series term-by-term over an interval, t, that is small compared to the period of the driving oscillation.

First, we rewrite the basic equation with the acceleration on the left-hand side. For example, the equation for the driven, damped oscillator with quadratic and cubic nonlinear terms given above can be rewritten

$$a = \frac{d^2x}{dt^2} = E \sin \omega t - \gamma \frac{dx}{dt} - \omega_0^2 \left[x + \frac{x^2}{X_q} + \frac{x^3}{X_c} \right]. \tag{A.52}$$

Note that we now have an explicit closed-form expression for the acceleration that can be differentiated as many times as we want to obtain the coefficients in the Taylor series. Thus, after the small time interval t, the new values for the acceleration, velocity, and position will be given by

$$a' = a + bt + c\frac{t^2}{2!} + d\frac{t^3}{3!} + Order(t^4) = \frac{dv}{dt} \tag{A.53}$$

$$v' = v + at + b\frac{t^2}{2!} + c\frac{t^3}{3!} + d\frac{t^4}{4!} + Order(t^5) = \frac{dx}{dt} \tag{A.54}$$

$$x' = x + vt + a\frac{t^2}{2!} + b\frac{t^3}{3!} + c\frac{t^4}{4!} + d\frac{t^5}{5!} + Order(t^6) \qquad (A.55)$$

where we have integrated the successive equations term-by-term and have included constants of integration (the values of the quantities x, v, and a at $t = 0$) in each case. The errors in a' are now of order t^4 and those for the new position of the particle x' are of order t^6. Hence, by making t sufficiently small, we can make the results as precise as we wish.

Putting these expressions into a reiterative computer loop where at each successive step one sets the initial values of a, v, and x to be the values computed for a', v', x' at the end of the last interval t yields x as a function of time to high accuracy. One could, of course, add more and more terms to the Taylor series for a, but terminating the series as shown above is more than adequate for the present examples.

The equations for a, v, and x are especially adaptable to inclusion in a computer program loop of the type used in BASIC or FORTRAN. One can simplify the numerical equations further by adopting a time scale in which the time increment t is defined to be unity. In that case the important program steps for solving the nonlinear oscillator may be written as follows. First, we define some constants dependent on the parameters of the oscillator.

Driving frequency	Resonant frequency
W = ω	W0 = ω_0
W2 = W * W	W02 = W0*W0
W3 = W * W2	
Damping constant per unit mass	
G = γ	

In terms of these computer variables, the equation for the acceleration becomes

$$AE * SIN(W * T)G * VW02 * (XX * X/XqX * X * X/Xc)dV/dT. \qquad (A.56)$$

For example, choosing the oscillator period to be 60 time units (or $\omega_0 = 2\pi/60$), means that a step size of $t = 1$ leads to an error ≈ 5 parts in 10^{10} in the calculation of x over one increment. One typically wants to do the calculation over a time interval $T_{max} \approx 5$ oscillator periods for the transient solution to settle down to a steady-state solution.

The initial conditions on position and velocity are:
$X = V = 0$
The reiterative loop equations then take the form in BASIC:[6]

---

[6]Or in FORTRAN, DO T $= 0$,Tmax,1 followed by END DO instead of NEXT T.

FOR T = 0   TO   T max   *'with implied steps of T = 1*
S1 = **SIN** (W ∗ T)   *'W = Driving Frequency. S1 and C1*
are defined to avoid
C1 = **COS**(W ∗ T)   *'recalculating the SIN and COS within*
each iteration
A = E∗S1–G∗V–W02∗(X+X∗X/Xq+X∗X∗X/Xc)
B = W∗E∗C1–G∗A–W02∗(V+2∗X∗V/Xq+3∗X∗X∗V/Xc)
C = –W2∗E∗S1–G∗B–W02∗(A+2∗(X∗A+V∗V)/Xq+3∗(2∗X∗V∗V
        +X∗X∗A)/Xc)
D = –W3∗E∗C1–G∗C–W02∗(B+2∗(V∗A+X∗B+2∗V∗A)/Xq)
D = D–W02∗3∗(2∗V∗V∗V+4∗X∗V∗A+2∗X∗V∗A+X∗X∗B)/Xc
X = X+V+A/2+B/6+C/24+D/120   *'Use V to find X before*
changing V
V = V + A + B / 2 + C / 6 + D / 24
REM Plot **or** Print results here **NEXT**   T

Note that $A = dV/dt$, $B = dA/dt$, $C = dB/dt$, $D = dC/dt$ and that the time
scale was defined so that $t = 1$ in the Taylor expansion for A and step size in the
loop on T.

A.13   Examples of Nonlinear Oscillator Solutions

Solutions of the nonlinear equation containing both quadratic and cubic force terms
are shown in Fig. A.8, where it was assumed that $Xq = Xc = 10$ and the driving
frequency was tuned to the normal resonance frequency, ω_0. A damping constant
$\gamma = 0.5\omega_0$ was chosen for this illustration.

Time increases from $t = 0$ at the left of each trace in Fig. A.8 and the full
transient response of the oscillator for the mass starting from rest at $x = 0$ is
shown for increasing values of the driving amplitude, E. The driving term is shown
in dotted lines and the oscillator response is in solid curves. For $E = 0.01$ (the
lowest curve), the response is nearly sinusoidal. However, as the driving amplitude
increases, various even and odd harmonics develop in the waveform. The behavior
of the oscillator becomes very complex when the excursion of the mass reaches
the critical region where $x \approx Xq = Xc = 10$. For the conditions assumed, that
point occurs at $E \approx 0.1$. Although the different Fourier amplitudes are continuous
in their dependence on E, there are discontinuities in slope that appear when new
harmonics cross threshold. Many of these discontinuities result in a reversal of
the direction of the dependence of a given harmonic on excitation level. As a
result, small changes in E can produce large changes in the spectral distribution.
Even harmonics are present that also oscillate with increasing excitation level, but
decrease in relative importance above $E \approx 0.05$. The cubic characteristic dominates
at large excitation levels where one sees a series of peaks in the different odd

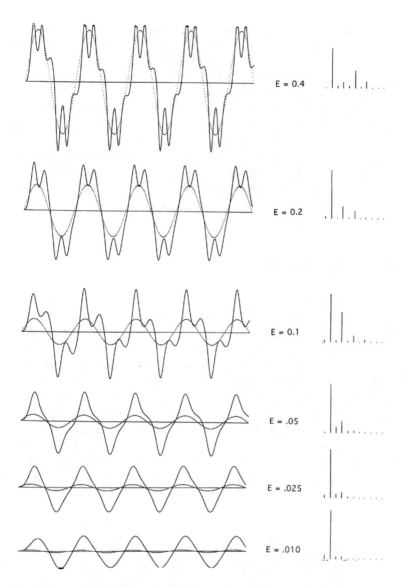

Fig. A.8 Solutions to the nonlinear oscillator equation when driven at the normal oscillator resonance frequency with $\gamma = 0.5\omega_0$, and $Xq = Xc = 10$. E increases from bottom to top. The driving wave (amplitude E) is shown dotted. A histogram of the relative spectral amplitudes is shown at the right in each case, normalized to the response at the driving frequency (the first term represents the response at DC)

harmonics. The waveform amplitude saturates above $E \approx 0.5$, although more and more ripples occur with increasing excitation. That behavior is characteristic of many reed and brass instrument, not to mention the human voice. Above the saturation level, the sensation of further loudness arises primarily due to the increase in harmonic content rather than output power.

The relative harmonic amplitudes for $N = 0, 1, 2, 3, \ldots, 10$ normalized to the fundamental component ($N = 1$) are shown at the right in Fig. A.8 as a histogram, as determined by Fourier analysis. ($N = 0$ corresponds to a DC rectification term.) Rectification, harmonics, and even subharmonics are an automatic consequence of the nonlinearities assumed. Although subharmonics are not shown in the Fourier spectrum of Fig. A.8, their production in the case of a strong quadratic nonlinear term is discussed in the chapter on the human voice.

A.14   Dynamic Chaos

At very high values of the driving force E, the nonlinear oscillator begins to exhibit characteristics of chaos in its motion. As an extreme example, consider the case where there is no damping force at all in the presence of large quadratic and cubic terms in the spring constant with high values of a periodic driving force. Here, even though the solutions are still bound and have strong spectral components at the driving frequency and its higher harmonics, the solutions are not periodic (i.e., they don't repeat themselves from one period to the next). Hence, the oscillator motion becomes unpredictable in detail from one cycle of the driving frequency to the next. An example of this behavior is shown in Fig. A.9. It is not clear that this regime of the nonlinear mechanical oscillator has any direct relevance to the behavior of the real human voice or other musical instruments. But some laryngologists have suggested that dynamic chaos may be present in vibrations of the vocal fold in the

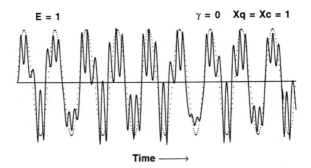

Fig. A.9 Example of chaotic behavior of the nonlinear oscillator. Here, the damping has been eliminated altogether ($\gamma = 0$) and the oscillator is driven at a very high excitation level (E = 1). As before, the driving term is at the normal resonance frequency and indicated by dotted lines. Note that the waveform does not repeat itself precisely from one period to the next

singing voice. (See Sataloff et al. 1998 and Hawkshaw et al. 2001.) However at the present time, these suggestions seem largely to be speculation drawn by analogy to other biomechanical systems where chaos has actually been observed.

A.15   Subharmonic Solutions

There is some indication that subharmonics may be present in some musical instruments. For example, the subtones produced by violinist Mari Kimura through use of very large bowing pressure were interpreted by her as subharmonics. (See Chap. 5.) Similarly, the extremely low frequencies produced by the Gyütö monks and the Tuva throat singers might also be due to subharmonic generation. (See Chap. 6 on the singing voice (Fig. A.10).)

Figure A.11 shows a subharmonic solution to the nonlinear oscillator when it only contains a quadratic nonlinear term in the spring constant. As discussed above, the oscillator is highly unstable under these conditions. A very weak potential well exists to trap the mass and the driving force has to be exceedingly small (or the

Fig. A.10 Variation of Fourier coefficients for nonlinear oscillator solutions of the type in Fig. A.9. As a function of driving amplitude

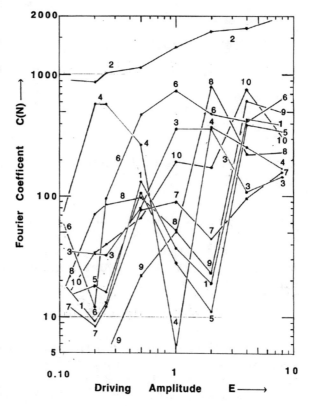

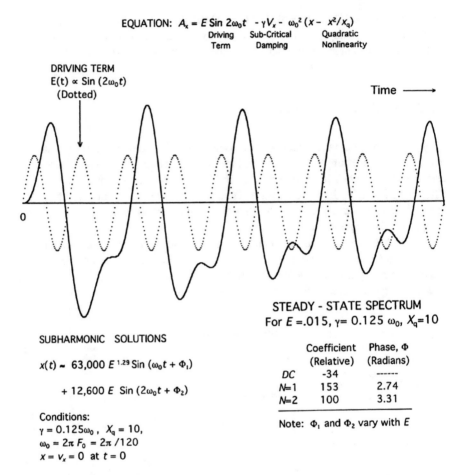

EQUATION: $A_x = E \sin 2\omega_0 t \; - \gamma V_x - \; \omega_0^2 \, (x - \; x^2/x_q)$

| | Driving Term | Sub-Critical Damping | Quadratic Nonlinearity |

DRIVING TERM
$E(t) \propto \sin(2\omega_0 t)$
(Dotted)

Time $\longrightarrow$

0

SUBHARMONIC   SOLUTIONS

$x(t) \sim 63{,}000 \, E^{\,1.29} \sin(\omega_0 t + \Phi_1)$

$+ 12{,}600 \, E \; \sin(2\omega_0 t + \Phi_2)$

Conditions:
$\gamma = 0.125\omega_0$, $X_q = 10$,
$\omega_0 = 2\pi \, F_0 = 2\pi \, /120$
$x = v_x = 0$ at $t = 0$

STEADY - STATE SPECTRUM
For $E = .015$, $\gamma = 0.125 \, \omega_0$, $X_q = 10$

	Coefficient (Relative)	Phase, Φ (Radians)
DC	-34	------
$N{=}1$	153	2.74
$N{=}2$	100	3.31

Note: Φ_1 and Φ_2 vary with E

Fig. A.11 Subharmonic solution to the forced nonlinear oscillator containing a quadratic term in the spring constant

damping very high) to prevent the oscillator from flying apart, which will happen if $|x| \geq X_q$.

To illustrate a subharmonic solution, it is useful to assume that the driving acceleration is applied at twice the resonant frequency of the normal linear oscillator. Hence, we will assume

$$E(t) = E \sin(2\omega_0 t) \tag{A.57}$$

as indicated by the dotted lines in Fig. A.11. The subharmonic then occurs at ω_0. The damping constant γ for the solution in the figure was chosen to be much smaller than the value $(2\omega 0)$ for critical damping in order to enhance production of the suboctave harmonic. (Although adding a large damping term would increase the stability of the oscillator, it would suppress the subharmonic component.) The solution shown includes the transient response when the driving force is initially turned on with the particle at rest at $x = 0$. The oscillator amplitude (solid curve) decays into the steady-state solution at the extreme right of the figure, where the fact that it is periodic in half the driving frequency is easily seen by eye. The first two Fourier coefficients and their phases are shown in Fig. A.11, where $C(1)$ is the subharmonic amplitude and $C(2)$ is the amplitude of the response at $2\omega_0$.

The phases of the two components (Φ_1, Φ_2) in Fig. A.11 vary in a complex manner with the driving amplitude. As previously noted, the oscillator with a quadratic nonlinearity acts as a rectifier. That is, a significant DC ("Direct Current") offset occurs in the solutions. With the human voice, the latter might correspond to a net average opening of the glottis which increases with driving amplitude. As discussed above, the presence of a cubic force term results in a stable potential well. A cubic (or at least an odd-symmetric force term) is also to be expected with most mechanical oscillators on a physical basis. As the spring in the mechanical oscillator compresses, the coils will eventually touch each other, creating a strong repulsive force back toward the equilibrium position ($x = 0$). An analogous situation occurs in the larynx when the glottis completely closes. At the other extreme, a large opening of the glottis would result in increased contact with other tissue in the larynx and provide a similar repulsive force back toward equilibrium. Hence, there is justification for an odd-symmetric force at large amplitudes and a cubic term is the simplest one to incorporate. However, the presence of the cubic term produces odd harmonics and reduces the subharmonic content.

A.16   The Relaxation Oscillator

A relaxation oscillator was used in the Dudley VODER discussed in the chapter on speech synthesis and a mechanical analog to the relaxation oscillator occurs in the grab-slip phenomenon in bowed strings discussed in the chapter on bowed strings. A very simple version is shown in Fig. A.12.

In Fig. A.12, a battery or other DC ("Direct Current") voltage supply at the left charges the capacitor C by means of a current $i = dq/dt$ flowing through the resistor R. Ignoring the neon tube at the right of the circuit for the moment, the charge q flowing to the capacitor would obey the equation,

Fig. A.12 A simple
relaxation oscillator circuit

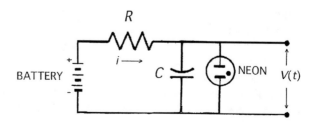

$$E = R\frac{dq}{dt} + \frac{q}{C}, \tag{A.58}$$

where E is the battery voltage (or "electromotive force"). Dividing the equation by R and using the integrating factor $e^{(t/RC)}$, it is seen that the voltage across the capacitor is

$$V(t) = \frac{q(t)}{C} = E\left[1 - e^{t/RC}\right], \tag{A.59}$$

where we have assumed that there was no initial charge on the capacitor before connecting the battery. Expanding the exponent,

$$V(t) = E\frac{t}{RC} + \text{Order}\left(\frac{t}{RC}\right)^2. \tag{A.60}$$

Hence, for $t \ll RC$, the voltage $V(t)$ rises linearly with time.

It is a characteristic of neon bulbs that once the voltage reaches a threshold value called the "ignition voltage," a value somewhat above the ionization potential of the neon atom (≈ 21.6 V) but dependent on the pressure and electrode geometry, a discharge occurs through the gas. The discharge current persists until the voltage drops below the "extinction voltage," a value somewhat below the first excited state of the atom (≈ 11.5 V), but again modified by the geometry and pressure. Hence, the capacitor, C, will be repeatedly charged up through the resistor R and then discharged through the neon bulb. The period for this process will be in the order of RC (a time constant characteristic of the circuit), but will be modified by the magnitude of the power supply voltage.

If the discharge through the neon bulb occurred instantaneously, the output voltage $V(t)$ in the limit that t«RC would approach an ideal sawtooth waveform, which has harmonic amplitudes that decrease as $1/n$, where n is the harmonic number; i.e., the power spectrum falls off as $1/n^2$. For example, if the RC time constant were adjusted so that the fundamental frequency of the oscillator were 120 Hz (typical of the vocal cord resonance in an adult male and the source frequency used in the Dudley VODER and VOCODER), the power spectrum of the oscillator would be distributed as shown by the solid line in Fig. A.13. In practice, the discharge time is not zero and is limited by drift mobilities of ions and electrons and other collision processes in the neon tube to values somewhat less than a millisecond (but also modified by pressure and tube geometry). The finite discharge time can reduce the harmonic output of the relaxation oscillator significantly, as shown by the histogram in Fig. A.13. The histogram was computed by Fourier analysis for an assumed discharge time of 2/3 of a millisecond. The minima at the harmonics for N = 13 and 26 correspond to multiples of the ratio of the oscillator period to the discharge time (≈ 13:1).

The limitations of the simple oscillator in Fig. A.12 may be overcome by using more complex circuitry such as that shown in Fig. A.14. For example, the capacitor

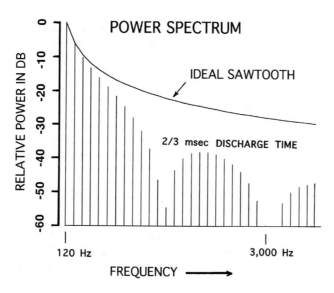

Fig. A.13 Power spectrum for a relaxation oscillator with finite discharge time compared with the spectrum from an ideal sawtooth waveform. The histogram was computed from the square of the Fourier coefficient amplitudes for the waveform

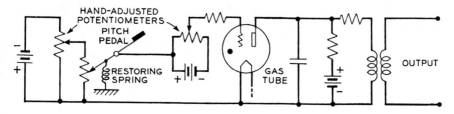

Fig. A.14 Relaxation oscillator used in the Dudley VODER (a manually operated synthetic speaker) first exhibited at the 1939 San Francisco Exhibition and New York World's Fair. Reproduced from Dudley (1939) by permission of Lucent Bell Laboratories, Jean Dudley Tintle and Richard Dudley. The spectrum produced by this circuit closely matched that for the Ideal Sawtooth shown in Fig. A.13

could be charged by a constant current generator to provide greater linearity during the charging cycle and the discharge of the capacitance could be achieved through a triggered, high- mutual-conductance tube or avalanche diode. Dudley's oscillator was actually the inverse of the circuit shown in Fig. A.12. He used a gas triode with a control grid and filament to charge the capacitance (in ≈ 0.3 ms). The capacitance was then allowed to discharge by itself with roughly a 0.8-ms time constant. The period of his oscillator (nominally ≈ 10 ms) was varied by biasing the control grid on the gas tube. Because the discharge pulse occupied a small fraction of the period, the power spectrum was very close to that for the ideal sawtooth shown in

Fig. A.13.[7] (Contemporary VOCODERS such as the MAM Model VF-11 use linear sawtooth generators that do not depend on gas discharge tubes.)

A.17 The Helmholtz Resonator

A number of acoustical problems can be analyzed in terms of Helmholtz Resonators. The diagram in Fig. A.15 is useful in deriving the resonant frequency of such a volume resonator.

We first want to calculate the change in pressure in the large volume that occurs when the gas (not necessarily just air) contained in the small cylinder is pushed into the sphere of initial volume V_0. Since a small volume of gas is added and we are ultimately going to consider changes in which that volume of gas goes in and out of the sphere, it is appropriate to think of the process as one in which there is no net heat flow in or out of the system. For such an "adiabatic" process involving an ideal gas, it is shown in thermodynamic texts that[8]

$$PV^\gamma = \text{constant, with } \gamma = C_p/C_V, \tag{A.61}$$

where P is the pressure of gas occupying volume V and γ is the ratio of specific heats for the gas at constant pressure C_P and constant volume C_V. The ratio γ varies appreciably depending on the molecular complexity of the gas. (See Table A.2.)

Now suppose a piston of area A is pushed through the cylinder of length L in Fig. A.14 forcing the gas from the small volume

$$\Delta V = A \times L \tag{A.62}$$

Fig. A.15 Schematic diagram of a Helmholtz oscillator of volume V_0

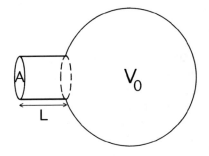

---

[7] See Dudley 1939; Figs, 7, 10, and 11.
[8] See, e.g., Zemansky (1951, Section 6.7.)

Table A.2 Values of γ for different common gases near room temperature

Gas	Wet air	Helium	Argon	Hydrogen	Nitrogen	Carbon dioxide	Methane
γ	1.37	1.67	1.67	1.40	1.40	1.29	1.30

Source: Zemansky (1951, pp. 129, 130). Zemansky notes that a method for measuring γ was devised by Rüchhardt in 1929 in which the mass of air in the Helmholtz cylinder was replaced by a small metal ball fitting snugly in the tube and measuring the oscillation frequency when the apparatus was oriented with the tube in the vertical direction

into the sphere. Differentiating Eq. (A.60) yields

$$\Delta P V^\gamma \gamma P V^{\gamma-1} \Delta V = 0, \qquad (A.63)$$

where ΔP is the change from the initial pressure P in the sphere due to the change in gas volume ΔV. Rearranging that equation,

$$\Delta P = \gamma P \Delta V / V. \qquad (A.64)$$

Hence, the change in pressure in the sphere is

$$\Delta P = \frac{\gamma P A L}{V_0} \qquad (A.65)$$

which results in a force pushing back on the hypothetical piston per unit length of the cylinder (i.e., the effective "spring constant" K for the system) given by

$$K = \frac{\gamma P A^2}{V_0}. \qquad (A.66)$$

One thus has a situation where a pressure fluctuation pushing in on the cylinder of gas with mass $\rho A L$ activates harmonic oscillation. Comparing these results with the basic harmonic oscillator equations [Eq. (A.1) through Eq. (A.4)], the oscillation frequency at resonance is seen to be

$$\omega_0 = 2\pi f_0 = \sqrt{\frac{\gamma P A^2}{V_0}} \Big/ \sqrt{\rho A L} = \sqrt{\frac{\gamma P A}{\rho L V_0}}. \qquad (A.67)$$

The resonant frequency can be rewritten in terms of the velocity of sound c in the gas as

$$f_0 = \frac{c}{2\pi} \sqrt{\frac{A}{L V_0}} \text{ where } c = \sqrt{\frac{\gamma P}{\rho}} \qquad (A.68)$$

Appendix B
Vibrating Strings and Membranes

B.1 The Wave Equation for the Vibrating String

Just as with the harmonic oscillator described in Appendix A, the differential equation describing the motion of a vibrating string can be derived by a simple application of Newton's law of motion, $\mathbf{F} = m\mathbf{a}$. Here, it is especially important to realize that both the force $\mathbf{F}$ and the acceleration $\mathbf{a}$ are vectors (i.e., have direction as well as magnitude).

Consider the element of string under tension T between positions x and $x + \Delta x$ along the horizontal axis as shown in Fig. B.1. We assume that the magnitude of the tension is constant throughout the length of the string and that the deflection $y(x)$ in the vertical direction is small. By small, we mean specifically that the angle θ that the curve makes with respect to the x-axis is small enough so that

$$\sin\theta = \tan\theta = \frac{\partial y}{\partial x} \tag{B.1}$$

at each point on the curve. The net force on the string in the vertical (y) direction is then given by the difference in y components (each of form $T\sin\theta$) of the tension between x and $x + \Delta x$. Defining μ to be the mass per unit length and using approximation B.1, the net vertical component of Newton's equation acting on the differential string element Δx with mass $\Delta m = \mu \Delta x$ becomes

$$F_y \approx T\left[\left(\frac{\partial y}{\partial x}\right)_{x+\Delta x} - \left(\frac{\partial y}{\partial x}\right)_x\right] = \Delta m a_y = \mu\Delta x \frac{\partial^2 y}{\partial t^2}. \tag{B.2}$$

(The horizontal components of force cancel within the present approximation.)

The original version of this chapter was revised: Pages 359, 366, 367, and 369 were corrected. The correction to this chapter is available at https://doi.org/10.1007/978-3-319-92796-1_8

© Springer Nature Switzerland AG 2018
W. R. Bennett, Jr., *The Science of Musical Sound*,
https://doi.org/10.1007/978-3-319-92796-1

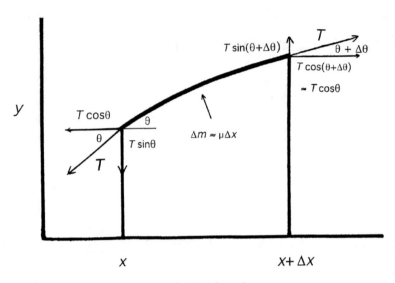

Fig. B.1 Forces on a differential element of a vibrating string

Next, we expand the function y/x in a Taylor series about the point x, noting[1]

$$\left(\frac{\partial y}{\partial x}\right)_{x+\Delta x} = \left(\frac{\partial y}{\partial x}\right)_x + \frac{\partial}{\partial x}\left(\frac{\partial y}{\partial x}\right)_x \Delta x^1 + \text{Order}(\Delta x^2). \tag{B.3}$$

Substituting Eq. (B.3) into the left side of Eq. (B.2) gives

$$F_y \approx T\frac{\partial}{\partial x}\left(\frac{\partial y}{\partial x}\right)_x \Delta x^1 + \text{Order}(\Delta x^2). \tag{B.4}$$

Substituting Eq. (B.4) into Eq. (B.2) and taking the limit as $\Delta x \to 0$ yields

$$T\frac{\partial^2 y}{\partial x^2} = \mu\frac{\partial^2 y}{\partial t^2} \tag{B.5}$$

which is the same as the wave equation,

$$\frac{\partial^2 y}{\partial x^2} = \frac{1}{c^2}\frac{\partial^2 y}{\partial t^2} \tag{B.6}$$

---

[1]The Taylor series permits evaluating any well-behaved function f(x) at a point displaced to $x+\Delta x$ by the infinite series

$f(x + \Delta x) = f(x) + \left(\frac{\partial y}{\partial x}\right)_x \Delta x^1 + \frac{1}{2}\left(\frac{\partial^2 y}{\partial x^2}\right)_x \Delta x^2 + \frac{1}{3!}\left(\frac{\partial^3 y}{\partial x^3}\right)_x \Delta x^3 + \dots$ Here, we simply

let $f(x) = \partial y/\partial x$.

given in Chap. 1, provided the velocity of the wave is

$$c = \sqrt{\frac{T}{\mu}} \qquad (B.7)$$

Note that μ is in units of mass per unit length.

B.2   General Solution of the Wave Equation

Equation (B.6) is linear and can be solved by separating the space- and time-dependent variables through a substitution of the type

$$y(x, t) = X(x)T(t) \qquad (B.8)$$

where $X(x)$ is only a function of x and $T(t)$ is only a function of t. Substituting this definition into Eq. (B.6) and dividing by $y(x, t)$ yields

$$\frac{1}{X(x)} \frac{\partial^2 X(x)}{\partial x^2} = \frac{1}{c^2 T(t)} \frac{\partial^2 T(t)}{\partial t^2} \equiv -k^2 = \text{constant.} \qquad (B.9)$$

That is, the only way the left side of the equation (which depends only on x) could equal the right side (which depends only on t) for all values of x and t is to have both sides equal to the same constant, chosen here to be $-k^2$ to insure that it is a negative value. (If we had chosen a positive constant at this point the solutions would not be oscillatory.) Equation (B.9) implies two separate differential equations,

$$\frac{\partial^2 X(x)}{\partial x^2} + k^2 X(x) = 0 \text{ and } \frac{\partial^2 T(t)}{\partial t^2} + k^2 c^2 T(t) = 0. \qquad (B.10)$$

These equations are both of the Harmonic Oscillator type discussed in Appendix A and have solutions of the form

$$X(x) \propto \sin kx \text{ and } T(t) \cos \omega t \text{ or } T(t) \propto \sin \omega t \qquad (B.11)$$

provided $k^2 c^2 = \omega^2$. (The solution $X(x) \propto \cos kx$, although perfectly valid, would not satisfy the boundary conditions on x in the present problem.) The requirement that $X(x) = 0$ at both $x = 0$ and $x = L$ in Eq. (C.11) is satisfied by

$$k_n = n\pi/L \text{ where } n = 1, 2, 3, \ldots, \text{ hence, } \omega_n = k_n c = n\pi c/L. \qquad (B.12)$$

The cyclical resonant frequencies are then given by $F_n = \omega n/2\pi = nc/2L$ as we showed in Chap. 1 by two different methods. Note that the spatial boundary conditions in Eq. (B.12) are what actually determine the resonant frequencies of the string. Again, because the wave equation is linear, any linear combination of solutions of the type described by Eq. (B.11) is a solution to Eq. (B.6). The most general solution is of the form originally given by Daniel Bernoulli,

$$y(x, t) = \sum_{n=1}^{\infty} A_n \sin(n\pi x/L) \cos(2\pi n F_0 t) \text{ or } \sum_{n=1}^{\infty} A_n \sin(n\pi x/L) \sin(2\pi n F_0 t).$$

(B.13)

The form on the left containing $\cos(2\pi n F_0 t)$ is most useful when the shape of the string is known at $t = 0$ (as in the case of the plucked string). The second form containing $\sin(2\pi n F_0 t)$ is most useful when the initial conditions involve the velocity at some point on the string, because there one needs to evaluate $\partial y/\partial t \propto \cos(2\pi n F_0 t)$ at $t = 0$ (as for a string struck by a piano hammer, or bowed on a violin).[2]

B.3   The Plucked String

Here, the first form of the solution given in Eq. (B.13) is the most useful. At $t = 0$, the string is distorted into a triangular shape of the type shown in Fig. B.2. [But keep in mind that the solutions given above apply for sufficiently small deflections of the string such that the approximation in Eq. (B.1) is satisfied. Figure B.2 greatly exaggerates the relative size of the deflection for the sake of illustration.]

Consider what happens when a string is plucked at $t = 0$ at some specific point $x = P_0$ along its length, L. If the amplitude of the string at $t = 0$ at the plucking point is A, we see from Eq. (1) that

Fig. B.2 Shape of a plucked string at t = 0 (the amplitude is greatly exaggerated)

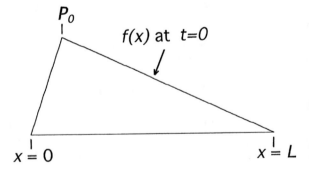

---

[2]The point in both cases is, of course, that $\cos(2\pi n F_0 t) = 1$ at $t = 0$, which leaves we with $y(x, 0) = \sum_{n=1}^{\infty} A_n \sin(n\pi x/L)$.

$$y(x, 0) = f(x) = \begin{cases} Ax/P0, & \text{for } 0 \leq x \leq P_0 \\ A(L-x)/(L-P_0), & \text{for } P_0 \leq x \leq L. \end{cases} \quad \text{(B.14)}$$

From Eq. (B.13) at t = 0,

$$f(x) = \sum_{n=1}^{\infty} A_n \sin(n\pi x/L). \quad \text{(B.15)}$$

Here, we can use the same orthogonality properties of the sine function summarized in Chap. 1 to obtain the values of the coefficients A_n, yielding

$$An = \frac{2}{L} \int_{x=0}^{L} f(x) \sin(n\pi x/L)dx. \quad \text{(B.16)}$$

The integral must be broken into two parts in accordance with the form of $f(x)$ given in Eq. (B.14). The spectral amplitudes are then determined from

$$A_n = \frac{2}{L} \int_{x=0}^{P_0} (A_x/P_0) \sin(n\pi x/L)dx + \frac{2}{L} \int_{P_0}^{L} [A(L-x)/(L-P_0)] \sin(n\pi x/L)dx. \quad \text{(B.17)}$$

Integrating by parts and collecting terms, one obtains

$$A_n = \frac{2AM^2}{(M-1)n^2\pi} \sin(n\pi/M) \quad \text{(B.18)}$$

for the spectral distribution, where $M = L/P_0$ is a measure of the plucking point and it is assumed that $n = 1, 2, 3, \ldots$ is an integer. (M does not have to be an integer.) Note that for $M = 2$ (plucking at the mid-point), one only gets odd harmonics and that $A_n = 0$ for n equal to integral multiples of M.

The time-dependent motion of the string is then obtained by substituting the amplitude coefficients A_n from Eq. (B.18) back into the original equation for $y(x, t)$ given at the left of Eq. (B.13). Examples of the motion are shown in Chap. 3.

B.4   The Struck String

Some instruments (especially, various forms of the piano and a few Hungarian instruments such as the Cembalom used by Kodaly) hit the string with a hammer. In this case, the right-hand solution in Eq. (B.13) is the appropriate form to start with because the boundary condition at $t = 0$ is one on velocity. Hence, we will start with a solution of the form,

$$y(x, t) \sum_{n=1}^{\infty} A_n \sin(n\pi x/L) \sin(2\pi n F_0 t). \tag{B.19}$$

The velocity distribution over the string is obtained by taking $\partial y/\partial t$, giving

$$v(x, t) = \frac{\partial y(x, t)}{\partial t} = 2\pi F_0 \sum_{n=1}^{\infty} n A_n \sin(n\pi x/L) \cos(2\pi n F_0 t). \tag{B.20}$$

In principle one could integrate the equation over some finite pulse duration during which the hammer was in contact with the string. However, we shall just assume here that a velocity distribution is suddenly imparted to the string by the hammer before the string has a chance to move. The approximation will be best for the low notes on the instrument where the vibrational periods are longest. In that approximation the velocity distribution at $t = 0$ is given by

$$v(x, 0) = V(x) = 2\pi F_0 \sum_{n=1}^{\infty} n A_n \sin(n\pi x/L). \tag{B.21}$$

Then, in analogy with Eq. (B.16), the spectral coefficients are given by[3]

$$A_n = \frac{1}{n F_0 L} \int_0^L V(x) \sin(n\pi x/L) dx. \tag{B.22}$$

As an example, consider a hammer that imparts a rounded velocity distribution to the string of the type shown in Fig. B.3, for which the velocity distribution is given by

$$V(x) = V_0 \left[1 - \left(\frac{P_0 - x}{R} \right)^2 \right] \quad \text{for } P_0 - R \le x \le P_0 + R \tag{B.23}$$

and $V(x)$ is assumed to be zero everywhere else. $V(x)$ has a rounded maximum value of V_0 at $x = P_0$, with an effective width of $2R$. (Roughly speaking, R corresponds to the hammer radius at the tip.) Although the shape was arbitrarily assumed, it is not unlike that found at the top of felt hammers currently used in grand pianos.

For the assumption in Eq. (B.23), the spectral coefficients in Eq. (B.22) become

---

[3]i.e., noting that for m, n integers, $\int_0^L \sin(mx/L) \sin(nx/L) dx = \begin{cases} L/2, & \text{for } m = n. \\ 0, & \text{for } m \ne n. \end{cases}$

Fig. B.3 Shape of the velocity pulse given by Eq. (B.23)

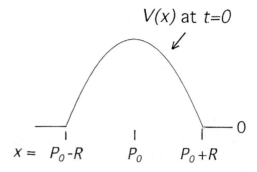

$$A_n = V_0 \frac{2H^2}{n^4 \pi^4 F_0} \left\{ -\cos n\pi \left(\frac{H+M}{MH} \right) + \cos n\pi \left(\frac{H-M}{MH} \right) \right.$$

$$\left. - \frac{n\pi}{H} \sin n\pi \left(\frac{H-M}{MH} \right) - \frac{n\pi}{H} \cos 2\pi \left(\frac{H+M}{MH} \right) \right\}$$

(B.24)

where n is the harmonic number, $H = L/R$ (the string length divided by the hammer radius) and $M = L/P_0$ (the ratio of the string length to the striking point, in analogy to the case of the plucked string). As in the case of the plucked string, Eq. (B.24) gives only odd harmonics when $M = 2$, or the string is struck in the middle. Also, $A_n = 0$ for $n = M$ (or the string is struck at a node for the Mth harmonic). In this approximation, voicing the hammer corresponds to adjusting the value of H.

The shape of the string as a function of time is obtained by putting the values of A_n from Eq. (B.24) back into the right-hand side of Eq. (B.13). Immediately after the hammer hits the string, a narrow pulse (of width 2τ at the left of Fig. B.5) pops up at the striking point ($x = P_0$, occurring at $t = 0$ in the figure). This narrow pulse consists of two equal-amplitude, oppositely-directed running waves. As time increases, the initial pulse broadens until the wave running to the left bounces off the support at $x = 0$, where it undergoes a "hard" reflection and changes sign. Now negative, it travels back in the $+x$ direction, canceling out its previous positive portion. Meanwhile, the running wave initially moving to the right has continued on its way. The result of adding these two running waves together is an isolated broader, positive pulse (at the right in Fig. B.4) that runs the length of the string.

The rise and fall times (τ) of this wider pulse are each equal to half of the initial narrow-pulse duration. However, the breadth of the wide pulse is determined by the time delay taken for that half of the initial pulse that bounces off the support at $x = 0$ to get back to the striking point at $x = P_0$. Hence, as indicated in the figure, the broad pulse time duration is $2P_0/c$, where c is the velocity of the wave. After reaching the point $x = L$, two "hard" reflections (in succession for each running wave) occur which send an inverted pulse back toward $x = 0$. Examples of these solutions, together with their spectral distribution, are shown in Chap. 4.

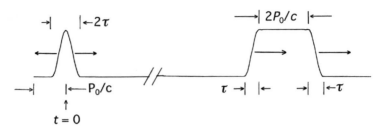

Fig. B.4 Pulses launched on the string by the striking process. Left: The initial pulse at the striking point. Right: The broader pulse running down the string after the first reflection

B.5   The Bowed String

As described in Chap. 5, the excitation of a violin string consists of a stick-slip process which is repeated at the fundamental round-trip frequency $(c/2L)$ and causes pulses to run back and forth on the string of length L. The problem is different from the plucked- and struck-string problems treated above in that the string at the contact point (which we will again take to be $x = P_0$) is forced to move at about the constant velocity of the bow until slipping occurs. When the string does slip, it returns rapidly to its initial point where it is again grabbed by the bow and the process repeats. The string displacement thus executes a sawtooth motion at the point P_0. Helmholtz (1885, pp.384–386) presented an approximate solution to the problem in which he assumed straight-line motion in the two halves of the stick-slip cycle. He then used Fourier analysis of this sawtooth motion and expressed the spectral amplitudes in the solution to the string equation in terms of those Fourier coefficients. The idealized motion is illustrated in Fig. B.5, where will assume the amplitude varies from -1 to $+1$ in the vertical direction and note that the vertical displacements are centered about the time axis.

As in the case of the struck string waveform, we will take the general solution for the shape of the string to be

$$y(x, t) = \sum_{n=1}^{\infty} A_n \sin(n\pi x/L) \sin(2\pi n F_0 t). \qquad (B.25)$$

But here, the boundary condition that determines the expansion coefficients An is on the time-dependent saw-tooth motion at the bowing point $x = P_0$ shown in Fig. B.5. In Helmholtz's formulation of the problem, he expressed one cycle of the motion in a general Fourier series including both sine and cosine terms in the time. He then shifted the time axis to obtain a result involving a series of sine terms only. Although that approach is perfectly valid and straightforward, there is a lot tedious algebra required to keep track of all the terms. It is simpler to note at the start from symmetry that, if we choose the origin of the time axis to be centered in the slip cycle as shown in Fig. B.6, one only needs sine terms in the Fourier series. Thus, the

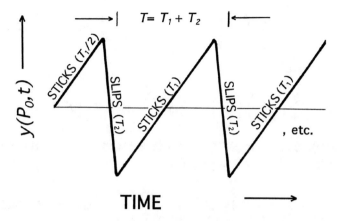

Fig. B.5 Slightly more than two idealized Stick-Slip cycles, corresponding to up-bowing. The total period is $T = T_1 + T_2$, where T_1 is the "Stick" time and T_2 is the "Slip" time

Fig. B.6 Choice of time origin to simplify the calculation of the Fourier series

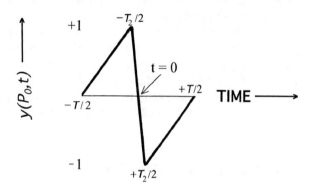

series for the motion at the bowing point may be written

$$y(P_0, t) = \sum_{n=1}^{\infty} C_n \sin(2\pi n F_0 t). \tag{B.26}$$

The Fourier coefficients are then given by

$$C_n = \frac{2}{T} \int_{T/2}^{+T/2} \sin(2\pi n F_0 t)\,dt$$

$$= \frac{2}{T} \int_{-T/2}^{-T_2/2} \left(t + \frac{T}{2}\right) \sin(2\pi n F_0 t)\,dt$$

$$+ \frac{2}{T} \int_{-T_2/2}^{+T_2/2} \left[\frac{T_2}{2} - \left(t + \frac{T_2}{2}\right)\right] \frac{2}{T_2} \sin(2\pi n F_0 t)\,dt$$

$$+\frac{2}{T}\int_{T_2/2}^{T/2}\left[-\frac{T_1}{2}+\left(t-\frac{T_2}{2}\right)\right]\frac{2}{T_1}\sin(2\pi n F_0 t)dt,$$

which simplifies to

$$C_n = -\frac{2}{n^2\pi^2}\left(\frac{T}{T_1}\right)\left(\frac{T}{T_2}\right)\sin\left(n\pi\frac{T_2}{T}\right) \quad \text{where } T = T_1 + T_2. \tag{B.27}$$

One then substitutes Eq. (B.27) in Eq. (B.26) and compares the result with Eq. (B.25) for $x = P_0$. It is then seen that

$$y(x,t) = -A\sum_{\substack{n\neq L/P_0}}^{\infty}\frac{2}{n^2\pi^2}\left(\frac{T}{T_1}\right)\left(\frac{T}{T_2}\right)\frac{\sin(n\pi T_2/T)}{\sin(n\pi P_0/L)}\sin(n\pi x/L)\sin(2\pi n F_0 t),$$

$$\tag{B.28}$$

where A has been introduced as an amplitude scaling factor and Eq. (B.28) was used to compute the figures illustrating the Helmholtz method in Chap. 5. The singularity that would occur if L/P_0 were an integer is avoided by setting the partial amplitude to zero for that harmonic. (That particular harmonic would not be excited because the bow would be at a node.) For more general behavior such as that reported by Pickering for real strings and discussed in Chap. 5, one must use the general form of the Fourier series including cosine terms in which the Fourier coefficients are computed numerically by methods equivalent to those discussed in Appendix C.

B.6   The Torsional Wave Equation[4]

As discussed in Chap. 5, the generation of torsional waves is important in the case of large diameter bowed strings under large bowing force. In this case, a form of the wave equation analogous to that in Eq. (B.5) also applies, but with very different wave velocity.

First, consider a hollow uniform cylinder of radius r and thickness dr stretched in the x-direction with the angular rotation at point x along the cylinder given by $\partial\varphi/\partial x$. The shear force on the cylinder material is given by $nr\partial\varphi/\partial x$ where n is defined as the "rigidity" given by

$$n \equiv \frac{Y}{2(\mu+1)} \quad \text{where } \frac{\text{Lateral Contraction}}{\text{Longitudinal Extension}}, \tag{B.29}$$

Y is Young's modulus, and by definition

---

[4]Nearly all treatises on mechanics ignore the torsional wave equation. The present derivation is based on one given by Lord Rayleigh (1877, pp. 243–254.)

$$\text{Longitudinal Extension} \equiv \frac{\text{Actual Length } - \text{ Natural Length}}{\text{Natural Length}}. \tag{B.30}$$

The quantity n lies between $Y/2$ and $Y/3$ for different materials.
The moment of inertia dI for a hollow cylinder of length dx is given by

$$dI = \rho 2\pi r^3 dr dx$$

where ρ is the mass density. The net change in resisting torque from shear over that length is

$$\Delta \text{Torque} = n2\pi r^3 dr dx \frac{\partial^2 \varphi}{\partial x^2} = n\frac{dI}{\rho}\frac{\partial^2 \varphi}{\partial x^2} dx$$

where we have made use of a Taylor expansion to show that

$$\left(\frac{\partial \varphi}{\partial x}\right)_{x+dx} - \left(\frac{\partial \varphi}{\partial x}\right)_{x} = \frac{\partial^2 \varphi}{\partial x^2} dx$$

Applying Newton's law relating torque and angular acceleration, we get the torsional wave equation,

$$n\frac{dI}{\rho}\frac{\partial^2 \varphi}{\partial x^2} dx = dI\frac{\partial^2 \varphi}{\partial t^2} dx \text{ or } \frac{\partial^2 \varphi}{\partial x^2} = \frac{1}{n/\rho}\frac{\partial^2 \varphi}{\partial t^2}. \tag{B.31}$$

Hence, the torsional waves have a velocity given by

$$c_T = \sqrt{\frac{n}{\rho}}. \tag{B.32}$$

Note that the radial dependence of the moment of inertia cancelled out in these equations. Therefore, the result for c_T is *independent of the radial mass distribution* as long as it has axial symmetry. One, of course, needs to know the values of the mass density ρ and of n from Eq. (B.30) for the string material in order to compute the velocity.

B.7   The Vibrating Membrane as a Two-Dimensional String

The extension of the wave equation to two-dimensional form follows in a straightforward manner from the derivation of the vibrating string equation given at the start of this appendix. Here, we will consider a thin membrane extending in the x and y directions with small vibrational amplitude in the z direction.

Fig. B.7 View looking down
on the membrane

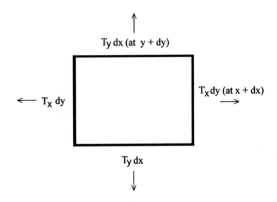

Here, we use a two-dimensional extension of the argument given in connection with Fig. B.1. (See Fig. B.7.) The constant tension T will now be per unit length along the x and y directions and we are concerned with the net restoring force in the z-direction acting on a differential mass element $\Delta m = \mu \Delta x \Delta y$, where σ is the mass density per unit area of the membrane. Applying Newton's Law to the motion in the z-direction for this differential membrane element, we get

$$F_z \approx T_x \Delta y \left[\left(\frac{\partial z}{\partial x} \right)_{x+\Delta x} - \left(\frac{\partial z}{\partial x} \right)_x \right] + T_y \Delta X \left[\left(\frac{\partial z}{\partial y} \right)_{y+\Delta y} - \left(\frac{\partial z}{\partial y} \right)_y \right] \Delta x$$

$$= \Delta m a_z = \sigma \Delta x \Delta y \frac{\partial^2 z}{\partial t^2}.$$

As with the one-dimensional case, we can get the component of force normal to the surface by expanding the functions dz/dx and dz/dy in a Taylor series about the point x, y noting that

$$\left(\frac{\partial z}{\partial x} \right)_{x+\Delta x} = \left(\frac{\partial z}{\partial x} \right)_x + \frac{\partial}{\partial x} \left(\frac{\partial z}{\partial x} \right)_x \Delta x^1 + \mathrm{Order}(\Delta x^2) \text{ and } \left(\frac{\partial z}{\partial y} \right)_{y+\Delta y}$$

$$= \left(\frac{\partial z}{\partial y} \right)_y + \frac{\partial}{\partial y} \left(\frac{\partial z}{\partial y} \right)_y \Delta y^1 + \mathrm{Order}(\Delta y^2)$$

Substituting in the force equation and neglecting quadratic terms in $\Delta x, \Delta y$, we get

$$F_z \approx T_y \frac{\partial}{\partial x} \left(\frac{\partial z}{\partial x} \right)_x \Delta x^1 + T_x \frac{\partial}{\partial y} \left(\frac{\partial z}{\partial y} \right)_y \Delta y^1 + \mathrm{Order}(\Delta x^2, \Delta y^2)$$

where we have assumed that T_x might not be equal to T_y for the sake of generality. (That permits having different wave velocities in the two orthogonal directions.) Substituting in the force equation and taking the limit as $\Delta x, \Delta y \to 0$, we get a non-isotropic form of the two-dimensional wave equation

$$\left(\frac{1}{T_y}\right)\frac{\partial^2 z}{\partial x^2} + \left(\frac{1}{T_x}\right)\frac{\partial^2 z}{\partial y^2} = \frac{\sigma}{T_x T_y}\frac{\partial^2 z}{\partial t^2}.$$

Of course, if $T_x = T_y$, this expression reduces to the more usual isotropic form of the two-dimensional wave equation,

$$\frac{\partial^2 z}{\partial x^2} + \frac{\partial^2 z}{\partial y^2} = \frac{1}{c^2}\frac{\partial^2 z}{\partial t^2} \tag{B.33}$$

for which the wave velocity, c, is the same in both directions. That is,

$$c = \sqrt{\frac{T}{\sigma}} \text{ for } T = T_x = T_y. \tag{B.34}$$

With the isotropic case, the variables are easily separable and we can write

$$Z(x, y, t) = X(x)Y(y)T(t) \tag{B.35}$$

which substituted in the wave equation yields

$$\frac{1}{X(x)}\frac{\partial^2 X(x)}{\partial x^2} + \frac{1}{Y(y)}\frac{\partial^2 Y(y)}{\partial y^2} = \frac{1}{c^2 T(t)}\frac{\partial^2 T(t)}{\partial t^2} \equiv -k^2 = \text{contant.} \tag{B.36}$$

The solutions must be valid for all values of x, y, and t and the only way that can happen is for the separate terms to equal constants. Here as in the one-dimensional case, we take a negative definite constant for the time-dependent part of the solution to insure stable oscillatory solutions. The first two terms on the left of the equation may be rewritten

$$\frac{1}{X(x)}\frac{\partial^2 X(x)}{\partial x^2} = -k_x^2, \ \frac{1}{Y(y)}\frac{\partial^2 Y(y)}{\partial y^2} = -k_y^2, \text{ where } k_x^2 + k_y^2 = k^2 \tag{B.37}$$

where k_x and k_y are constants. This substitution gives rise to solutions of the form

$$X(x) \propto \sin k_x x \text{ and / or } \cos k_x x$$
$$Y(y) \propto \sin k_y y \text{ and / or } \sin k_y y \tag{B.38}$$
$$T(t) \propto \cos \omega t \text{ and / or } \sin \omega t$$

where $Z(x, y, t)$ is given by the product of the three.

Spatial boundary conditions determine the resonant frequencies. For example, if the membrane is clamped on all edges, the solutions must have zero amplitude there and the appropriate spatial functions are

$$X(x) \propto \sin k_x x \text{ and } Y(y) \propto \sin k_y y \tag{B.39}$$

with

$$k_x = n_x\pi/L_x \text{ and } k_y = n_y\pi/L_y \text{ where } n_x, n_y = 1, 2, 3, \ldots \tag{B.40}$$

The oscillation frequencies of these modes are then given by

$$\omega_n = c\sqrt{k_x^2 + k_y^2} = \pi c\sqrt{\left(\frac{n_x}{L_x}\right)^2 + \left(\frac{n_y}{L_y}\right)^2}. \tag{B.41}$$

The equations for the homogeneous membrane give rise to modes of the type illustrated in Fig. B.8.

Solutions for the nonisotropic membrane are of interest because of their relationship to those in thin soundboards (for example, in harpsichords and violins) in which the velocity of wave propagation is quite different in the two principal orthogonal directions. The solutions in that case may be obtained in exactly the same way using the principle of separability. However, it is easier to note that the solutions for the non-isotropic case may be obtained from those for the isotropic case by a simple coordinate transformation of the type

$$x' = T_y x \text{ and } y' = T_x y. \tag{B.42}$$

Hence, the case of nonequal velocities in the two orthogonal directions has solutions that are equivalent to the isotropic case with different relative dimensions of the rectangular membrane.

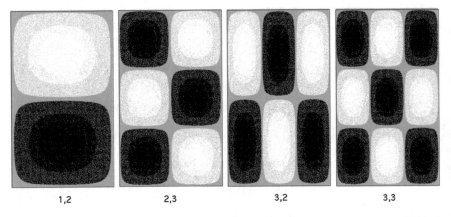

| 1,2 | 2,3 | 3,2 | 3,3 |

Fig. B.8 Modes for a rectangular membrane (or drumhead) in which the tensions (wave velocities) are equal in both coordinate directions and the sides are in the ratio of 2:3

As noted above, the equation of motion in this case is of the form

$$\left(\frac{1}{T_y}\right)\frac{\partial^2 z}{\partial x^2} + \left(\frac{1}{T_x}\right)\frac{\partial^2 z}{\partial y^2} = \left(\frac{\sigma}{T_x T_y}\right)\frac{\partial^2 z}{\partial t^2} \tag{B.43}$$

Again the equation is separable in the time and space variables by the substitution

$$Z(x, y, t) = X(x)Y(y)T(t). \tag{B.44}$$

Hence

$$\left(\frac{1}{T_y X(x)}\right)\frac{\partial^2 X(x)}{\partial x^2} + \left(\frac{1}{T_x Y(y)}\right)\frac{\partial^2 Y(y)}{\partial y^2} = \frac{\sigma}{T_x T_y}\left(\frac{1}{T(t)}\right)\frac{\partial^2 T(t)}{\partial t^2}$$

$$= -\omega^2 = \text{constant} \tag{B.45}$$

where a negative constant was again chosen to assure a stable oscillatory solution with angular frequency ω. For the two terms on the left of the equation to add up to a constant, each must separately be constant. Hence,

$$\frac{\partial^2 X(x)}{\partial x^2} + \omega_x^2 T_y X(x) = 0 \text{ and } \frac{\partial^2 Y(y)}{\partial y^2} + \omega_y^2 T_x Y(y) = 0 \tag{B.46}$$

with solutions

$$X(x) \propto \sin \omega_x \sqrt{T_y} x \text{ and } Y(y) \propto \sin \omega_y \sqrt{T_x} y, \tag{B.47}$$

where

$$\omega_x \sqrt{T_y} = n_x \pi / L_x \text{ and } \omega_y \sqrt{T_x} = n_y \pi / L_y \text{ where } n_x, n_y = 1, 2, 3, \ldots \tag{B.48}$$

That is, the modes are determined by the spatial boundary condition that the vibrational amplitudes are zero on the boundaries; that is, we assume the membrane is clamped on the edges. The oscillatory frequency for a given mode is then given by

$$\omega = \sqrt{\omega_x^2 + \omega_y^2} = \pi \sqrt{n_x^2 / L_x^2 T_y + n_y^2 / L_y^2 T_x} \tag{B.49}$$

and they, in general, will not be harmonically related.

B.8   Circular Membranes

Drums utilize circular membranes in most cases. These are easiest to treat using circular (or cylindrical) coordinates. Consider a coordinate system where the radius vector r is in the xy plane at angle ϕ with respect to the x axis (Fig. B.9). (The new and old z-axes are identical.)

It may be shown that the two-dimensional wave-equation operator transforms as[5]

$$\frac{\partial^2 A(x, y)}{\partial x^2} + \frac{\partial^2 A(x, y)}{\partial y^2} = \frac{1}{r}\frac{\partial}{\partial r}\left(r\frac{\partial A(r, \varphi)}{\partial r}\right) + \frac{1}{r^2}\frac{\partial^2 A(r, \varphi)}{\partial \varphi^2}. \tag{B.50}$$

Hence, the wave equation in circular coordinates becomes

$$\frac{1}{r}\frac{\partial}{\partial r}\left(r\frac{\partial A}{\partial r}\right) + \frac{1}{r^2}\frac{\partial^2 A}{\partial \varphi^2} = \frac{1}{c^2}\frac{\partial^2 A}{\partial t^2} \quad \text{where } c = \sqrt{\frac{T}{\sigma}} \tag{B.51}$$

c is the running wave velocity (assumed to be the same in all directions), T is the constant tension per unit length around the circumference, and σ is the mass density per unit area.

Again, we can separate the variables in the form

$$A(r, \varphi, t) = R(r)\Phi(\varphi)T(t), \tag{B.52}$$

where $T(t)\cos(\omega t)$ or $T(t) = \sin(\omega t)$.

Fig. B.9 Relation of rectangular and cylindrical coordinates

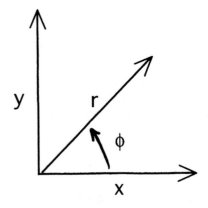

---

[5]See, e.g., Weatherburn (1951), pp. 15 and 16.

Then we get

$$\frac{\partial^2 \Phi}{\partial \varphi^2} + m^2 \Phi = 0 \text{ where } \Phi(\varphi) = \cos(m\varphi) \text{ or } \Phi(\varphi) = \sin(m\varphi). \quad \text{(B.53)}$$

Because $\Phi(\varphi)$ must be periodic in 2π so that the function closes on itself in one revolution about z-axis, we must have $m = 1, 2, 3, 4, \ldots$ This condition means that

$$\frac{\partial^2 R}{\partial r^2} + \frac{1}{r}\frac{\partial R}{\partial r} + \left(\frac{\omega^2}{c^2} - \frac{m^2}{r^2}\right) R = 0, \quad \text{(B.54)}$$

which is known as Bessel's Equation and has solutions of the form

$$J_m(x) = \sum_{p=0}^{\infty} (-1)^p \frac{(x/2)^{m+2p}}{p!(m+p)!}. \quad \text{(B.55)}$$

where $J_m(x)$ is a Bessel function of real argument of order m.[6] Hence, the radial solutions for the circular membrane are of the form

$$R(r) \propto J_m\left(\frac{\omega}{c}r\right) = J_m(kr). \quad \text{(B.56)}$$

A particular time-dependent solution is then

$$A(r, \varphi, t) \propto \cos(m\varphi) J_m(k_{m,n}r) \cos(2\pi f_{m,n}t). \quad \text{(B.57)}$$

The nodal diameter for a particular allowed solution is determined by the value of m and arises from the periodicity requirement on $\Phi(\varphi)$. The allowed values of the parameter k are then determined by the requirement that the rim of the circular drum corresponds to a root n of the Bessel function of that particular order, m. Thus, the resultant values of k depend on two integers, m and n. The frequency of the mode of vibration then is given by

$$f_{m,n} = \frac{k_{m,n}c}{2\pi} \quad \text{(B.58)}$$

To illustrate a particular mode for a circular kettle drum of radius, R_0, it is necessary to scale the argument of the Bessel function so that it goes through zero for a particular root when $r = R_0$. Thus, a given kettle drum mode distribution would be proportional to

---

[6]During the Great Depression, mathematicians were hired by the WPA to compute tables of Bessel Function up to order 600 or more and in high precision. These tables, which are now totally obsolete if any rudimentary computer is available, were stored in the basement of Low Library at Columbia University.

Table B.1 Values of $k_{m,n}$ (roots) for the Bessel function J_m

m, n	1	2	3	4	5
0	2.40483	5.52008	8.65373	11.7915	14.9309
1	3.83171	7.01559	10.1735	13.3237	16.4706
2	5.13562	8.41724	11.6198	14.796	17.9598
3	6.38016	9.76102	13.0152	16.2235	19.4090
4	7.58834	11.0647	14.3725	17.616	20.8269

Computed by the author

$$\cos(m\varphi)\, J_m\left(k_{m,n}\frac{r}{R_0}\right). \qquad (B.59)$$

Approximate values for the first few roots of the Bessel function J_m are given in Table B.1. (Such modes are sometimes called "eigen functions" and the quantities $k_{m,n}$ are "eigen values.") Representative kettledrum mode shapes are shown in Fig. B.10 in 3-D.

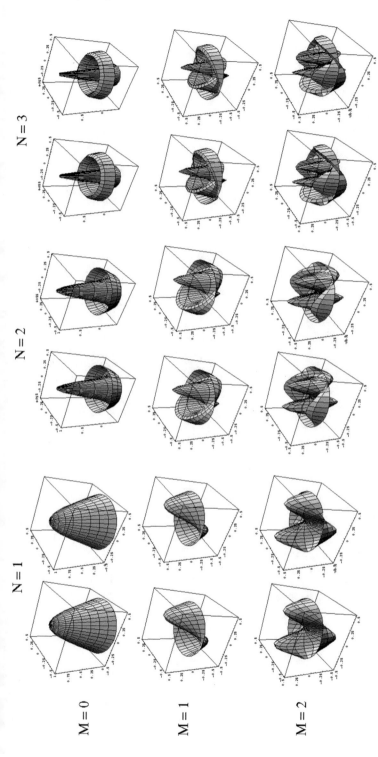

Fig. B.10 The first few kettledrum modes from Eq. (B.59) and Table B.1 are shown in stereoscopic pairs. Note that for $M = 0$, there are no nodal diameters; for $M = 1$, there is one nodal diameter; and for $M = 2$, there are two nodal diameters. Similarly, for $N = 1$ there is one circular node (at the rim of the drum); for $N = 2$ there are two circular nodes; and for $N = 3$ there are three such nodal circles. To see these figures stereoscopically, hold the paper as near to your eyes as permits comfortable focusing. For each pair, you should see four images—two with each eye. Make the two center images coincide in your brain. The amplitudes are exaggerated for clarity (The figures were computed by the author)

Appendix C
Fourier Analysis

We will include here some of the mathematical details dependent on calculus that were omitted in Chap. 2, together with a working program for discrete Fourier analysis. We will restrict the discussion to "well-behaved" periodic functions which obey

$$V(\theta + 2\pi) = V(\theta), \tag{C.1}$$

for which Fourier (1822) showed that the function $V(\theta)$ may be written

$$V(\theta) = C_0 + \sum_{n=1}^{\infty} A_n \sin n\theta + \sum_{n=1}^{\infty} B_n \cos n\theta. \tag{C.2}$$

We wish to evaluate the DC constant C_0 together with the harmonic coefficients A_n and B_n.

Determining the DC constant in Eq. (C.2) is particularly easy. Recalling the fact that the definite integral is just equal to the area under the curve,[1] it is apparent that the integrals of all the sine and cosine terms in Eq. (C.2) over one period will identically vanish; i.e., the sine and cosine functions have equal areas above and below the horizontal axis. Hence, in averaging Eq. (C.2) over one fundamental period, all of the $\sin n\theta$ and $\cos n\theta$ terms will drop out, leaving

$$C_0 = \frac{1}{2\pi} \int_0^{2\pi} V(\theta) d\theta \tag{C.3}$$

---

[1] The fact that the area under the curve in the integrand is given by the definite integral between two points on the horizontal axis is sometimes called "The Fundamental Theorem of Calculus."

© Springer Nature Switzerland AG 2018
W. R. Bennett, Jr., *The Science of Musical Sound*,
https://doi.org/10.1007/978-3-319-92796-1

where, as indicated using a notation invented by Fourier himself, the definite integral runs from 0 to 2π radians.

Determining the coefficients A_n and B_n for the waveform is a little harder. Here, we will make use of something known as the "orthogonality" of the sine and cosine functions over one period.[2] You can show from trigonometric identities that[3]

$$\int_0^{2\pi} \sin m\theta \cos n\theta d\theta = 0 \tag{C.4}$$

for all integral values of m and n.

Similarly, it can be shown that[4]

$$\int_0^{2\pi} \sin m\theta \sin n\theta d\theta = \int_0^{2\pi} \cos m\theta \cos n\theta d\theta = \begin{cases} \pi, & \text{for } m = n. \\ 0, & \text{for } m \neq n. \end{cases} \tag{C.5}$$

where m and n are integers.

To evaluate the general coefficient A_m, multiply both sides of Eq. (C.2) by $\sin m\theta$ and integrate from 0 to 2π. All the terms in the Fourier series then vanish except for

$$A_m = \frac{1}{\pi} \int_0^{2\pi} V(\theta) \sin m\theta d\theta. \tag{C.6}$$

Similarly to get B_m, multiply Eq. (C.2) by $\cos m\theta$ obtaining

$$B_m = \frac{1}{\pi} \int_0^{2\pi} V(\theta) \cos m\theta d\theta. \tag{C.7}$$

Then, as shown in Chap. 2, we may rewrite $V(\theta)$ as

$$V(\theta)C_0 + \sum_{n=1}^{\infty} C_n \sin(n\theta + \varphi_n), \tag{C.8}$$

---

[2] The term arises because the sines and cosines can be regarded as projections of orthogonal "vectors" in a multi-dimensional space in the sense that their generalized dot- (or scalar-) products vanish.

[3] Since
$$2 \sin m\theta \cos n\theta = \sin(m+n)\theta + \sin(m-n)\theta$$
and m and n are integers, the integrals of the sine functions on the right side of this identity over one period must vanish.

[4] E.g., consider the identity
$$2 \sin m\theta \sin n\theta = \cos(m+n)\theta \cos(m-n)\theta.$$
If $m \neq n$, integrals over one period on the right side both vanish. But if $m = n$, the second term on the right side of the trig identity is $\cos(0) = 1$. Hence, for $m = n$,
$$\int_0^{2\pi} \sin m\theta \sin n\theta d\theta = \frac{1}{2} \int_0^{2\pi} d\theta = \pi$$

where the constants are given by

$$C_n = \sqrt{A_n^2 + B_n^2} \text{ for } n \geq 1 \text{ and } \varphi_n = \arctan(B_n/A_n). \tag{C.9}$$

The procedure for determining the Fourier coefficients then consists of computing the integrals in Eqs. (C.3), (C.6), and (C.7), which generally must be done numerically from digital samples of the waveform over one period. Then one uses Eq. (C.9) to determine the net harmonic amplitude and phase. There is one pitfall to avoid in computing the phase from the arctangent relation in Eq. (C.9). Computer programming languages and pocket calculators do not always have provision for automatically putting the angle whose tangent is (B/A) in the correct quadrant. If the computer only deals with the numerical result from evaluating $x = B/A$, it cannot tell what the quadrant should be from the statement $\phi = \text{ATN}(x)$. If A is positive, the answer for ϕ will be correct and the angle will fall in either the first or fourth quadrant. However, if A is negative, the answer will be wrong and should fall in the second or third quadrant. The problem is solved in BASIC by including statements such as

$$\text{ATN}(B/A)$$
$$\text{IF } A < 0 \text{ THEN } \phi = \phi + \pi. \tag{C.10}$$

The FORTRAN programming language has a single arctangent statement of the type ATAN2(B/A) that takes care of the problem. If any phase angle were off by π, reconstruction of the waveform from the Fourier coefficients would give the wrong shape. Of course, the spectrum of harmonic amplitudes wouldn't be altered. Reconstruction of the original waveform is, of course, obtained by substituting the coefficients C_n and ϕ back into Eq. (C.8). If the reconstructed waveform does not match the original, you have either made a mistake or have not taken enough harmonics into account. (See Fig. 2.12.)

C.1   Energy Distribution in a Fourier Series

It is generally true in classical physics (i.e., non quantum-mechanical problems) that the energy in a vibrating system is proportional to the square of the amplitude of the disturbance. That fact is shown in detail for the harmonic oscillator in Appendix A. What we are interested in here is the average value of the square of the amplitude over the fundamental period of the Fourier series. The relative values are usually of main concern rather than the absolute values. Suppose we have a waveform of the type

$$y(t) = \sum_{n=1}^{\infty} C_n \sin(n\omega t + \varphi_n). \tag{C.11}$$

Then, the average energy over one cycle is[5]

$$Energy \propto \frac{1}{T} \int_0^{2\pi} y(t)^2 dt = \sum_{n=1}^{\infty} C_n^2 \tag{C.12}$$

where $T = 1/f = 2\pi/\omega$ is the fundamental period of the waveform. Hence, the relative energy in the nth harmonic is simply proportional to C_n^2.

C.2   Program for Discrete Fourier Analysis[6]

The following program is written in BASIC, but could easily be rewritten in FORTRAN or another language. In what follows, REM means a remark that is ignored by the computer. Explanatory comments after the apostrophe (') are also ignored when the computer runs.

REM Dimension statements[7]

         DIM A(50), B(50), C(50), P(50)   ' *For Fourier coefficients*
         DIM V(255)   ' *To store the Data*
         Pi = 4***ATN**(1.0)   ' *Pi = Greek \pi*

REM Enter data **for** one period = P[8]
READ P

---

[5]To prove the statement, note that $y^2(t)$ could be written
    $y(t)^2 = \sum_{n=1}^{\infty} C_n \sin(n\omega t + \varphi_n) \sum_{m=1}^{\infty} C_m \sin(m\omega t + \varphi_m)$.
Hence,
    Energy $\propto \frac{1}{T} \sum_{n=1}^{\infty} \sum_{m=1}^{\infty} C_n C_m \int_0^T \sin(n\omega t + \varphi_n) \sin(m\omega t + \varphi_m) dt$
where the terms for $m \neq n$ vanish because of orthogonality of the sine functions.

[6]Note: In order to save running time, it really pays to compile the program.

[7]Arrays are just subscripted variables; e.g., A(N) is the same as the mathematical variable AN. Dimension(DIM) statements tell the computer to set aside the maximum number of elements that could appear in the array; e.g., DIM A(50) means that there could be 50 elements (harmonics) in array A(N).

[8]The READ statement assigns the next previously unread number within DATA statements in the program to a given variable. E.g., here P = 109, which is the total number of points in the following DATA statements and V(1) = 3. The loop structure, FOR I = 1 TO P permits using the same statement, READ V(I), within the loop over and over for P times. The loop is closed by the NEXT I statement. Note that the waveform is periodic. (The first and last points are both equal

```
FOR   I=1  TO  P
READ  V( I )  NEXT  I
```

REM Mode–Locked Garden Hose (normalized to ±1000)
REM P= Number of points in Period = the first datum
```
DATA 109
DATA 3,20,43,68,111,176
DATA 273,426,634,861,1000,989
DATA 864,705,563,452,358,273
DATA 199,136,85,45,17,−14
DATA −40,−63,−85,−105,−125,−129
DATA −151,−162,−165,−165,−165,−165
DATA −170,−165,−170,−170,−170,−173
DATA −173,−170,−162,−159,−156,−153
DATA −151,−142,−139,−136,−131,−128
DATA −128,−128,−125,−119,−114,−108
DATA −102,−94,−91,−91,−91,−91
DATA −91,−91,−91,−91,−91,−91
DATA −85,−82,−82,−80,−74,−74
DATA −74,−80,−80,−80,−80,−80
DATA −82,−82,−82,−85,−88,−91
DATA −91,−91,−91,−91,−91,−85
DATA −82,−80,−74,−71,−68,−65
DATA −60,−57,−45,−37,−28,−14
DATA    3
```

REM Fundamental frequency in Hz
```
DATA   307.692     ' Not used in this program
```

REM Display of the Waveform[9]
```
Xmax=550          'Xmax and Ymax are the maximum number
Ymax=500          'of pixels available and vary with the
computer
```
REM Plot Horizontal Axis
```
        Pen=0
    X=0
        Y=Ymax/2
    GOSUBS PLOT
```

---

to 3.) The DATA were all taken with an A- to-D (Analog-to-Digital) converter with 1 part in 1000 resolution, hence are in the range 1000.

[9]Xmax and Ymax are the maximum number of points ("pixels") used on the plotting device. Pen is a "Pen- Lift" variable. (Imagine plotting on an old-fashioned xy-recorder.) See Subroutine PLOT: for its meaning.

```
        Pen=1
    X=Xmax
    GOSUB PLOT
REM Plot Waveform
        Pen=0
        FOR I=1 TO P
                Y=0.5*Ymax*(1+V(I)/1000)
        X=(I-1)*Xmax/(P-1)
        GOSUB PLOT
                Pen=1
    NEXT I
```

REM Chosing the maximum number of harmonics[10]

```
PRINT ''Nmax''
INPUT  Nmax
```
REM Harmonic Analysis[11]

```
A0=2*Pi/(P-1)     ' A0 is a constant used in the loop
    on N.
FOR N=0 TO Nmax   ' N=0 corresponds to the constant or
DC term
        A(N)=0
        B(N)=0
        FOR I=1 TO P
                A(N)=A(N)+V(I)*SIN(N*A0*(I-1))
                B(N)=B(N)+V(I)*COS(N*A0*(I-1))
    NEXT I
        A(N)=A(N)*2/(P-1)
        B(N)=B(N)*2/(P-1)
```

---

[10]Normally one could simply estimate the maximum number of harmonics (Nmax) from the number of wiggles in one period of the waveform and INPUT (enter) that number from the keyboard. The present case is tricky because the waveform is a sharp pulse without wiggles. One way is to make a guess (e.g., Nmax = 20) and see how closely the waveform reconstructed from the Fourier coefficients agrees with the original. Then increase (or decrease) Nmax as needed. See Fig. 2.12.

[11]This part of the program takes most of the running time. A(N) and B(N) correspond to the Fourier amplitudes given by the integrals in Eq. (C.6) and (C.7). They are done here numerically by a method equivalent to drawing straight lines between the successive points and adding up the areas (the "trapezoidal Rule"). Each variable A(N) and B(N) is initially set to zero and the increments repeatedly added within the loop FOR I = 1 TO P-1 in a series of the type $A(N)0.5(y_1 + y_2)\Delta x 0.5(y_2 + y_3)\Delta x + \ldots + 0.5(y_{P-1} + y_P)\Delta x = (y_1 + y_2 + y_3 + \ldots + y_{P-1})\Delta x + 0.5(y_P - y_1)\Delta x$ which simplifies because $y_P = y_1$ (i.e., the series is periodic.) Here, e.g., $y_i = V(I) * SIN(N * A0 * (I - 1))$ and $\Delta x = 1$. The net Fourier amplitude $C(N)$ and phase $P(N)$ are computed using Eqs. (C.6), (C.7), and (C.8).

```
  C(N)=SQR(A(N)*A(N)+B(N)*B(N))
      IF  N>0  THEN  P(N)=ATN(B(N)/A(N))    ' avoid DC term
      IF  A(N)<0  THEN  P(N)=P(N)+Pi        'arctangent
  problem
NEXT N
REM DC or constant term next
      C(0)=B(0)/2
REM Plot Histogram of the Coefficients
      REM Find Maximum Harmonic coefficient , Cmax
      Cmax=-1E30           'A negative number below the
smallest likely C(N)
      FOR N=1 TO Nmax
                 IF C(N)>Cmax THEN Cmax=C(N)
      NEXT N
      FOR N=1 TO Nmax
                 X=(N-1)*Xmax/Nmax
      Y=0
                 Pen=0
      GOSUB PLOT
                 Y=C(N)*Ymax/Cmax
      Pen=1
                 GOSUB PLOT
      NEXT N
```

C.3   Additional Waveforms

Additional musical instrument waveforms are presented below in the same format
as that given above for the garden hose for those who might like to study them
and for use elsewhere in this book. These data were taken in Davies Auditorium
at Yale, using a high-quality Sennheiser MKH104 omnidirectional microphone
with uniform response (± 1 dB) over the range from 50 Hz to 20 kHz and the
Hewlett-Packard equipment shown in Fig. 2.9. The 10-bit A-to-D converter used
had a dynamic range of about 60 dB. The instruments were played without vibrato
by professional musicians to whom the author is greatly indebted. The brass
instruments were played by James Undercoffler; the violins, by Syoko Aki; the
oboe, heckelphone, krummhorn and rohr schalmei by James Ryan; and flute and
piccolo by Leone Buyse. I am indebted to the late William Liddell for the use of
his krummhorns, to Richard Rephann for the loan of the historic brass instruments
from the Yale Instrument Collection, and to Robert Sheldon of the Smithsonian
Collection in Washington, DC for playing the serpent. The heckelphone was loaned
by the Yale Concert Band and the rohr schalmei was borrowed from the author's
personal pipe organ.

F-Cornet

DATA 107
DATA −4,44,114,163,212,249
DATA 280,301,311,319,321,321
DATA 311,290,259,223,166,111
DATA 57,5,−47,−83,−106,−122
DATA −124,−119,−98,−83,−67,−52
DATA −41,−47,−62,−83,−109,−150
DATA −192,−231,−259,−285,−301,−303
DATA −293,−269,−228,−166,−96,−13
DATA 78,174,275,370,477,578
DATA 681,777,870,948,995,1000
DATA 974,886,764,617,433,244
DATA 67,−101,−238,−345,−412,−448
DATA −461,−448,−415,−383,−337,−301
DATA −272,−259,−251,−269,−301,−332
DATA −365,−399,−425,−446,−453,−456
DATA −446,−425,−399,−383,−368,−355
DATA −347,−337,−329,−301,−262,−241
DATA −210,−166,−117,−65,−4
REM Cornet Frequency in Hz
DATA 316.075

Piccolo Trumpet

DATA 168
DATA −5,97,195,280,367,450
DATA 530,610,680,755,807,857
DATA 898,935,967,982,997,1000

Piccolo Trumpet DATA −755,−720,−673,−623,−563,−500
DATA   168   DATA −433,−355,−272,−193,−110,−5
DATA   −5,97,195,280,367,450   **REM** Piccolo Trumpet
Frequency in Hz
DATA   530,610,680,755,807,857 DATA   597.672
DATA   898,935,967,982,997,1000
DATA 995,987,965,940,910,875
DATA 835,797,742,695,637,585
DATA 527,475,417,357,302,247
DATA 187,130,85,37,−10,−50
DATA −80,−110,−140,−160,−180,−193
DATA −203,−200,−200,−193,−170,−155
DATA −130,−93,−55,−12,35,77
DATA 125,177,227,277,325,367
DATA 408,450,490,517,545,567

```
DATA  580,587,590,590,580,560
DATA  540,515,485,450,408,367
DATA  327,285,240,190,147,97
DATA  55,15,−35,−65,−100,−133
DATA  −155,−180,−200,−215,−230,−233
DATA  −245,−245,−242,−245,−242,−240
DATA  −238,−235,−238,−235,−235,−242
DATA  −253,−260,−265,−282,−300,−320
DATA  −343,−363,−390,−413,−440,−463
DATA  −490,−513,−535,−557,−583,−603
DATA  −623,−643,−660,−680,−690,−715
DATA  −733,−753,−773,−793,−810,−830
DATA  −845,−857,−873,−875,−875,−882
DATA  −873,−870,−857,−835,−810,−785
DATA  −755,−720,−673,−623,−563,−500
DATA  −433,−355,−272,−193,−110,−5
```
REM Piccolo Trumpet Frequency in Hz
```
DATA  597.672
```

REM Visual reconstruction from the Fourier coefficients[12]

```
Pen=0
FOR  I=1  TO  P
        V( I )=C( 0 )            ' constant or DC term
      FOR  N=1  TO  Nmax
                V( I )=V( I )+C(N)*SIN(N*A0*I+P(N))
      NEXT  N
      Y=0.5*Ymax*(1+V( I )/1000)
    X=( I −1)*Xmax/( P−1)
    GOSUB  PLOT
        Pen=1
NEXT  I
```

REM Dummy Input to retain display
```
INPUT  Q$
```

---

[12]Here, we reconstruct the Fourier Series to compare with the original waveform initially stored in V(I). (The contents of the array V(I) are changed when the program runs.) Note that without the constant or DC term, the reconstructed waveform may be offset vertically from the original waveform. An effective DC offset can sometimes arise from the presence of low-frequency background noise that is unrelated to the waveform of interest. Several languages also permit constructing the sound of the waveforms, in which case the fundamental frequencies contained in the last DATA statement is important.

END

SUBROUTINE[13]

PLOT

RETURN

REM Pen=0 plots points at (X,Ymax–Y), with origin at (0,Ymax)
REM Pen=1 Draws line from last point to coordinates (X, Ymax–Y)
IF Pen=0 **THEN** POINT (X, Ymax–Y)
IF Pen=1 **THEN** LINE (X, Ymax–Y)

Note: Some programming languages permit constructing the sound of the waveform. The fundamental frequencies for each waveform are included in Hz in the DATA statements for that purpose.

French Horn (Loud)

```
DATA    151
DATA    21,112,188,242,273,318
DATA    356,387,399,394,385,371
DATA    349,309,264,200,133,76
DATA    38,21,17,17,26,52
DATA    67,95,121,147,159,171
DATA    176,166,157,143,140,138
DATA    128,124,124,124,133,143
DATA    147,147,143,140,135,133
DATA    119,105,93,76,62,26
DATA    -12,-52,-95,-138,-181,-214
DATA    -252,-283,-299,-314,-309,-295
DATA    -268,-249,-240,-233,-214,-195
DATA    -166,-128,-76,-10,64,121
DATA    176,216,245,273,285,295
DATA    292,292,302,314,333,340
```

---

[13] The instructions POINT (X,Y) and LINE (X,Y) are intended to plot a point at coordinates X, Y whenPen = 0 and a line from the previously plotted point to coordinates X, Y if Pen = 1, respectively, assuming the origin is in the lower left-hand corner. Plotting commands of this type are contained in most languages, but with varying designations. Most languages now put the origin in the upper left hand corner of the plotting device, in which case one needs to replace Y by Ymax-Y as has been done in the subroutine here. One may also need scaling factors dependent on the maximum number of pixels available in the x and y directions. The convention on naming subroutines varies from one version of BASIC to another. In some versions the term PLOT has a different, specific meaning built into the operating system.

```
DATA    349,347,347,349,375,428
DATA    508,575,627,656,672,679
DATA    684,684,689,684,670,641
DATA    596,551,508,451,380,304
DATA    214,128,45,-43,-164,-314
DATA    -461,-589,-670,-703,-717,-743
DATA    -791,-846,-895,-931,-960,-988
DATA    -1000,-971,-924,-886,-855,-838
DATA    -836,-841,-838,-829,-774,-698
DATA    -613,-527,-437,-335,-219,-95
DATA    21
```
REM French Horn Frequency in Hz DATA   222.488

French Horn (Soft)

```
DATA    151
DATA    4,46,87,121,154,183
DATA    212,231,262,272,291,297
DATA    306,297,297,291,283,268
DATA    264,247,237,229,216,206
DATA    200,191,179,166,158,141
DATA    131,116,121,112,106,104
DATA    100,98,96,83,87,79
DATA    73,71,62,56,48,40
DATA    40,42,46,54,58,62
DATA    64,71,67,73,71,71
DATA    75,81,96,104,114,133
DATA    156,166,183,212,225,247
DATA    272,297,322,349,372,399
DATA    422,441,462,484,499,516
DATA    528,541,555,563,563,572
DATA    570,563,555,545,528,511
DATA    486,464,424,391,356,308
DATA    262,218,164,116,56,-10
DATA    -71,-135,-210,-279,-345,-407
DATA    -482,-541,-607,-661,-719,-773
DATA    -825,-867,-911,-940,-965,-979
DATA    -996,-1000,-994,-983,-973,-950
DATA    -925,-892,-857,-811,-767,-717
DATA    -659,-603,-536,-468,-403,-343
DATA    -285,-233,-175,-127,-71,-25
DATA    4
```
REM Soft Fr. Horn Frequency in Hz DATA   223.294

Ophicleide

```
DATA    196
```

```
DATA   -4,-117,-186,-235,-283,-324
DATA   -348,-393,-421,-421,-377,-324
DATA   -247,-170,-73,40,126,170
DATA   158,142,113,45,-32,-81
DATA   -138,-202,-259,-287,-312,-340
DATA   -360,-364,-364,-377,-364,-348
DATA   -316,-308,-287,-275,-291,-324
DATA   -348,-377,-393,-413,-405,-389
DATA   -364,-340,-328,-324,-324,-312
DATA   -291,-259,-259,-259,-267,-287
DATA   -291,-283,-279,-247,-170,-73
DATA   49,158,259,372,462,526
DATA   579,607,611,607,575,518
DATA   462,417,401,385,360,332
DATA   308,300,267,235,178,121
DATA   77,32,24,32,61,81
DATA   89,93,109,97,97,105
DATA   134,194,251,255,186,77
DATA   -24,-85,-101,-97,-85,-101
DATA   -101,-40,40,105,142,142
DATA   154,186,223,271,316,381
DATA   421,421,405,364,328,304
DATA   316,364,421,478,518,514
DATA   486,413,304,142,-89,-348
DATA   -632,-911,-1219,-1510,-1741
DATA   -1915,-1960,-1911,-1781,-
DATA   1603,-1381,-1154
DATA   -927,-700,-462,-219,28,235
DATA      429,595,749,879,955,996
DATA      1000,1000,972,964,935,866
DATA      777,688,599,514,437,364
DATA      304,255,211,170,158,142
DATA      126,121,130,126,130,142
DATA      142,126,89,-4
REM Ophicleide  Frequency  in  Hz
DATA      85.103
```

Serpent

```
DATA 203
DATA   -5,53,113,167,207,237
DATA 233,223,207,170,140,90
DATA 47,-7,-47,-87,-107,-113
DATA   -107,-80,-43,10,73,133
DATA 187,223,250,263,263,250
DATA 223,200,157,130,73,20
```

DATA −40,−97,−147,−200,−240,−267
DATA −287,−280,−267,−240,−203,−180
DATA −163,−133,−123,−127,−120,−123
DATA −127,−123,−130,−133,−127,−133
DATA −133,−113,−107,−73,−30,37
DATA 103,160,210,237,240,250
DATA 237,207,170,127,90,73
DATA 60,50,40,23,20,−13
DATA −47,−80,−120,−150,−187,−193
DATA −193,−173,−150,−133,−107,−80
DATA −57,−50,−43,0,60,143
DATA 237,323,440,533,620,687
DATA 740,747,640,460,253,20
DATA −353,−737,−1053,−1333,−1497
DATA −1527,−1440,−1273,−993,−667
DATA −327,−20,210,340,340,260
DATA 160,80,−7,−20,50,147,277
DATA 420,553,667,773,853,943
DATA 997, 1000,970,927,810
DATA 673,500,273,40,−193
DATA −387,−507,−560
DATA −540,−447,−337,−217,−93,23
DATA 83,130,147,143,140,140
DATA 147,153,167,173,153,117
DATA 77,23,−40,−60,−100,−133
DATA −173,−203,−240,−267,−280,−280
DATA −300,−287,−273,−257,−257,−230
DATA −227,−213,−207,−213,−200,−193
DATA −163,−150,−110,−70,−5
REM Serpent Frequency in Hz
DATA 61.702

Andreas Amati Violin (G String)

DATA 171
DATA 14,−153,−330,−507,−641,−732,−
DATA 813,−900,−967,−1000
DATA −967,−880,−737,−603,−498,−450,−
DATA 426,−431,−440,−426
DATA −402,−354,−306,−330,−335,−306,−
DATA 273,−177,53,211,364
DATA 474,545,593,608,641,641,603
DATA 574,493,431,416,431
DATA 512,603,651,651,584,459,316,
DATA 187,86,0,−100,−177
DATA −234,−273,−239,−177,−100

DATA 33,134,201,220,249,278
DATA 335,392,469,507,469,383,297
DATA 191,124,72,91,124
DATA 148,225,354,550,775,895
DATA 880,689,397,144,−10,29
DATA 144,239,311,278,206,163
DATA 234,335,383,440,392
DATA 258,33,−62,−24,57,115
DATA 96,29,−124,−234,−249
DATA −182,−38,57,110,115,124
DATA 144,230,340,440,474
DATA 402,301,196,67,−62,−201
DATA −344,−445,−464,−388
DATA −273,−144,−38,129,244,392
DATA 507,531,512,435
DATA 378,354,344,378,435,536
DATA 646,737,766,699,627
DATA 545,512,478,474,459,402
DATA 378,354,335,340,344
DATA 340,301,249,167,77,14
REM Amati Violin Frequency in Hz
DATA 196.71

Ordinary Violin (G String)

DATA 130
DATA 76,389,686,854,945,980
DATA 872,600,290,27
DATA −75,7,168,325,456,566
DATA 662,748,796,761
DATA 644,491,336,226,137,111
DATA   157,254,332,403
DATA   496,606,690,690,597,440
DATA   254,77,−102,−215
DATA −296,−354,−407,−460,−513,−566
DATA −637,−746,−878,−976
DATA −1000,−912,−743,−584,−454
DATA −341,−186,49,270,440
DATA   531,518,458,389,350,305
DATA   237,166,115,53
DATA −18,−62,−82,−111,−146,−146
DATA −122,−58,−2,31
DATA   53,53,27,0,7,49
DATA   124,212,281,369
DATA 467,549,566,451,283,53
DATA −104,−170,−146,−86

DATA −53,−33,31,119,181,243
DATA 310,296,184,−31
DATA −283,−442,−515,−546,−540
DATA −558,−580,−617,−642,−666
DATA −639,−577,−584,−692,−808,−878
DATA −816,−569,−199,76
REM Violin Frequency in Hz
DATA 194.41

Oboe (Lauré)

DATA 194
DATA 2,−181,−397,−581,−717,−809
DATA −839,−829,−777,−697,−596,−486
DATA −370,−261,−159,−79,−10,45
DATA 82,107,127,149,161,174
DATA 179,179,174,169,156,149
DATA 134,124,104,84,55,15
DATA −35,−79,−112,−144,−159,−164
DATA −161,−159,−154,−151,−159,−161
DATA −164,−169,−159,−144,−114,−84
DATA −50,−10,37,84,127,179
DATA 218,246,261,253,223,176
DATA 127,69,10,−40,−77,−122
DATA −159,−189,−223,−251,−270,−283
DATA −288,−280,−273,−263,−258,−258
DATA −258,−258,−258,−258,−251,−243
DATA −241,−231,−218,−208,−191,−169
DATA −154,−139,−124,−122,−122,−129
DATA −141,−154,−159,−159,−156,−151
DATA −139,−129,−122,−124,−129,−134
DATA −151,−169,−179,−199,−208,−231
DATA −243,−253,−258,−258,−251,−238
DATA −213,−191,−161,−139,−104,−74
DATA −40,−5,27,55,77,89
DATA 97,97,94,87,74,60
DATA 37,15,−20,−60,−104,−156
DATA −213,−280,−347,−409,−474,−526
DATA −566,−588,−586,−558,−499,−422
DATA −330,−221,−109,15,136,256
DATA 377,501,620,739,831,913
DATA 960,993,1000,995,983,963
DATA 938,913,883,856,821,789
DATA 752,715,663,600,496,355
DATA 176,2
REM Oboe Frequency in Hz

DATA 259.581

Heckelphone

DATA 130
DATA −8,142,250,308,300,283
DATA 262,196,58,33,42,42
DATA 29,−8,−54,−104
DATA −225,−321,−417,−475,−488
DATA −417,−242,0
DATA 217,225,346,608,867,1000
DATA 942,642,25,−50,−108,−267
DATA −475,−650,−667,−471,−217,−204
DATA −167,−25,171,317,417,425
DATA 329,308,229,92,−50,−142
DATA −208,−267, −358,−425,−458,−454
DATA −425,−388,−304,−221,13,142
DATA 262,346,392,425,442,425
DATA 238,158,75,−33,−125,−242
DATA −367,−450, −388,−317,−233
DATA −142,−75,13,112,192,225
DATA 258,292,292,250,175,46,−104
DATA −254,−292,−321,−317,−292,−25
DATA −188,−121,50,196,396,612
DATA 746,725,608,479, 262,167
DATA 50,−117,−283,−400,−417,−350
DATA −67,108,258,367,442,475
DATA 500,458,−17,−321,−600,−817
DATA −892,−808,−671,−538,−192,−8
REM Heckelphone Frequency in Hz
DATA 129.777

Krummhorn

DATA 174
DATA −59,−91,−157,−216,−100,−78
DATA −162,−176,333,1000,623,−130
DATA −412,−355,−100,422,853,850
DATA 490,−100,−623,−760,−760,−603
DATA −150,255,157,−169,−277,−196
DATA −250,−309,−150,113,206,279
DATA 377,331,108,−74,−51,15
DATA 145,353,510,507,333,164
DATA 93,152,243,341,431,439
DATA 419,373,257,230,186,59
DATA 64,135,201,206,86,−78
DATA −194,−196,−142,−83,47,103

DATA 39,−32,−91,−167,−199,−120
DATA 0,34,27,54,83,118
DATA 125,137,196,230,174,115
DATA 83,20,−120,−316,−471,−412
DATA −277,−81,152,211,47,−137
DATA −267,−353,−319,−176,−69,−25
DATA 10,47,93,100,132,142
DATA 142,162,157,105,27,−44
DATA −78,−110,−118,−110,−78,−44
DATA −32,−51,−74,−98,−103,−137
DATA −137,−110,−78,−88,−113,−123
DATA −78,−34,−5,−20,−88,−196
DATA −353,−451,−453,−512,−534,−578
DATA −534,−446,−397,−324,−306,−358
DATA −306,−245,−140,−61,69,96
DATA 162,152,142,172,213,289
DATA 294,294,309,225,145,−59
REM Krummhorn Frequency in Hz DATA 192.52

Rohrschalmei (Laukuft)

DATA   129
DATA   4,−94,−132,−186,−172,−224
DATA −256,−202,−146,−116,−68,−20
DATA −36,−64,−64,−104,−168,−136
DATA −98,−16,82,126,212,158
DATA   158,152,20,−80,−188,−282
DATA −206,720,692,444,1000,648
DATA   694,230,68,−228,−938,−826
DATA −726,−830,−474,−168,126,322
DATA   376,562,408,280,316,206
DATA   40,16,−44,−96,−178,−168
DATA −170,−194,−242,−402,−568,−488
DATA −584,−754,−604,−410,−370,−32
DATA   144,476,542,678,848,816
DATA   638,272,146,−176,−320,−420
DATA −482,−450,−400,−202,−43,−132
DATA −40,56,−16,0,−36,88
DATA   172,246,414,404,366,430
DATA   364,276,88,−48,−124,−282
DATA −272, −238,−314,−306,−282,−306
DATA −266,−136,−124,12,52,128
DATA   212,166,262,220,192,164
DATA   152,78,4
REM Rohrschalmei Frequency in Hz
DATA 259.674

Flute

DATA 192
DATA 18,−54,−75,−101,−134,−165
DATA −178,−247,−281,−291,−335,−376
DATA −428,−425,−464,−438,−446,−425
DATA −402,−412,−330,−299,−250,−160
DATA −82,21,93,180,216,317
DATA 325,358,320,289,255,222
DATA 144,119,111,85,5,26
DATA 5,−34,−52,−57,−106,−67
DATA −64,−75,−34,−34,8,−5
DATA 36,59,101,111,90,119
DATA 119,98,39,57,15,−26
DATA 3,15,0,8,8,−41
DATA −62,−77,−106,−126,−183,−232
DATA −260,−240,−253,−247,−191,−157
DATA −103,−67,−26,28,108,131
DATA 250,340,430,503,595,634
DATA 660,698,675,619,526,448
DATA 369,340,206,124,88,−26
DATA −82,−178,−250,−291,−302,−371
DATA −407,−410,−436,−464,−521,−559
DATA −582,−652,−765,−881,−920,−972
DATA −912,−910,−838,−678,−577,−479
DATA −353,−247,−134,−21,62,121
DATA 155,216,204,193,222,201
DATA 204,216,237,291,376,428
DATA 585,799,840,892,959,1000
DATA 907,747,588,430,222,−46
DATA −289,−376,−518,−598,−655,−655
DATA −621,−526,−459,−327,−188,−26
DATA 90,170,268,361,454,454
DATA 443,485,495,423,407,348
DATA 291,227,144,77,62,18
REM Flute Frequency in Hz
DATA 262.05

Piccolo

DATA 170
DATA −11,34,101,100,218,276
DATA 333,391,451,508,556,590
DATA 631,643,679,695,706,707
DATA 706,695,667,652,631,604
DATA 590,542,525,487,460,422

```
DATA    388,362,341,326,307,295
DATA    276,269,247,240,247,237
DATA    225,218,192,189,168,151
DATA    134,106,91,74,53,36
DATA    36,14,-10,-29,-34,-62
DATA    -77,-86.-115,-115,-125,-129
DATA    -139,-146,-149,-146,-153,-146
DATA    -146,-137,-137,-127,-127,-118
DATA    -108,-89,-72,-55,-38,-10
DATA    24,48,94,125,170,206
DATA    247,285,317,353,388,408
DATA    436,458,460,484,494,477
DATA    482,468,458,451,429,412
DATA    388,369,345,331,321,293
DATA    266,249,216,199,173,161
DATA    115,86,46,22,-14,-70
DATA    -120,-182,-245,-317,-386,-480
DATA    -556,-616,-686,-751,-815,-859
DATA    -897,-935,-959,-981,-993,-990
DATA    -990,-1000,-993,-993,-964,-954
DATA    -959,-930,-914,-887,-887,-856
DATA    -811,-763,-731,-671,-631,-568
DATA    -499,-444,-367,-317,-261,-175
DATA    -118,  -62,-11
REM Piccolo Frequency in Hz
DATA 593.877
```

C.4   Program for Discrete Fourier Analysis

The following program is written in MATLAB using much of the same code from the original BASIC program by Bennett.

C.4.1   Contents

- Load waveform data for one period
- Variables
- Compute the Fourier Coefficients
- Plot the Waveform
- Plot Histogram of the Coefficients
- Reconstruct the Waveform from the Fourier Coefficients

- Plot the Waveform

```
clear all;
```

C.4.2 Load waveform data for one period

Call load_waveforms.m

```
load_waveforms;
```

C.4.3 Variables

```
% Choose the maximum number of harmonics
Nmax=20;

% Select which waveform to analyze
waveform = 'gardenhose';

switch waveform
    case 'gardenhose'
        V=gardenhose;
        strLegend=sprintf('Garden Hose\nf_{n=1}
        =307.692 Hz');
    case 'fcoronet'
        V=fcoronet;
        strLegend=sprintf('F-Cornet\nf_{n=1}
        =316.075 Hz');
    case 'piccolotrumpet'
        V=piccolotrumpet;
        strLegend=sprintf('Piccolo Trumpet\nf_{n=1}
        =597.672 Hz');
    case 'frenchhornloud'
        V=frenchhornloud;
        strLegend=sprintf('French Horn (Loud)\nf_{n=1}
        =222.488 Hz');
    case 'frenchhornsoft'
        V=frenchhornsoft;
        strLegend=sprintf('French Horn (Soft)\nf_{n=1}
        =223.294 Hz');
    case 'ophicleide'
        V=ophicleide;
```

```
            strLegend=sprintf('Ophicleide\nf_{n=1}
            =85.103 Hz');
    case 'serpent'
        V=serpent;
        strLegend=sprintf('Serpent\nf_{n=1}
        =61.702 Hz');
    case 'amativiolin'
        V=amativiolin;
        strLegend=sprintf('A.Amati Violin (G)\nf_{n=1}
        =196.71 Hz');
    case 'violin'
        V=violin;
        strLegend=sprintf('Ordinary Violin
        (G)\nf_{n=1}=194.41 Hz');
    case 'oboelaure'
        V=oboelaure;
        strLegend=sprintf('Oboe (Laure)\nf_{n=1}
        =259.581 Hz');
    case 'heckelphone'
        V=heckelphone;
        strLegend=sprintf('Heckelphone\nf_{n=1}
        =129.777 Hz');
    case 'kurmhorn'
        V=krummhorn;
        strLegend=sprintf('Krummhorn\nf_{n=1}
        =192.52 Hz');
    case 'rohrschalmei'
        V=rohrschalmei;
        strLegend=sprintf('Rohrschalmei (Laukuft)\
        nf_{n=1}
        =259.674 Hz');
    case 'flute'
        V=flute;
        strLegend=sprintf('Flute\nf_{n=1}=262.05 Hz');
    case 'piccolo'
        V=piccolo;
        strLegend=sprintf('Piccolo\nf_{n=1}
        =593.877 Hz');
end
```

C.4.4   Compute the Fourier Coefficients

Call fouriercoef_trapezoid.m

```
[A0 A B0 B C0 C PHI]=fouriercoef_trapezoid(V,Nmax);
```

C.4.5   Plot the Waveform

```
figure;
plot(V./max(abs(V)),'Color','k','LineWidth',3)
line([0 length(V)],[0 0],'Color','k','LineWidth',1)
set(gca,'LineWidth',2,'FontSize',14)
set(gca,'YLim',[-1.1 1.1],'YTick',[0]);
set(gca,'XLim',[0 length(V)],'XTick',[]);
xlabel('Time','FontSize',14);
ylabel('Microphone Signal','FontSize',14);
pbaspect([2 1 1])
```

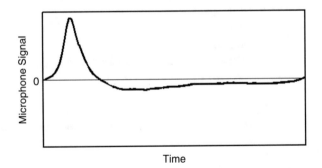

C.4.6   Plot Histogram of the Coefficients

```
figure;
stem([1:Nmax],C./max(C),'Color','k','LineWidth',3,
 'Marker','none')
set(gca,'YLim',[0 1],'YTickLabel',[]);
set(gca,'XLim',[0 Nmax],'XTick',[0:2:Nmax]);
set(gca,'LineWidth',2,'FontSize',14)
xlabel('Harmonic Number, (n)','FontSize',14);
```

```
ylabel('Fourier Coefficients, C(n)','FontSize',14);
pbaspect([2 1 1])
text(12,.8,strLegend,'FontSize',14,
  'HorizontalAlignment','left');
```

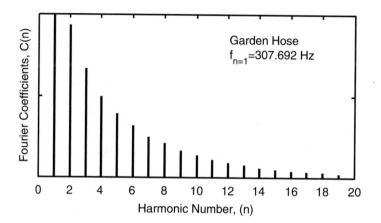

C.4.7   Reconstruct the Waveform from the Fourier Coefficients

```
P=length(V);
for I=1:P
    V2(I)=C0;
    for N=1:Nmax
        V2(I)=V2(I)+C(N)*sin(N*A0*(I-1)+PHI(N));
    end
end
```

C.4.8   Plot the Waveform

```
figure; hold on;
plot(V./max(abs(V)),'Color','k','LineWidth',3)
plot(V2./max(abs(V2)),'--','Color','b','LineWidth',3)
line([0 length(V)],[0 0],'Color','k','LineWidth',1)
set(gca,'LineWidth',2,'FontSize',14)
set(gca,'YLim',[-1.1 1.1],'YTick',[0]);
set(gca,'XLim',[0 length(V)],'XTick',[]);
```

```
xlabel('Time','FontSize',14);
ylabel('Microphone Signal','FontSize',14);
clear strLegend;
strLegend{1}='Recorded Waveform';
strLegend{2}=sprintf('Reconstructed Waveform,
 Nmax = %u', Nmax);
legend(strLegend);
pbaspect([2 1 1])
```

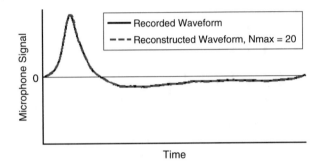

C.5   Additional Waveforms

Additional musical instrument waveforms are presented below in the same format as that given above for the garden hose for those who might like to study them and for use elsewhere in this book. These data were taken in Davies Auditorium at Yale, using a high-quality Sennheiser MKH104 omnidirectional microphone with uniform response (±1 dB) over the range from 50 Hz to 20 kHz and the Hewlett-Packard equipment shown in Fig. 2.9. The 10-bit A-to-D converter used had a dynamic range of about 60 dB. The instruments were played without vibrato by professional musicians to whom the author is greatly indebted. The brass instruments were played by James Undercoffler; the violins, by Syoko Aki; the oboe, heckelphone, krummhorn and rohr schalmei by James Ryan; and flute and piccolo by Leone Buyse. I am indebted to the late William Liddell for the use of his krummhorns, to Richard Rephann for the loan of the historic brass instruments from the Yale Instrument Collection, and to Robert Sheldon of the Smithsonian Collection in Washington, DC for playing the serpent. The heckelphone was loaned by the Yale Concert Band and the rohr schalmei was borrowed from the author's personal pipe organ.

C.5.1   *Contents*

- Mode-Locked Garden Hose (normalized to +/- 1000)
- F-Cornet
- Piccolo Trumpet
- French Horn (Loud)
- French Horn (Soft)
- Ophicleide
- Serpent
- Andreas Amati Violin (G String)
- Ordinary Violin (G String)
- Oboe (Laure)
- Heckelphone
- Krummhorn
- Rohrschalmei (Laukuft)
- Flute
- Piccolo

C.5.2   *Mode-Locked Garden Hose (normalized to +/- 1000)*

```
%DATA 109 % P = Number of points in Period = the first
datum gardenhose=[...
3,20,43,68,111,176,...
273,426,634,861,1000,989,...
864,705,563,452,358,273,...
199,136,85,45,17,-14,...
-40,-63,-85,-105,-125,-129,...
-151,-162,-165,-165,-165,-165,...
-170,-165,-170,-170,-170,-173,...
-173,-170,-162,-159,-156,-153,...
-151,-142,-139,-136,-131,-128,...
-128,-128,-125,-119,-114,-108,...
-102,-94,-91,-91,-91,-91,...
-91,-91,-91,-91,-91,-91,...
-85,-82,-82,-80,-74,-74,...
-74,-80,-80,-80,-80,-80,...
-82,-82,-82,-85,-88,-91,...
-91,-91,-91,-91,-91,-85,...
-82,-80,-74,-71,-68,-65,...
-60,-57,-45,-37,-28,-14,...
3];
% Fundamental frequency in Hz
%DATA  307.692
```

C.5.3   F-Cornet

```
%DATA    107
fcoronet=[...
-4,44,114,163,212,249,...
280,301,311,319,321,321,...
311,290,259,223,166,111,...
57,5,-47,-83,-106,-122,...
-124,-119,-98,-83,-67,-52,...
-41,-47,-62,-83,-109,-150,...
-192,-231,-259,-285,-301,-303,...
-293,-269,-228,-166,-96,-13,...
78,174,275,370,477,578,...
681,777,870,948,995,1000,...
974,886,764,617,433,244,...
67,-101,-238,-345,-412,-448,...
-461,-448,-415,-383,-337,-301,...
-272,-259,-251,-269,-301,-332,...
-365,-399,-425,-446,-453,-456,...
-446,-425,-399,-383,-368,-355,...
-347,-337,-329,-301,-262,-241,...
-210,-166,-117,-65,-4];
% Cornet Frequency in Hz
%DATA    316.075
```

C.5.4   Piccolo Trumpet

```
%DATA    168
piccolotrumpet=[...
-5,97,195,280,367,450,...
530,610,680,755,807,857,...
898,935,967,982,997,1000,...
995,987,965,940,910,875,...
835,797,742,695,637,585,...
527,475,417,357,302,247,...
187,130,85,37,-10,-50,...
-80,-110,-140,-160,-180,-193,...
-203,-200,-200,-193,-170,-155,...
-130,-93,-55,-12,35,77,...
125,177,227,277,325,367,...
408,450,490,517,545,567,...
580,587,590,590,580,560,...
```

```
540,515,485,450,408,367,...
327,285,240,190,147,97,...
55,15,-35,-65,-100,-133,...
-155,-180,-200,-215,-230,-233,...
-245,-245,-242,-245,-242,-240,...
-238,-235,-238,-235,-235,-242,...
-253,-260,-265,-282,-300,-320,...
-343,-363,-390,-413,-440,-463,...
-490,-513,-535,-557,-583,-603,...
-623,-643,-660,-680,-690,-715,...
-733,-753,-773,-793,-810,-830,...
-845,-857,-873,-875,-875,-882,...
-873,-870,-857,-835,-810,-785,...
-755,-720,-673,-623,-563,-500,...
-433,-355,-272,-193,-110,-5];
% Piccolo Trumpet Frequency in Hz
%DATA    597.672
```

C.5.5   French Horn (Loud)

```
%DATA    151
frenchhornloud=[...
21,112,188,242,273,318,...
356,387,399,394,385,371,...
349,309,264,200,133,76,...
38,21,17,17,26,52,...
67,95,121,147,159,171,...
176,166,157,143,140,138,...
128,124,124,124,133,143,...
147,147,143,140,135,133,...
119,105,93,76,62,26,...
-12,-52,-95,-138,-181,-214,...
-252,-283,-299,-314,-309,-295,...
-268,-249,-240,-233,-214,-195,...
-166,-128,-76,-10,64,121,...
176,216,245,273,285,295,...
292,292,302,314,333,340,...
349,347,347,349,375,428,...
508,575,627,656,672,679,...
684,684,689,684,670,641,...
596,551,508,451,380,304,...
214,128,45,-43,-164,-314,...
-461,-589,-670,-703,-717,-743,...
```

```
-791,-846,-895,-931,-960,-988,...
-1000,-971,-924,-886,-855,-838,...
-836,-841,-838,-829,-774,-698,...
-613,-527,-437,-335,-219,-95,...
21];
% French Horn Frequency in Hz
%DATA   222.488
```

C.5.6   French Horn (Soft)

```
%DATA   151
frenchhornsoft=[...
4,46,87,121,154,183,...
212,231,262,272,291,297,...
306,297,297,291,283,268,...
264,247,237,229,216,206,...
200,191,179,166,158,141,...
131,116,121,112,106,104,...
100,98,96,83,87,79,...
73,71,62,56,48,40,...
40,42,46,54,58,62,...
64,71,67,73,71,71,...
75,81,96,104,114,133,...
156,166,183,212,225,247,...
272,297,322,349,372,399,...
422,441,462,484,499,516,...
528,541,555,563,563,572,...
570,563,555,545,528,511,...
486,464,424,391,356,308,...
262,218,164,116,56,-10,...
-71,-135,-210,-279,-345,-407,...
-482,-541,-607,-661,-719,-773,...
-825,-867,-911,-940,-965,-979,...
-996,-1000,-994,-983,-973,-950,...
-925,-892,-857,-811,-767,-717,...
-659,-603,-536,-468,-403,-343,...
-285,-233,-175,-127,-71,-25,...
4];
% Soft Fr. Horn Frequency in Hz
%DATA   223.294
```

C.5.7   *Ophicleide*

```
%DATA   196
ophicleide=[...
-4,-117,-186,-235,-283,-324,...
-348,-393,-421,-421,-377,-324,...
-247,-170,-73,40,126,170,...
158,142,113,45,-32,-81,...
-138,-202,-259,-287,-312,-340,...
-360,-364,-364,-377,-364,-348,...
-316,-308,-287,-275,-291,-324,...
-348,-377,-393,-413,-405,-389,...
-364,-340,-328,-324,-324,-312,...
-291,-259,-259,-259,-267,-287,...
-291,-283,-279,-247,-170,-73,...
49,158,259,372,462,526,...
579,607,611,607,575,518,...
462,417,401,385,360,332,...
308,300,267,235,178,121,...
77,32,24,32,61,81,...
89,93,109,97,97,105,...
134,194,251,255,186,77,...
-24,-85,-101,-97,-85,-101,...
-101,-40,40,105,142,142,...
154,186,223,271,316,381,...
421,421,405,364,328,304,...
316,364,421,478,518,514,...
486,413,304,142,-89,-348,...
-632,-911,-1219,-1510,-1741,...
-1915,-1960,-1911,-1781,-1603,-1381,-1154,...
-927,-700,-462,-219,28,235,...
429,595,749,879,955,996,...
1000,1000,972,964,935,866,...
777,688,599,514,437,364,...
304,255,211,170,158,142,...
126,121,130,126,130,142,...
142,126,89,-4];
% Ophicleide Frequency in Hz
%DATA   85.103
```

C.5.8 Serpent

```
%DATA   203
serpent=[...
-5,53,113,167,207,237,...
233,223,207,170,140,90,...
47,-7,-47,-87,-107,-113,...
-107,-80,-43,10,73,133,...
187,223,250,263,263,250,...
223,200,157,130,73,20,...
-40,-97,-147,-200,-240,-267,...
-287,-280,-267,-240,-203,-180,...
-163,-133,-123,-127,-120,-123,...
-127,-123,-130,-133,-127,-133,...
-133,-113,-107,-73,-30,37,...
103,160,210,237,240,250,...
237,207,170,127,90,73,...
60,50,40,23,20,-13,...
-47,-80,-120,-150,-187,-193,...
-193,-173,-150,-133,-107,-80,...
-57,-50,-43,0,60,143,...
237,323,440,533,620,687,...
740,747,640,460,253,20,...
-353,-737,-1053,-1333,-1497,...
-1527,-1440,-1273,-993,-667,...
-327,-20,210,340,340,260,...
160,80,-7,-20,50,147,277,...
420,553,667,773,853,943,...
997, 1000,970,927,810,...
673,500,273,40,-193,...
-387,-507,-560,...
-540,-447,-337,-217,-93,23,...
83,130,147,143,140,140,...
147,153,167,173,153,117,...
77,23,-40,-60,-100,-133,...
-173,-203,-240,-267,-280,-280,...
-300,-287,-273,-257,-257,-230,...
-227,-213,-207,-213,-200,-193,...
-163,-150,-110,-70,-5];
% Serpent Frequency in Hz
%DATA   61.702
```

C.5.9 Andreas Amati Violin (G String)

```
%DATA   171
amativioling=[...
14,-153,-330,-507,-641,-732,-813,-900,-967,-1000,...
-967,-880,-737,-603,-498,-450,-426,-431,-440,-426,...
-402,-354,-306,-330,-335,-306,-273,-177,53,211,364,...
474,545,593,608,641,641,603,...
574,493,431,416,431,...
512,603,651,651,584,459,316,...
187,86,0,-100,-177,...
-234,-273,-239,-177,-100,...
33,134,201,220,249,278,...
335,392,469,507,469,383,297,...
191,124,72,91,124,...
148,225,354,550,775,895,...
880,689,397,144,-10,29,...
144,239,311,278,206,163,...
234,335,383,440,392,...
258,33,-62,-24,57,115,...
96,29,-124,-234,-249,...
-182,-38,57,110,115,124,...
144,230,340,440,474,...
402,301,196,67,-62,-201,...
-344,-445,-464,-388,...
-273,-144,-38,129,244,392,...
507,531,512,435,...
378,354,344,378,435,536,...
646,737,766,699,627,...
545,512,478,474,459,402,...
378,354,335,340,344,...
340,301,249,167,77,14];
% Amati Violin Frequency in Hz
%DATA   196.71
```

C.5.10 Ordinary Violin (G String)

```
%DATA   130
violin=[...
76,389,686,854,945,980,...
872,600,290,27,...
-75,7,168,325,456,566,...
```

```
662,748,796,761,...
644,491,336,226,137,111,...
157,254,332,403,...
496,606,690,690,597,440,...
254,77,-102,-215,...
-296,-354,-407,-460,-513,-566,...
-637,-746,-878,-976,...
-1000,-912,-743,-584,-454,...
-341,-186,49,270,440,...
531,518,458,389,350,305,...
237,166,115,53,...
-18,-62,-82,-111,-146,-146,...
-122,-58,-2,31,...
53,53,27,0,7,49,...
124,212,281,369,...
467,549,566,451,283,53,...
-104,-170,-146,-86,...
-53,-33,31,119,181,243,...
310,296,184,-31,...
-283,-442,-515,-546,-540,...
-558,-580,-617,-642,-666,...
-639,-577,-584,-692,-808,-878,...
-816,-569,-199,76];
% Violin Frequency in Hz
%DATA 194.41
```

C.5.11   Oboe (Laure)

```
%DATA    194
oboelaure=[...
2,-181,-397,-581,-717,-809,...
-839,-829,-777,-697,-596,-486,...
-370,-261,-159,-79,-10,45,...
82,107,127,149,161,174,...
179,179,174,169,156,149,...
134,124,104,84,55,15,...
-35,-79,-112,-144,-159,-164,...
-161,-159,-154,-151,-159,-161,...
-164,-169,-159,-144,-114,-84,...
 -50,-10,37,84,127,179,...
 218,246,261,253,223,176,...
 127,69,10,-40,-77,-122,...
-159,-189,-223,-251,-270,-283,...
```

```
-288,-280,-273,-263,-258,-258,...
-258,-258,-258,-258,-251,-243,...
-241,-231,-218,-208,-191,-169,...
-154,-139,-124,-122,-122,-129,...
-141,-154,-159,-159,-156,-151,...
-139,-129,-122,-124,-129,-134,...
-151,-169,-179,-199,-208,-231,...
-243,-253,-258,-258,-251,-238,...
-213,-191,-161,-139,-104,-74,...
-40,-5,27,55,77,89,...
97,97,94,87,74,60,...
37,15,-20,-60,-104,-156,...
-213,-280,-347,-409,-474,-526,...
-566,-588,-586,-558,-499,-422,...
-330,-221,-109,15,136,256,...
377,501,620,739,831,913,...
960,993,1000,995,983,963,...
938,913,883,856,821,789,...
752,715,663,600,496,355,...
176,2];
% Oboe Frequency in Hz
%DATA 259.581
```

C.5.12   Heckelphone

```
%DATA   130 [JLR probably a typo, should be 138]
%[JLR] Note: There are two typos in the manuscript
which are fixed here heckelphone=[...
-8,142,250,308,300,283,...
262,196,58,33,42,42,...
29,-8,-54,-104,...
-225,-321,-417,-475,-488,...
-417,-242,0,...
217,225,346,608,867,1000,...
942,642,25,-50,-108,-267,...
-475,-650,-667,-471,-217,-204,...
-167,-25,171,317,417,425,...
329,308,229,92,-50,-142,...
-208,-267, -358,-425,-458,-454,...
-425,-388,-304,-221,13,142,...
262,346,392,425,442,425,...
238,158,75,-33,-125,-242,...
-367,-450,-388,-317,-233,...
```

```
-142,-75,13,112,192,225,...
258,292,292,250,175,46,-104,...
-254,-292,-321,-317,-292,-250,...   %[JLR] last point
 changed to -250
-188,-121,50,196,396,612,...
746,725,608,479,262,167,...
50,-117,-283,-400,-417,-350,...
-67,108,258,367,442,475,...
500,458,-17,-321,-600,-817,...
-892,-808,-671,-538,-192,-8];
% Heckelphone Frequency in Hz
%DATA 129.777
```

C.5.13   Krummhorn

```
%DATA 174
krummhorn=[...
-59,-91,-157,-216,-100,-78,...
-162,-176,333,1000,623,-130,...
-412,-355,-100,422,853,850,...
490,-100,-623,-760,-760,-603,...
-150,255,157,-169,-277,-196,...
-250,-309,-150,113,206,279,...
377,331,108,-74,-51,15,...
145,353,510,507,333,164,...
93,152,243,341,431,439,...
419,373,257,230,186,59,...
64,135,201,206,86,-78,...
-194,-196,-142,-83,47,103,...
39,-32,-91,-167,-199,-120,...
0,34,27,54,83,118,...
125,137,196,230,174,115,...
83,20,-120,-316,-471,-412,...
-277,-81,152,211,47,-137,...
-267,-353,-319,-176,-69,-25,...
10,47,93,100,132,142,...
142,162,157,105,27,-44,...
-78,-110,-118,-110,-78,-44,...
-32,-51,-74,-98,-103,-137,...
-137,-110,-78,-88,-113,-123,...
-78,-34,-5,-20,-88,-196,...
-353,-451,-453,-512,-534,-578,...
-534,-446,-397,-324,-306,-358,...
```

```
-306,-245,-140,-61,69,96,...
162,152,142,172,213,289,...
294,294,309,225,145,-59];
% Krummhorn Frequency in Hz
%DATA 192.52
```

C.5.14   Rohrschalmei (Laukuft)

```
%DATA   129
rohrschalmei=[...
4,-94,-132,-186,-172,-224,...
-256,-202,-146,-116,-68,-20,...
-36,-64,-64,-104,-168,-136,...
-98,-16,82,126,212,158,...
158,152,20,-80,-188,-282,...
-206,720,692,444,1000,648,...
694,230,68,-228,-938,-826,...
-726,-830,-474,-168,126,322,...
376,562,408,280,316,206,...
40,16,-44,-96,-178,-168,...
-170,-194,-242,-402,-568,-488,...
-584,-754,-604,-410,-370,-32,...
144,476,542,678,848,816,...
638,272,146,-176,-320,-420,...
-482,-450,-400,-202,-43,-132,...
-40,56,-16,0,-36,88,...
172,246,414,404,366,430,...
364,276,88,-48,-124,-282,...
-272, -238,-314,-306,-282,-306,...
-266,-136,-124,12,52,128,...
212,166,262,220,192,164,...
152,78,4];
% Rohrschalmei Frequency in Hz
%DATA 259.674
```

C.5.15   Flute

```
%DATA 192
flute=[...
18,-54,-75,-101,-134,-165,...
-178,-247,-281,-291,-335,-376,...
```

```
    -428,-425,-464,-438,-446,-425,...
    -402,-412,-330,-299,-250,-160,...
    -82,21,93,180,216,317,...
    325,358,320,289,255,222,...
    144,119,111,85,5,26,...
    5,-34,-52,-57,-106,-67,...
    -64,-75,-34,-34,8,-5,...
    36,59,101,111,90,119,...
    119,98,39,57,15,-26,...
    3,15,0,8,8,-41,...
    -62,-77,-106,-126,-183,-232,...
    -260,-240,-253,-247,-191,-157,...
    -103,-67,-26,28,108,131,...
    250,340,430,503,595,634,...
    660,698,675,619,526,448,...
    369,340,206,124,88,-26,...
    -82,-178,-250,-291,-302,-371,...
    -407,-410,-436,-464,-521,-559,...
    -582,-652,-765,-881,-920,-972,...
    -912,-910,-838,-678,-577,-479,...
    -353,-247,-134,-21,62,121,...
    155,216,204,193,222,201,...
    204,216,237,291,376,428,...
    585,799,840,892,959,1000,...
    907,747,588,430,222,-46,...
    -289,-376,-518,-598,-655,-655,...
    -621,-526,-459,-327,-188,-26,...
    90,170,268,361,454,454,...
    443,485,495,423,407,348,...
    291,227,144,77,62,18];
    % Flute Frequency in Hz
    %DATA 262.05
```

C.5.16   Piccolo

```
    %DATA   170
    piccolo=[...
    -11,34,101,100,218,276,...
    333,391,451,508,556,590,...
    631,643,679,695,706,707,...
    706,695,667,652,631,604,...
    590,542,525,487,460,422,...
    388,362,341,326,307,295,...
```

```
276,269,247,240,247,237,...
225,218,192,189,168,151,...
134,106,91,74,53,36,...
36,14,-10,-29,-34,-62,...
-77,-86.-115,-115,-125,-129,...
-139,-146,-149,-146,-153,-146,...
-146,-137,-137,-127,-127,-118,...
-108,-89,-72,-55,-38,-10,...
24,48,94,125,170,206,...
247,285,317,353,388,408,...
436,458,460,484,494,477,...
482,468,458,451,429,412,...
388,369,345,331,321,293,...
266,249,216,199,173,161,...
115,86,46,22,-14,-70,...
-120,-182,-245,-317,-386,-480,...
-556,-616,-686,-751,-815,-859,...
-897,-935,-959,-981,-993,-990,...
-990,-1000,-993,-993,-964,-954,...
-959,-930,-914,-887,-887,-856,...
-811,-763,-731,-671,-631,-568,...
-499,-444,-367,-317,-261,-175,...
-118, -62,-11];
% Piccolo Frequency in Hz
%DATA 593.877
```

C.6 Harmonic Analysis

This part of the program takes most of the running time. A(N) and B(N) correspond to the Fourier amplitudes given by the integrals in Eqs. (C.6) and (C.7). They are done here numerically by a method equivalent to drawing straight lines between the successive points and adding up the areas (the "trapezoidal Rule").

```
function [A0 A B0 B C0 C PHI]
  =fouriercoef_trapezoid(V,Nmax)

% The net Fourier amplitude C(N) and phase PHI(N)
  are computed using
% Eqs (C.6), (C.7), and (C.8).

% Check for input errors
error(nargchk(1, 2, nargin));
if (nargin < 2); Nmax = 20; end;
```

```
P=length(V);

A0=2*pi/(P-1); % A0 is a constant used in the loop on N.

for N=1:Nmax % FOR N=0 TO Nmax
% N=0 corresponds to the  constant or DC term,
 we will calculate it later

    A(N)=0;
    B(N)=0;

    for I=1:P % FOR I=1 TO P
        A(N)=A(N)+V(I)*sin(N*A0*(I-1));
        B(N)=B(N)+V(I)*cos(N*A0*(I-1));
    end; % NEXT I

    A(N)=A(N)*2/(P-1);
    B(N)=B(N)*2/(P-1);
    C(N)=sqrt(A(N)*A(N)+B(N)*B(N));

    %IF N>0 THEN P(N)=ATN(B(N)/A(N)) % avoid DC term
    PHI(N)=atan(B(N)/A(N));
    %IF A(N)<0 THEN P(N)=P(N)+Pi 'arctangent problem
    if (A(N) < 0); PHI(N)=PHI(N)+pi; end;

end; %NEXT N

% DC or constant term (A0 calculation above)
B0=sum(V)*2/(P-1);
C0=B0/2;
```

Published with MATLAB®7.12

Appendix D
The Well-Tempered Scale

The numbers here were computed to 0.06 ppm and rounded off to the nearest 0.01 Hz assuming the 1936 international convention that $A_4 = 440.0000$ Hz, using $2^{1/12} = 1.05946310$. Note as a check: $1.05946310^{12} = 2.00000012$

Frequency in Hz

A	27.50	55.00	110.00	220.00	440.00	880.00	1760.00	3520.00	7040.00	14080.00
A♯	29.14	58.27	116.54	233.08	466.16	932.33	1864.66	3729.31	7458.62	14917.25
B	30.87	61.74	123.47	246.94	493.88	987.77	1975.53	3951.07	7902.13	15804.27
C	32.70	65.41	130.81	261.63	523.25	1046.50	2093.00	4186.01	8372.02	16744.04
C♯	34.65	69.30	138.59	277.18	554.37	1108.73	2217.46	4434.92	8869.85	17739.69
D	36.71	73.42	146.83	293.66	587.33	1174.66	2349.32	4698.64	9397.28	18794.55
D♯	38.89	77.78	155.56	311.13	622.25	1244.51	2489.02	4978.03	9956.07	19912.13
E	41.20	82.41	164.81	329.63	659.26	1318.51	2637.02	5274.04	10548.08	21096.17
F	43.65	87.31	174.61	349.23	698.46	1396.91	2793.83	5587.65	11175.31	22350.61
F♯	46.25	92.50	185.00	369.99	739.99	1479.98	2959.96	5919.91	11839.83	23679.65
G	49.00	98.00	196.00	392.00	783.99	1567.98	3135.96	6271.93	12543.86	25087.72
G♯	51.91	103.83	207.65	415.30	830.61	1661.22	3322.44	6644.88	13289.75	26579.51

© Springer Nature Switzerland AG 2018
W. R. Bennett, Jr., *The Science of Musical Sound*,
https://doi.org/10.1007/978-3-319-92796-1

Solutions

Problems of Chap. 1

1.1 The oil would spread out very rapidly over the water, reducing its surface tension and thereby preventing wave motion at the short wavelengths characterized by the ripples.

1.2 (a) 3.11 m. (b) 615 THz.

1.3 Answers: The frequency is $233.08/(1.441 \times 10^{17}) = 1.62 \times 10^{-15}$ Hz. The period is 6.183×10^{14} s = 19.59 million years.

1.4 (a) 633 nm. (b) 43.6 ft.

1.5 (a) 3.8. (b) 2.7

1.6

From the original length of the string, one expects frequencies at $nc/2L$ where n = 1,2,3,.... But because of reflections from the kink you also get odd harmonics of $3c/8L$ and of $3c/4L$. Note that the string is free to move up and down at the kink, so that the frequencies on either side of the kink are determined by only one "hard" phase shift per round trip, and the resonances have the same form as those for a closed pipe.

1.7 (a) Multiples of 110 Hz. (Neglecting phase shift and time delays in the amplifier, the running wave phase shift would be $2\pi f_L/c$ per trip around the loop.) (b) It would oscillate at some other frequency if there were significant phase shift in the amplifier (e.g., from tone controls or a large gain variation with frequency.)

1.8 Answer: The pipe is closed at the top, initially by the flap valve, and then by the water flowing down the pipe. The fundamental frequency of the pipe starts out at $c/4L = 46$ Hz, with odd harmonics at 138, 230,... Hz. As the water goes down the pipe, the acoustic length shortens and the pitch for each harmonic goes up.

© Springer Nature Switzerland AG 2018
W. R. Bennett, Jr., *The Science of Musical Sound*,
https://doi.org/10.1007/978-3-319-92796-1

1.9 The main frequencies are 523 Hz, 622 Hz, 740 Hz, 932 Hz, 1570 Hz, 1865 Hz, 2217 Hz, and 2797 Hz. Starting at C above A=440 Hz, the notes are C, E♭, G♭, B♭, G, B♭, D♭, and F. (See figure for musical notation.) The first two intervals are minor thirds, giving the whistle its characteristic, mournful sound. The bottom half might be regarded as a half-diminished seventh. The second half is just the first chord raised by a twelfth. (The lower chord would resolve on a diminished seventh and then on a B♭ minor chord) (Fig. S.1).

1.10 Answers: Hard phase shifts occur at both the ceiling and the floor, hence one might expect the main resonances to be given approximately by $f_n = nc/2L$. Resonances do occur at those frequencies (multiples of 55 Hz.) But because the floor is in the focal plane of the concave ceiling, resonances also occur at $f_n = nc/8L$. (The running wave from a point source on the floor makes eight transits to the ceiling and back before it closes on itself.) Assuming the radius of curvature of the ceiling is 20 ft and twice the ceiling height, the main resonances would occur every 13.75 Hz throughout the audio band. (Speech is totally unintelligible, but the mumbling sound is impressive.)

1.11 One reason is that the open end of a closed pipe radiates as a monopole source (equally in all directions), whereas an open pipe acts like a dipole with the maximum radiated intensity going in the vertical direction (unless the pipe is turned 90°).

1.12 By listening to the full Doppler shift on the bells at grade crossings, he or she could determine the median pitch and the total fractional change in frequency on the Well Tempered Scale. From that, one could calculate the speed.

1.13 Answers: (a) 38.2 mph. (b) About one whole step.

1.14 The Doppler effect from sound reflected by the rotating blades creates a warbling effect.

1.15 (a) 676 Hz. (b) 786 Hz. (c) 738 Hz.

1.16 +715 Hz. (Note that the velocity of the image of the radar gun is twice the speed of the car.)

1.17 Answer: $c/4L = 1040/(4 \times 6.9) = 37.7$ Hz.

1.18 Answer: Since $f = 65.4$ Hz, $\lambda = c/f = 1087/65.4 = 16.6$ ft.

Fig. S.1 The steam engine whistle

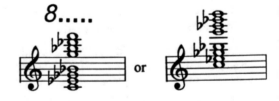

Problems of Chap. 2

2.1 $\Delta F \approx 2\pi \, \Delta t \approx 0.032 \, \text{Hz}$.

2.2 (a) To go through 50 Hz would require at least
$\Delta t \approx \frac{1}{2\pi \Delta F} = \frac{1}{2\pi 50} = 318$ ms.
Therefore, to go through 10,000 Hz, it would take a minimum of about 64 s, or (b) a scanning rate of 156 Hz/s.

2.3 See Eqs. (2.7) and (2.8)

2.4 10 dB

2.5 $20 Log_{10}(80 \times 5280/10) + 60 = 152 \, \text{dB}$

2.6 Serber was about 15.4 miles away. Los Alamos was about 250 miles from the test site. The blast would have been about 24 dB louder at Serber's position.

2.7 $20 Log_{10}[3000/1] + 60 \approx 129 \, \text{dB}$ (or about as loud as a rock band in a typical concert setting).

2.8 About 60 dB ($\approx 20 Log_{10} 1024$.)

2.9 About 96 dB ($\approx 20 Log_{10} 2^{16} = 320 Log \, 102$). But as shown by Bennett (1948), one gets an extra 3 dB by averaging the quantization errors in most systems.

2.10 Answer: The sound intensity per unit area would be attenuated by a factor of 36, which corresponds to a decrease of 16 dB. Hence, the sound level at the neighbor's house should be about 64 dB (something between normal conversation and shouting at 4 ft).

2.11 Answer: (a) $20 Log_{10}(4) \approx 12 \, \text{dB}$. (b) $10 Log_{10}(4) \approx 6 \, \text{dB}$.

2.12

Frequency (Hz)	31.5	63	125	250	500	1000	2000
Signal (dB)	65	78	88	80	70	67	50
Amplitudes ($\div 1000$) $\approx$	1.8	7.9	25	10	3.2	2.2	0.3

2.13

Frequency (Hz)	31.5	63	125	250	500	1000	2000	4000	8000
Signal (dB)	50	52.5	52.5	55	60	67.5	60	57.5	52.5
Amplitudes ($\div 100$) $\approx$	3.2	4.2	4.2	5.6	10	23	10	7.5	4.2

2.14

$$
\begin{array}{ccccccc}
\text{n} = & 1 & 3 & 5 & 7 & 9 & 11 \\
\text{dB} & 0 & -9.5 & -14 & -17 & -19 & -21
\end{array}
$$

2.15 A closed organ pipe.

2.16 The spectral components would have side bands at ± 6 Hz.

Problems of Chap. 3

3.1 If only one coil located at x_0 were used, nulls in the spectrum would occur at $\frac{n\pi x_0}{L} = m\pi$ where $m = 1, 2, 3, \ldots$ hence, at the harmonics $n = L/x_0 \approx 9, 18, 27, \ldots$

3.2 If the output voltages from the two coils in Fig. 3.10 were added, the result would be

$V(t) \propto \sum_{n=1}^{\infty} A_n \{\sin[n\pi(x_0 + a)/L] + \sin[n\pi(x_0 - a)/L)]\} f_n(t)$
$= 2 \sum_{n=1}^{\infty} A_n \sin\left(\frac{n\pi x_0}{L}\right) \cos\left(\frac{n\pi a}{L}\right) f_n(t)$ and the spectrum would have minima at $n = m\frac{L}{x_0}$ and $n = (2m - 1)\frac{L}{2a}$ where $m = 1, 2, 3 \ldots$ which for the dimensions given in the text would occur at $n \approx 9, 18, 26, 27, 36, 43$, etc.

3.3 The separation between the negative and positive peaks for the 56.8 Hz resonance should correspond to half the wavelength of the surface wave along the grain direction. Hence, for the long dimension assumed above, $\lambda/2 \approx 28.1$ inches and the surface wave velocity would be about

$c_{\text{Surface}} = \lambda f \approx (56.2)(56.8) = 3192$ in./s ≈ 266 ft/s, or about 1/4 of that for the velocity of sound in air.

3.4 The frequency is F = 261.6 Hz. The tension = 6.64 106 dynes = 14.9 lbs.

3.5 We want the density per unit length to be the same. Hence, the cross-sectional area of the wire should be $1.4/7.83 = 0.1788$ times that of the gut string, or the diameter of the steel wire should be 0.423×0.029 in. $= 0.012$ in.

Problems of Chap. 4

4.1 Answer: About 9.4 times for the Stein prellmechanik action, 9.2 times for the Broadwood action, and 9.5 times for the Streicher action.

4.2 Answer: Remove all the dampers for notes on that chord before dropping it. (Don't try that with a good piano!!! If the cast iron frame breaks, it is virtually impossible to repair it.)

4.3 Answer: The highest loss (symmetric) mode occurs when all three strings are in phase. Intermediate loss modes occur when one string is 180° out of phase with the other two. The lowest loss (odd-symmetric) mode occurs when the outer strings are 180° out of phase and the middle string is not vibrating.

4.4 Answer. The strength of the wire increases with its cross-sectional area, but so does its density per unit length. Hence, the ratio of the breaking tension to the density per unit length is about constant and that ratio determines the wave velocity, hence the pitch, at the breaking point.

4.5 The wavelength is $\lambda \approx 2 \times 72 = 144$ in.; hence, the wave velocity $c = \lambda f = 144 \times 87.3 = 12{,}571$ in./s ≈ 1048 ft/s.

4.6 The tension is about 9×10^7 dynes ≈ 202 pounds of force.

4.7 About 29 lbs, or about 6 lbs more than to bring the string up to normal pitch. (It takes about $1.26 \times 202 \approx 255$ lbs of force to break the string. The mechanical advantage of the tuning hammer is $10/0.1125 \approx 8.9$)

4.8 About 388 ft/s. (The maximum and minimum are separated by half a wavelength.)

4.9 For F2, about 3.001; for C9, about 6.8.

Problems of Chap. 5

5.1 Answer: E= $3 \times 440/2 = 660$ Hz, D= $2 \times 440/3 = 293.33$ Hz, G = $2 \times 293.33/3 = 195.55$ Hz. On the WTS, E = 659.25 Hz, D = 293.66 Hz, and G = 196.00 Hz.

5.2 For the A string, c = $2 \times 33 \times 440 = 29{,}040$ cm/s; for the G string, c = 12,906 cm/s.

5.3 $\mu = T/c^2 = 2.838 \times 10^{-5}$ g/cm^3.

5.4 195.55 Hz.

5.5 $391.1 - 330 = 61.1$ Hz, or B two octaves below middle C.

5.6 The difference frequency is an octave below the D, so the "Tartini Tone" doesn't stand out the way it would for a double third or fourth.

5.7 $1490/8 = 186$ strokes per second.

Problems of Chap. 6

6.1 $c/4L \approx 43\,\text{Hz}$.

6.2 Open pipe resonances would occur at about 92 Hz, 184 Hz, 276 Hz, and so on.

6.3 (a) The voice frequency is proportional to $1/\sqrt{M}$, where M is the effective molecular weight of the gas in the lungs. Therefore our SCUBA diver wants M to be about $29 \times (120/77.8)^2 \approx 69$.

(b) Let X = the fraction of oxygen in the tank and Y = the fraction of krypton in the tank. Then $X + Y = 1$ $32X + 84Y = 69$

Solving these two equations yields $X \approx 0.288$ for the fraction of oxygen, and $Y \approx 0.712$ for the fraction of krypton.

6.4 No. He'll only be able to reach 2725 Hz (about halfway between E and F.)

Problems of Chap. 7

7.1 Answer: The total pressure from a 2-inch difference in the height of water per unit area in a U-tube manometer is $5.08\,\text{g/cm}^2$. The air-pressure regulator area is $6\,\text{ft}^2 = 864\,\text{in.}^2 = 5574\,\text{cm}^2$. Hence the total weight needed is about $28.3\,\text{kg} \approx 62\,\text{lbs.}$

7.2 As a first approximation, the air velocity producing the edge tone in Fig. 7.13 is simply proportional to the air pressure. Hence the cut-up (L in the figure) should be reduced by $2/5 = 0.4$ on each pipe.

7.3 Up to some point, the pressure in the toe of the pipe would increase proportionally to the area of the toe-hole after the pipe has been turned on.

7.4 From Appendix D, the fundamental pitch should be 261.6 Hz. Since the Rohr Schalmei is an open pipe, the overall length should be $L = 1100/(2261.6) = 2.10\,\text{ft}$ $= 25.2\,\text{in.}$ Since the large cavity is tuned to the second harmonic of the pipe, both it and the copper tube should be $25.2/2 = 12.6\,\text{in.}$ long.

7.5 Answer: The orchestral oboe is made from a narrow-scale conical piece of wood. (Of course, the reed is placed at the vertex where the pressure is a maximum.)

7.6 Use two pipes for each note, chosen to produce successive harmonics of a 16-ft open pipe. The ear will then interpret the combination as having 16-ft pitch. An 8-ft closed pipe would provide the fundamental pitch and its odd harmonics. Adding an open 8-ft pipe would provide the needed even harmonics. If the two are on the same note channel, they will also tend to lock in phase.

7.7 The total length of the large diameter semi-closed pipe should be $L = 1100/(4440) = 0.625\,\text{ft} = 7.5\,\text{in.}$ The length of the short open pipe at the end should be tuned to the fifth harmonic of 440 Hz = 2200 Hz. Hence, the short length should be about $1100/(2 \times 2200) = 0.25\,\text{ft} = 3\,\text{in.}$

7.8 From Eq. (7.3),

$\frac{8.20}{5.80} = 12k$ or $k = 0.346/12 = 0.0289$ We want N such that $\log_e(2) = 0.0289N$ or $N = 24$ steps.

7.9 Answer: The 32-ft diapason is an open pipe with harmonics at 32-, 16-, and 8-ft pitch (the second, fourth, sixth, and eighth harmonic of the non-existent 64-ft pipe). The bordon is a closed pipe with a fundamental that is a fifth higher (i.e., third harmonic) than its nominal 64 foot pitch. Hence, the combination has frequencies at the second, third, fourth, sixth, and eighth harmonic of the phantom 64-ft pipe and the human ear will conclude that it actually has 64-ft pitch, even though the fundamental is completely missing. Assuming the velocity of sound is 1100 ft/sec, the frequencies would be:

64-ft open 8.59, 17.18, 25.7, 34.4, 42.9, 51.5, 60.1, 68.7, 77.3 Hz 32-ft Diapason— 17.18, —, 34.4, —, 51.5, —, 68.7, — Hz 32-ft Bordon Quint —, 25.7, —, —, —, —, 77.3 Hz Effective harmonic 2 3 4 6 8 9

References

Abele H (1905) The violin and its story. John Leng & Co, New York

Allen EL, Holien H (1973) A laminagraphic study of pulse (vocal fry) register phonationİ. Folia Phoniatr 25:241–250

Apel W (1972) Harvard dictionary of music. Harvard University Press, Cambridge

Appelman DR (1967) The science of vocal pedagogy. Indiana University Press, Bloomington

Arms WT (1959) History of leyden Massachusetts. The Enterprise and Journal, Orange; Colrain, MA Library, call number 9744 A73

Arnold GE (1973) Physiology of speech. The new encyclopedia Britannica, vol 17. Encyclopedia Britannica, London, pp 477–492

Atkins H, Newman A (1978) Beecham stories. Futura Publications, London

Attenborough D (1998) The life of birds. BBC Video Co, London

Atwood W (1997) A bridge collection examined. Catgut Acoust Soc J 3(4):38–43

Audsley GA (1905) The art of organ building. Dodd, Mead, and Co, New York; Reprinted by Dover, New York in 1965 in two volumes. Emphasizes the history of organ building, vol 1 and Emphasizes the constructional details, vol 2

Backus J (1969) The acoustical foundations of music. WW Norton, New York

Baken RJ (1998) An overview of laryngeal functions for voice production. In: Sataloff RT (ed) Vocal health and pedagogy. Singular Publishing Group, San Diego, pp 27–45

Ballot B (1845) Akustische Versuche auf der Niederländischen Eisenbahn, nebst gelentlichen Bemerkungen zur Theorie des Hrn. Prof. Dopperİ. Poggendorff's Ann (also known as Ann Phys Chem, Leipzig) 66(11):321–351

Bartoli C (2002) Interview on "All things considered". National Public Radio, Washington, D.C.; Wednesday, September 25

Beare C (1980a) Dominique Peccatte. In: Sadie S (ed) The new groves dictionary of music and musicians, vol 14. MacMillan, London, p 323

Beare, C (1980b) Eugène Sartory.İ In: Sadie S (ed) The new groves dictionary of music and musicians, vol 16. MacMillan, London, p 512

Benade AH (1975) The Wolf tone on violin family instruments. Catgut Acoust Soc Newslett 24:21–23; Reprinted in Hutchins and Benade (1997) 1:117–120. This material was also reproduced in Benade, 1976, pp 567–575

Benade, AH (1976) Fundamentals of musical acoustics. Oxford University Press, New York

Bennett WR (1948) Spectra of quantized signals. Bell Syst Tech J 27:446–472

Bennett WR (1983) Secret telephony as a historical example of spread-spectrum communication. IEEE Trans Commun COM-31:98–104

© Springer Nature Switzerland AG 2018

W. R. Bennett, Jr., *The Science of Musical Sound*,

https://doi.org/10.1007/978-3-319-92796-1

Bennett WR Jr (1962) Gaseous optical masers. In: Heavens OS Applied optics supplement no.1 on optical masers. Optical Society of America, Washington, D.C.

Bennett WR Jr (1976) Scientific and engineering problem-solving with the computer. Prentice Hall, Englewood Cliffs

Bennett WR Jr (1977) The physics of gas lasers. Gordon and Breach, New York

Bennett WR Jr (1990) Heart sounds and murmurs. Lecture with digitally recorded illustrations in compact disc format. MCG International Inc., New Haven; CD No. DIDX 009454 (December)

Bennett WR Jr, Bennett J (1990) Dynamic spectral phonocardiograph. U.S. Patent No. 4,967,760 (November 6)

Berlyn GP, Richardson AD (2001) Wood: its properties in relation to its use in turning. In: Wood turning in North America since 1930. Amilcare Pizzi, Milan, pp 152–160

Blot E (2001) Liuteria Italiana IV (1800–1950), Eric Bloc editioni, Cremona

Boalch D (1974) Makers of harpsichords and clavichords. Oxford University Press, London

Bouase H (1929) Tuyaux et Resonators - Introduction a l'Etude des Instruments da Vent. Libraire Delagrave, Paris

Bouasse H, Fouché M (1929) Instruments à vent, par H. Bouasse avec la collaboration expérimentale de M. Fouché. Delagrave, Paris

Boyden DD, Schwarz B (1980) Sadie S (ed) Violin in the new groves dictionary of music and musicians, vol 20. MacMillan, London, pp 819–855

Braunwald E (1984) Editor of heart disease. W.B. Saunders Company, Philadelphia

Brigham EO, Murrow RE (1967) The fast Fourier transform. IEEE Spectr 4:63–70

Brookes K (2002) Craft masters, Tempo (Oct/Nov/Dec), vol 6. Temple University Radio, Philadelphia, pp 4–11

Broos B (1995) Un celebre Peijntre nommé Verme[e]rİ. In: Johannes V National Gallery of Art, Washington, DC, pp 47–65

Browne, MW (1996) Computer recreates call of dinosaur sound organ. The New York times, New York, p C7; Tuesday, March 12

Bucur V (1988) Wood structural, anistropy estimated by acoustic invariants.İ IAWA Bull 9(1):67–74; Also, see reproduction in Hutchins and Benade (1997) 2:795–802

Bucur V (2006) Acoustics of wood. Springer, Berlin

Campbell GA (1911) Cisoidal oscillations. Trans Am Inst Electr Eng 30:873–909

Campbell GA, Foster RM (1920) Maximum output networks for telephone substations and repeater circuits. Trans Am Inst Electr Eng 39:231–280

Cazden J (1993) Vocal acoustics in electronic musician (February) pp 97–100

Cencié ME (1990) Boy soprano, in Schubert, Der Hirt auf dem Felsen, with Roman Ortner (piano), and Erwin Monschein (clarinet), in Wiener Sängerknabe Philips CD Recording No. 426307-2; Band 7

Chladni EFF (1809) Traite d'Acoustique. Courcier, Paris

Conan-Doyle SA (1981) The adventure of the cardboard box. In: The original illustrated Sherlock Holmes; Reprinted from The Strand Magazine of 1892 to 1893. Castle Books, Seacaucus, p 207

Cooke P (1980) Pygmy music. In: Sadie S (ed) The new groves dictionary of music and musicians, 92 vol 15. MacMillan, London, pp 482–483 and "Yodel", In: Sadie S (ed) The new groves dictionary of music and musicians, 92, vol 20, MacMillan, London, p 574

Cooley JW, Tukey JW (1965) An algorithm for the machine calculation of complex Fourier series, vol 19. Math Comput, pp 297–301

Crandall IB (1926) Theory of vibrating systems. Van Nostrand, New York

Cremer L (1981) Physik der Geige. S. Hirzel Verlag, Stuttgart; Translated into English by John S. Allen as The physics of the violin. MIT Press, Cambridge

Cuesta C, Valette C (1988) Evolution temporelle de la Vibration des Cordes de Clavecin. Acustica 66:37–45

Davis PJ, Fletcher NH (eds) (1996) Vocal fold physiology. Singular Publishing Group, San Diego

Dick Heinrich Company (2003–2005) Quality products for musical instruments. Dick GMBH, Germany

Diegert CF, Williamson TE (1998) A digital acoustic model of the lambeosaurine hadrosaur Parasaurolophus tubicen. J Vertebr Paleontol 18:38A

Dolge A (1911) Pianos and their makers: a comprehensive history of the development of the piano, new edn, 1972. Dover, New York

Dom Bédos de Celles F (1766–1768) L'art du facteur d'orgues, vol I text, vol II plates. Delatour, Paris

Donnington R (1980) Vibrato. In: Sadie S (ed) The new groves dictionary of music and musicians, vol 19. MacMillan, London, pp 697–698

Drower MS (1974) Thebes (Egypt).í In: The new encyclopedia Britannica, vol 18. Encyclopedia Britannica, London, p 264

Dudley H (1939) Remaking speech. J Acoust Soc Am 11:169–177

Dudley H (1940) The carrier nature of speech. Bell Syst Tech J 19:495–515

Dudely H (1955) Fundamentals of speech synthesis. J Audio Eng Soc 3:170–185

Dudley H, Tarnoczy TH (1950) The speaking machine of Wolfgang von Kempelen. J Acoust Soc Am 22:151–166

Dudley H, Riesz RR, Watkins SSA (1939) A synthetic speaker. J Franklin Inst 227:739–764

Elder D (1982) Pianists at play—interviews, master lessons, and technical regimes. The Instrumentalist Company Evanston, Illinois

Eulenstein K (1892) Eulensteins musikalische Laufbahn (edited by his daughter, Fanny Roodenfels, Stuttgart). Stadtarchiv, Heilbronn

Faber T (2005) Stradivari's genius. Random House, New York

Farga F (1950) Violins & violinists, translated by Egon Larsen and Bruno Raikin. Barrie & Rockliff, The Cresset Press, London

Farinelli (1994) Film biography of Carlo Broschi, directed by Gerard Corbiau (collaboratively produced in France, Belgium and Italy for Sony Pictures Classics)

Farnsworth BW (1940) High speed motion pictures of the human vocal tract. Bell Lab Rec 18(7):203–208

Feynman RP, Robert BL, Matthew LS (1963) The Feynman lectures on physics. Addison-Wesley, Reading, MA

Firth I (1976/77a) The nature of the tap tone in stringed instruments. Acustica 36(1):36–41

Firth I (1976/77b) A method for adjusting the pitch of top and back plates of the cello. Acustica 36(4):307–312

Flanagan JL (1972) Speech analysis synthesis and perception. Springer, Heidelberg

Forte M (2002) Chopin played by Madeleine Forte on the Erard Piano Paris, 1881—Yale collection of musical instruments, Roméo Records No. 7214

Fostle DW (1993) Henry Zeigler Steinway, a grand traditioní, interview in audio magazine (January), pp 52–58

Fourier BJv (1822) La Theorie analytique de la chaleur (Paris). See Freeman's translation. Cambridge University Press, Cambridge, 1878

Friedlander FG (1953) On the oscillations of the bowed string. Proc Camb Philos Soc 49(3):516

Gaillard F (1939) Bach six suites for Violoncello Solo.í Schirmer's library of musical classics, vol 1565; Preface

Gill D (ed) (1981) The book of the piano. Phaidon, Oxford

Giordano N (1998a) Mechanical impedance of a piano soundboard. J Acoust Soc Am 103:2128–2133

Giordano N (1998b) Sound production by a vibrating piano soundboard: experiment. J Acoust Soc Am 104:1648–1653

Giordano A, Dilworth J (2004) A fitting conclusion (Paganini's" Cannon" violin). STRAD 115, no. 1374: 1046-+

Giordano N, Winans JP II (1999) Plucked strings and the harpsichord. J Sound Vib 224:456–473

Giordano N, Winans II JP (2000) Piano hammers and their force compression characteristics: does a power law make sense? J Acoust Soc Am 107:2248–2255

Giordano N, Millis JP (2001) Hysteretic behavior of piano hammers. In: Proceedings of the international symposium on musical acoustics, Perugia

Gleason JB (1997) The development of language. Allyn and Bacon, Boston

Gold J (1995) Paganini: virtuoso, collector and dealer. J Violin Soc Am XIV(1):67–87

Gough CE (1981) The theory of string resonances on musical instruments. Acustica 49:124–141

Gray H (1918) Anatomy of the human body. Lea and Febiger, Philadelphia

Gray H, Warwick R, Williams PL (1980) Gray's anatomy. Saunders, Philadelphia

Grissino-Mayer H, Burckle L (2003) Stradivari violins, tree rings and the maunder minimum: a hypothesis. Dendrochronologia 21(1):41–45

Grissino-Mayer HD, Sheppard PR, Cleaveland MK (2004) A Dendroarchaeological re-examination of the 'Messiah' violin and other instruments attributed to Antonio Stradivari. J Archaeol Sci 31:167–174

Gruberova E (1988) Soprano, in Opera collection. Mozart, Die Zauber Flöte, Querschnite, Nikolaus Harnoncourt, CD 063-13810-9, Teldec Classics

Guettler K (2002) On playing 'harmonics' (flageolet tones). Catgut Acoust Soc J 4(5):5–7

Guhr K (1831) Uber Paganinis Kunst die Violin zu Spielen (Main, 1831). Translated from the original German by Sabilla Novello in 1915, revised by C. Egerton Lowe. Novello & Co., London; Also see the translation by Gold J (1982) Paganini's art of violin playing. Teresa Parker Associates, San Francisco

Gyütö M (1986) The Gyütö monks: Tibetan tantric choir WD-2001. 360Publishing, ASCAP and Windham Hill Records, Stanford

Hannings L, Chin Y (2006) J Violin Soc XX:146–162

Hanon CL (1900) The virtuoso pianist in sixty exercises for the piano translated from the French by Dr. Theodore Baker. G. Schirmer, Milwaukee

Hansell KK (1980) Sadie S (ed) Manzuoli, Giovanni̇ in the new groves dictionary of music and musicians, vol 11. MacMillan, London, p 638

Hanson RJ, Schneider AJ, Hagedahl FW (1994) Anomalous low-pitched tones from a bowed violin string. Catgut Acoust Soc 2(6):1–7

Harrison JM, Thompson-Allen N (1996) Loudness level survey of the Newberry Memorial Organ, Yale University. J Acoust Soc Am 100:3909–3916

Hartley RVL (1939) Oscillations in systems with non-linear reactance. Bell Syst Tech J 15:424–440

Hawkshaw M, Sataloff RT, Bhatia R (2001) Chaos in voice research. In: The professional voice—the science and art of clinical care. Singular Publishing Group, San Diego, pp 185–190; Chapter 9

Hayes C (1951) The ape in our house. Harper, New York

Hayes JW (1982) String behavior and scale design. In: Travis JW (ed) A guide to restringing. JohnTravis, Takoma Park, pp 31–64

Hedley A, Brown M (1980) Chopin, Fryderyk Franciszek. In: Sadie S (ed) The new groves dictionary of music and musicians, vol 4. Oxford University Press, Oxford, pp 292–312

Helmholtz HLF (1885), Die Lehre von den Tonempfindungen. Longmans & Co, Harlow; Translated into English by Ellis (1885) and republished as "On the sensations of tone". Dover, New York, 1954

Heriot A (1975) The Castrati in Opera. Da Capo Press, Cambridge

Herken G (2002) Brotherhood of the bomb. Henry Holt & Co, New York

Heron-Allen E (1885) Violin-making: as it was and is. Ward Lock & Co., London; Reprinted in 2000 by Algrove, Ottawa

Hill A, Hill A, Hill WH (1902) Antonion Stradivari his life & work. William E. Hill & Sons, London, pp 1644–1737; Reprinted by Dover Publications, New York in 1963

Hill WH, Arthur F, Ebsworth A (1931) The violin-makers of the Guarneri family. William F. Hill & Sons, London, pp 1626–1762; Reprinted by Dover Publications, New York in 1989

Hirano M (1970) Regulation of register, pitch and intensity of the voice. Folia Phoniatr 22:1–20; Also see, "The regulatory mechanisms of the voice in singing"İ, instructional video tape produced by The Voice Foundation, Philadelphia, p 19103

Hubbard F (1965) Three centuries of harpsichord making. Harvard University Press, Cambridge

Hussey LW, Wrathall LR (1936) Oscillations in an electromechanical system. Bell Syst Tech J 15:441–445

Hutchins CM (1962) The physics of violins. Sci Am 207(5):78–93

Hutchins CM (1981) The acoustics of violin plates. Sci Am 245(4):170–187

Hutchins CM (1983) A history of violin research. J Acoust Soc Am 73:1421–1440

Hutchins CM (2004) The catgut acoustical society story. Catgut Acoust Soc J 5:16–30

Hutchins C, Schelleng JC (1967) A new concert violin. J Audio Eng 15(4)

Hutchins C, Benade V (eds) (1997) Research papers in violin acoustics, vols 1, 2. Acoustical Society of America, Woodbury, pp 1975–1993

Jansson E (1973) On higher air modes in the violin. Catgut Acoust Soc NL 19:13–15

Jansson E, Molin NE, Sundin H (1970) Resonances of a violin body studied by hologram interferometry and acoustical methods. Phys Scr 2:243–256

Jansson EV, Niewczyk BK, Fryden L (1997) On body resonance C3 and violin construction. Catgut Acoust Soc J 3(3):9–14

Javan A, Bennett WR Jr, Herriott DR (1961) Population inversion and continuous optical maser oscillation in a gas discharge containing a He-Ne mixtureİ. Phys Rev Lett 6:106–110

Jeans SJ (1937) Science & music. Cambridge University Press, Cambridge

Jonas F (1968) Translation of selected essays on music by Vladimir Vasilevich Stasov. Praeger, New York

Kaplan HM (1971) Anatomy and physiology of speech. McGraw Hill, New York

Katz M (2005) Capturing sound: how technology has changed music. University of California Press, Berkeley

Kempelen WRv (1791) Mechanismus der menschlicken Sprache nebst der Beschreibung seiner sprechenden Maschine. Available in the Yale University Beinecke Library. Also see, Dudley and Tarnoczy, 1950

Kent RD (1997) The speech sciences. Singular Publishing Group, San Diego

Kimura M (2001) The world below G—subharmonics extend the violin's range. STRINGS (August/ September), pp 24–29

Kindel J (1989) Modal analysis and finite element analysis of a piano soundboardİ. M.S. thesis, University of Cincinnati

Kindlmann PJ (1966) Measurement of excited state lifetimes. Ph.D. dissertation in Engineering and Applied Science, Yale University

Kolbert E (2005) The climate of man. The New Yorker, New York; April 25, pp 56–71; May 2, pp 64–73; and May 9, pp 52–63

Kottick EL, Marshall KD, Hendrickson TJ (1991) The acoustics of the harpsichord. Sci Am 110–115

Kroodsma D (2005) The singing life of birds. Houghton Mifflin, New York

Kuperman WA, Lynch JF (2004) Shallow-water acoustics. Phys Today 57(10) 55–61

Lebrecht N (2001) The maestro myth. Citadel Press, New York

Leipp E (1963a) Un 'Vocoder' Mécanique: La Guimbarde.İ Ann Télécommun 18(5–6):82–87

Leipp E (1963b) Étude Acoutsique de la Guimbarde. Acustica 13(6):382–396

Leipp E (1967) La Guimbarde. Bulletin du Groupe d'Acoustique Musicale, No. 25. Laboratoire d'Acoustique Faculté des Sciences, Paris

Lenehan M (1982) Building steinway grand piano K2521—the quality of the instrument. Atl Mon 1–27

Levin TC, Edgerton ME (1999) The throat singers of Tuva. Sci Am 281:70–77

Litten A (1963) Soprano, in Schubert, Der Hirt auf dem Felsen with Stanley Udy (piano), and William R. Bennett, Jr. (clarinet), recorded at Yale, December 1963

Loen JS (2001) Thickness graduation mapping: methods and goals. Catgut Acoust Soc J 4(4):5–6

Loen JS (2002) Fingerprints under the varnish: comparing thickness graduations of the 'Messiah' violin to golden age strads. Catgut Acoust Soc J 4(6):14–16

Loen J (2003) Thickness graduation mapping: surprises and discoveries. J Violin Soc Am XIX(2):41–66; Proceedings of the thirty-first annual convention

Loen JS (2004) Thickness graduation maps of classic violins, violas, and cellos, 1st edn. Jeffery S Loen, Kenmore

Luke JC (1971) Measurement and analysis of body vibrations of a violin. J Acoust Soc Am 49(4)(part 2):1264–1274

Mackay J (1997) Alexander Graham Bell—a life. Wiley, New York

Marshall KD (1985) Modal analysis of a violin. J Acoust Soc Am 77(2):695–709

Martin WH (1924) The transmission unit and telephone reference systems. Bell Syst Tech J 3:400–408

Martin DW, Ward WD (1961) Subjective evaluation of musical scale temperament in pianos. J Am Acoust Soc 33:582–585

McIntyre ME, Schumacher RT, Woodhouse J (1982) Aperiodicity in bowed-string motion. Acoustica 50:294–295

Mersmann H (1972) Editor of letters of Wolfgang Amadeus Mozart. Dover, New York; Reprinted from an original edition by J.M.Dent & Sons, London, 1928

Michelson AA (1903) Light waves and their uses. University of Chicago Press, Chicago

Millant R (1972) J. B. Vuillaume. W. E. Hill & Sons, London; This book is in English, French and German

Milstein N, Solomon V (1990) From Russia to the west—musical reminiscences of Nathan Milstein, translated by Antonina W. Bouis. Limelight Editions, New York

Moreschi A (1902, 1904) Vatican cylinder recordings reproduced on LP disc by opal records circa 1984; Record no. 823

Morrison M (1983) Video highlights from the voice clinic, vol 1. The Voice Foundation, Philadelphia, pp 19103

Neumayr A (1994) Music and medicine—Haydn, Mozart, Beethoven, Schubert—notes on their lives, works, and medical histories, translated by Bruce Cooper Clarke. Medi-Ed Press, Bloomington

Neuwirth R (1994) From sciarrino to subharmonics. STRINGS, September/October, pp 60–63

Nilsson B (1984) Master class, vol 60.İ The New Yorker, New York, pp 44–45

Nyquist H (1924) Certain factors affecting telegraph speed. Bell Syst Tech J 3:324–346

Odiaga L (1974) Johann Sebastian Bachİ, Edition del Patronato Popular y Porvenir Pro Música Clásica; CBS Discos del Perù; Recorded in New Haven by William R. Bennett; P.E.002-A and P.E.002-B

Olson HF (1952) Musical engineering. McGraw-Hill, New York; Republished as music, physics and engineering. Dover, New York, 1967

Pedersen PO (1934) Subharmonics in forced oscillations in dissipative systemsİ, Part I [Theoretical]. J Acoust Soc Am VI:227–238; Part II [Experimental], J Acoust Soc Am VII:64–70

Pepperberg IM (1991) Referential communication with an African grey parrot. Harvard Graduate Society News Letter (Spring), pp. 1–4. Also see, "Brain Parrot Dies..."İ by Benedict Carey, N.Y. Times, Tuesday, September 11, 2007, p A23 and "Alex Wanted a Cracker..."İ by George Johnson, N.Y. Times, September 16, 2007, pp 1,4 wk

Perrins CM, Middleton ALA (1985) The encyclopedia of birds. Facts on file, New York

Peschel ER, Peschel RE (1987) Medical insights into the Castrati in opera. Am Sci 75:578–583

Peterson (1999) Skubic M (ed) AutoStrobe 490-ST strobe tuner instruction manual. Peterson Electro-Musical Products, Alsip

Pickering, NC (1991) The bowed string. Amereon Ltd, Mattituck

Pickering NC (1993) Problems in string making. Catgut Acoust Soc J 2(3):1–4

Pierce JR, alias, Coupling JJ (1983) The science of musical sound. Scientific American Books, New York

Pleasants H (1981) The great singers from the dawn of opera to Caruso, Callas, and Pavarotti. Simon and Schuster, New York

Potter RK, George AK, Green HC (1947) Visible speech. D. Van Nostrand Company, New York

Railsback OL (1938) J Acoust Soc Am 9:274

Raman CV (1918) On the mechanical theory of vibrations of bowed strings. Bull Indian Assoc Cultiv Sci (Calcutta) 15:1–158

Raman CV (1988) Ramaseshan S (ed) Scientific papers of C. V. Raman volume II acoustics. Indian Academy of Sciences, Bangalore

Ramanujan S (1914) Modular equations and approximations to p. Q J Pure Appl Math 45:350–373

Rayleigh L, alias, Strutt JW (1877) Theory of sound, vols I and II. Macmillan Co. Reprinted by Dover, New York, 1945

Riesz RR (1932) The relationship between loudness and the minimum perceptible increment of intensity. J Am Acoust Soc 4:211–216

Ripin EM (1980) Groves dictionary of music, vol 14. Macmillan, London, pp 682–714

Ripin EM, Schott H, Barnes J, O'brien G (1980) Groves dictionary of music, vol 8. Macmillan, London, pp 216–246

Rioul O, Vetterli M (1991) Wavelets and signal processing. IEEE Signal Process Mag 8:14–38

Robbins WW, Weier TE (1950) Botany: an introduction to plant science. Wiley, New York

Rosen C (2002) Piano notes—the world of the pianist. The Free Press, New York

Rossing T (1990) The science of sound. Addison-Wesley, Reading

Rossing T, Fletcher N (1991) The physics of musical instruments. Springer, New York

Rothstein E (1994) A violinist tests music of her time. New York Times, New York, p C21; April 21

Rothstein E (1995) Emblems of mind—the inner life of music and mathematics. Avon Books, New York

Runge C (1903) Z Math Phys 48:443

Rymer R (2004) Saving the music tree. Smithsonian 35(1):52–63

Sacconi SF (2000) The "secrets"ï of Stradivari, translated by Andrew Dipper and Cristina Rivaroli with corrections to the first English edition of 1975 by Andrew Dipper. Eric Blot Edizioni, Cremona

Sadie S (1980) Mozart (3), Wolfgang Amadeus. In: Sadie S (ed) New groves dictionary of music and musicians, vol 12. MacMillan, London, pp 680–752

Salgo S (2000) Thomas Jefferson, musician and violinist. Thomas Jefferson Foundation, Moticello

Sataloff RT (1992) The human voice. Sci Am 267:108–115

Sataloff RT, Hawkshaw M, Bhatia R (1998) Medical applications of chaos theory,ï in: Davis PJ, Fletcher NH (eds) Vocal fold physiology. Singular Publishing Group, San Diego, pp 369–387

Saunders FA (1962) Violins old and new – an experimental study. Sound: Its Uses Control 1(4):7–15 (1962)

Savart F (1819) Rapport sur un mémoire relatif a la construction des instruments àä cordes et àä archet. Ann Chim Phys 2:225

Schawlow AL (1996) Charles townes as i have known him. In: Chiao RY (ed) Amazing light, a volume dedicated to Charles Hard Townes on his 80th birthday. Springer, New York, pp 1–6

Schelleng JC (1963) The violin as a circuit. J Accoust Soc Am 35:326–338; Especially, see Section VI

Schleske M (1998) On the acoustical properties of violin varnish. Catgut Acoust Soc J 3(6):27–43

Schneider D (2004) Living in sunny times. Am Sci 93:22, 23

Schonberg HC (1984) History's last castrato is heard again. The New York Times, New York, p H25; September 16

Schonberg HC (1985) The glorious ones: classical music's legendary perfomers. Times Books, New York, p 3

Schroeder MR (1999) Computer speech. Springer, Berlin

Schuck OH, Young RW (1943) Observations on the vibrations of piano strings. J Am Acoust Soc 15:1–11

Schwartz D (1996) Snatching scientific secrets from the Hippo's gaping jaws. Smithsonian 26(12):90–102

Segré G (2002) A matter of degrees—what temperature reveals about the past and future of our species. In: Planet and universe. Viking, New York

Serber R, Crease RP (1998) Peace & war. Columbia University Press, New York

Shakespeare W, Craig WJ (1966) Shakespeare: complete works. Oxford University Press, London. Print

Shanks D, Wrench JW (1962) Calculation of PTO 100,000 decimals. Math Comput 16:76; poccupies a 20-page table

Shorto R (2004) The island at the center of the world. Doubleday, New York

Silverman WA (1957) The violin hunter. Paganiniana, Neptune City

Sitwell S (1967) Liszt. Dover Publications, New York

Solomon M (1995) Mozart—a life. Harper Collins, New York

Sonnaillon B (1985) King of the instruments—a history of the organ. Rizzoli, New York

Stasov, V. V. Selected essays on music. London: Cresset P, 1968. Print

Steadman P (2001) Vermeer's camera. Oxford University Press, Oxford

Steinhardt A (1998) Indivisible by four—a string quartet in pursuit of harmony. Farrar Straus Giroux, New York

Steinhardt A (2006) Violin dreams. Houghton Mifflin, New York

Stetson KA, Powell RI (1965) Interferometric vibration analysis by wave front reconstruction. J Opt Soc Am 55:1593–1598

Stetson KA, Powell RI (1966) Hologram interferometry.İ J Opt Soc Am 56:1161–1166

Stevens KN (1998) Acoustic phonemes. The MIT Press, Cambridge

Sundberg J (1987) The science of the singing voice. Northern Illinois University Press, DeKalb

Synge JL, Griffith BA (1949) Principles of mechanics. McGraw-Hill Book Co., New York

Tammet D (2007) Born on a blue day. Free Press, New York

Taub R (2002) Playing the Beethoven piano sonatas. Amadeus Press, Portland

Taylor DA (1965) Paderewski's piano, Smithsonian, Maryland, pp 30–34

Thayer AW, Forbes E (eds) (1973) Thayer's life of Beethoven. Princeton University Press, Princeton

Travis JW (1982) A guide to restringing. John W. Travis, Takoma Park

Trent HM, Stone DE (1957) Elastic constants, hardness, strength, and elastic limits of solids.İ In: American institute of physics handbook. McGraw Hill, New York; Chapter 2f

TUVA (1990), Voices from the center of Asia, Smithsonian Fokways CD 40017

Valente B (1960) Schubert, Der Hirt auf dem Felsen, in Recital No. 2 by Harold Wright (clarinet), Rudolf Serkin (piano), Boston Records BR1024CD; Band 10

Van Dyke M (1982) An album of fluid motion. The Parabolic Press, Stanford

Walker T (1980) Castratoİ. In: Sadie S (ed) The new grove dictionary of music and musicians, vol 3. Macmillan, London, pp 875–876

Wang LM (1999) Radiation mechanisms from bowed violinsİ. Doctoral Thesis in Acoustics, Pennsylvania State University, State College, Pennsylvania

Wang LM, Burroughs CB (2001) Acoustic radiation from bowed violins. J Acoust Soc Am 110:543–555

Weatherburn CE (1951) Advanced vector analysis. Bell and Sons, London

Weinreich G (1977) Coupled piano strings. J Acoust Soc Am 62:1474–1484

Weinreich G (1997) Directional tone color. J Acoust Soc Am 101:2338–2346

Weinreich G (2002) Sound radiation from the violin—as we know it today. J Acoust Soc Am 5:37–42

West MJ, King AP (1990) Mozart's starling. Am Sci 78:106–114

Wheatstone C (1828) Erkläring der vermittels des Brummeisens hervorgebrachten Töne. Allg Musik Z (Leipzig) 30:626

Wheatstone C (1879) The scientific papers of Sir Charles wheatstone. Physical Society of London, London, pp 348–367

Whittaker E, Watson G (1902) Modern analysis, Chapter 12. Cambridge University Press, Cambridge

Whittaker ET, Watson GN (1920) A course of modern analysis, 3rd edn. Cambridge University Press, Cambridge

Winchester S (2003) Krakatoa—the day the world exploded: August 27, 1883. HarperCollins, New York

Wood AB (1955) A textbook of sound. Bell, London

Woodhouse J (2002) Body vibrations of the violin—what can a maker expect to control? J Acoust Soc Am 5:43–49

Wright J (1980) Jew's Harp.İ In: Sadie S (ed) The new grove dictionary of music and musicians, vol 9. MacMillan Publishers, London, pp 645–646

Yoshikawa S (1997, in Japanese) Bowing positions to play cello tones on the violin. J Acoust Soc Jpn (Nihon Onk yo Gak kai-shi) 53(12):955–963

Young RW (1957) Frequencies of simple vibrators. Musical scales. American institute of physics handbook. McGraw-Hill, New York, pp 3-100–3-107

Yung BN (1980) Chin23 part IV. Theoryİ. In: Sadie S (ed) The new grove dictionary of music and musicians, vol 4. MacMillan Publishers, London, pp 260–283

Zemansky MW (1951) Heat and thermodynamics. McGraw-Hill, New York

Index

© Springer Nature Switzerland AG 2018
W. R. Bennett, Jr., *The Science of Musical Sound*,
https://doi.org/10.1007/978-3-319-92796-1

Printed in the United States
By Bookmasters